OPTIMAL CONTROL OF DISTRIBUTED SYSTEMS WITH CONJUGATION CONDITIONS

Nonconvex Optimization and Its Applications

Volume 75

Managing Editor:

Panos Pardalos
University of Florida, U.S.A.

Advisory Board:

J. R. Birge
University of Michigan, U.S.A.

Ding-Zhu Du
University of Minnesota, U.S.A.

C. A. Floudas
Princeton University, U.S.A.

J. Mockus
Lithuanian Academy of Sciences, Lithuania

H. D. Sherali
Virginia Polytechnic Institute and State University, U.S.A.

G. Stavroulakis
Technical University Braunschweig, Germany

H. Tuy
National Centre for Natural Science and Technology, Vietnam

OPTIMAL CONTROL OF DISTRIBUTED SYSTEMS WITH CONJUGATION CONDITIONS

By

IVAN V. SERGIENKO
Glushkov Institute of Cybernetics, Ukraine

VASYL S. DEINEKA
Glushkov Institute of Cybernetics, Ukraine

Editor

NAUM Z. SHOR
National Academy of Science of Ukraine

Kluwer Academic Publishers

Library of Congress Cataloging-in-Publication Data

A C.I.P. record for this book is available from the Library of Congress.

ISBN 978-1-4419-5477-0
e-ISBN 978-0-387-24256-9

Printed on acid-free paper.

©2010 Kluwer Academic Publishers

Printed in the United States of America.

9 8 7 6 5 4 3 2 1

springeronline.com

CONTENTS

5   CONTROL OF A SYSTEM DESCRIBED BY A PARABOLIC EQUATION UNDER CONJUGATION CONDITIONS

6   CONTROL OF A SYSTEM DESCRIBED BY A PARABOLIC EQUATION IN THE PRESENCE OF CONCENTRATED HEAT CAPACITY

x

At present, in order to resolve problems of ecology and to save mineral resources for future population generations, it is quite necessary to know how to maintain nature arrangement in an efficient way.

It is possible to achieve a rational nature arrangement when analyzing solutions to problems concerned with optimal control of distributed systems and with optimization of modes in which main ground medium processes are functioning (motion of liquids, generation of temperature fields, mechanical deformation of multicomponent media). Such analysis becomes even more difficult because of heterogeneity of the region that is closest to the Earth surface, and thin inclusions/cracks in it exert their essential influence onto a state and development of the mentioned processes, especially in the cases of mining.

Many researchers, for instance, A.N. Tikhonov – A.A. Samarsky [121], L. Luckner – W.M. Shestakow [65], Tien-Mo Shih, K.L. Johnson [47], E. Sanchez-Palencia [94] and others stress that it is necessary to consider how thin inclusions/cracks exert their influences onto development of these processes, while such inclusions differ in characteristics from main media to a considerable extent (moisture permeability, permeability to heat, bulk density or shear strength may be mentioned).

An influence exerted from thin interlayers onto examined processes is taken into account sufficiently adequately by means of various constraints, namely, by the conjugation conditions [4, 8, 10, 15, 17–20, 22–26, 38, 44, 47, 52, 53, 68, 76, 77, 81, 83, 84, 90, 95, 96–100, 112–114, 117, 123].

The mathematical models include the (partial differential) equations that describe states of components in multicomponent media and have boundary (object-medium interaction) and initial conditions. And the conjugation conditions, specified on median surfaces of thin inclusions and based on the main laws of conservation, are added to them. Such an approach generates the new mathematical problem classes, and a problem solution makes it possible for first-type discontinuities to be present on conjugation condition specification surfaces.

It should be noted that, in 1980s, the problem of construction of computation algorithm with a higher-order accuracy was resolved in general for elliptic, parabolic and hyperbolic equations and for elasticity theory equation systems with boundary and initial conditions [see 16, 43, 54, 55, 71, 78, 79, 91, 92, 119, 124 and other ones]. However, the correctness of these problem classes with conjugation conditions was not investigated and the efficient algorithms, used to solve them numerically, were also absent.

Some simple problems from the above-mentioned families were solved analytically. From the mechanical point of view, the energy functionals were obtained for deformed solids with inclusions of a low rigidity. When the conjugation conditions were considered, the penalty method was used by some authors. There are also the works, where an equation of a state is extended to a solution discontinuity surface by means of the Dirac function.

Unlike these works, the authors of the present monograph propose to use the respective classes of the discontinuous functions in order to investigate boundary-value and initial boundary-value problems with partial derivatives and conjugation conditions [18, 19, 21, 96–100, 112]. This circumstance allows to create the classical energy functionals and weakly stated problems specified on such function classes. The computation algorithms with an enhanced problem discretization accuracy order are developed for the mentioned group of problem classes with conjugation conditions. This is done when proceeding from the application of the finite-element method functions that allow discontinuity. As for a discretization step order, the accuracy of such algorithms is not worse than the accuracy of the similar ones and known for the respective problem classes with smooth solutions.

The authors of the present monograph show the existence of a unique generalized solutions for such problem classes, and a unique solution on a subspace is demonstrated for the Neumann problem. Such unique solutions continuously depend on the disturbances including the right-hand sides of equations, conjugation conditions, boundary conditions. Therefore, it is possible to prove the existence of the unique optimal controls as for the J.L. Lions' quadratic cost functionals.

The contents of the proposed authors' monograph is given mainly in their works [101–111].

It should also be noted that the basic fundamental results were obtained in the theory of optimal control in the works by L.S. Pontryagin, V.P. Boltyansky, R.V. Gamkrelidze, E.F. Mishchenko [85, 42], J. Warga [126], A.A. Feldbaum [40], R. Bellman [5], N.N. Krasovski [51], B.N. Pshenichnyi [87, 88], V.M. Tikhomirov [120] and by other authors.

States of objects (i.e. of systems with distributed parameters) are described on the basis of the laws of conservation by the classical and non-classical equations of mathematical physics in many technical applications and when nature arrangement and ecology problems are investigated and resolved. The works by F. Bensousanne [7], B.N. Bublik [11], A.G. Butkovsky [12–14], F.P. Vasilyev [125], A.I. Egorov [29–31],

Yu.M. Ermoliev [33–37], V.I. Ivanenko, V.S. Mel'nik [46], J.L. Lions [56–64], S.I. Lyashko [67], K.A. Lurje [66], Yu.S. Osipov [80], Yu.I. Samoylenko [14], A.M. Samoylenko [93], T.K. Sirazetdinov [118], R.P. Fedorenko [39], V.A. Dykhta [27] and other works are devoted to resolution of the problems concerned with control of systems with distributed parameters.

Chapter 1 considers new problems concerned with optimal control of distributed systems described by an elliptic equation with conjugation conditions and by a quadratic cost functional. Computation schemes are made up that have an increased order of problem discretization. This is done for the case when a feasible control set $\mathcal{U}_\partial$ of feasible controls coincides with a complete control Hilbert space $\mathcal{U}$.

Chapter 2 discusses optimal control of a conditionally correct system that is described by the Neumann problem for an elliptic equation with conjugation conditions. The aspects of how to create equivalent correct problems and of how to find optimal controls for conditionally correct systems on the basis of such problems are studied.

The problems of optimal control of one- and two-dimensional quartic equations with conjugation conditions are dealt with, respectively, in Chapters 3 and 4.

Motion of a liquid in an elastic medium and non-stationary heat diffusion in multicomponent media are described by initial boundary-value problems for parabolic-type equations with conjugation conditions. Chapter 5 is devoted to optimal control of such systems.

The presence of a concentrated heat capacity on thin inclusions generates classes of initial boundary-value problems for parabolic-type equations with conjugation conditions that contain the first-order time derivative of a solution [91]. Chapter 6 considers optimal control of such systems.

Chapter 7 is concerned with new problems of optimal control of distributed systems described by initial boundary-value problems for a pseudoparabolic equation with conjugation conditions and by a quadratic cost functional.

The initial boundary-value problems for the pseudoparabolic equations were previously considered [1–3, 6, 28, 69, 74, 75, 89, 116].

Chapter 8 studies optimal control of systems described by initial boundary-value problems for hyperbolic equations with conjugation conditions.

The initial boundary-value problems for the pseudohyperbolic equations were also previously considered [45, 48, 50, 70, 82, 115].

Chapter 9 deals with optimal control of systems described by initial boundary-value problems for pseudohyperbolic equations with conjugation conditions.

Chapter 10 discusses optimal control of stress-deformed states of solid bodies that contain thin and not very rigid inclusions.

The authors want to express their gratitude to Mr. Naum Z. Shor, the Scientific Editor, Academician of National Academy of Sciences of Ukraine, for his valuable remarks, useful advises and attention paid to the work.

The authors wish to express their gratitude to Mr. D.A. Kondrashov and Mrs. I.I. Riasnaia, who work at V.M. Glushkov Institute of Cybernetics of National Academy of Sciences of Ukraine, for the translation of the monograph manuscript into English and to Mrs. G.A. Sakhno and Mrs. N.N. Siyanitsa, who also work at V.M. Glushkov Institute of Cybernetics, for the preparation of the computer monograph version.

The authors express their thanks to Mr. John Martindale, the Senior Publisher, and to Ms. Angela Quilici for the fruitful cooperation during the preparation of the book.

Ivan V. Sergienko and
Vasyl S. Deineka

CONTROL OF SYSTEMS DESCRIBED BY ELLIPTIC-TYPE PARTIAL-DIFFERENTIAL EQUATIONS UNDER CONJUGATION CONDITIONS

Chapter 1 considers new problems concerned with optimal control of distributed systems. Such systems are described by an elliptic equation with conjugation conditions and a quadratic cost functional. Computation schemes are made up that have an increased order of problem discretization. This is done for the case when a feasible control set $\mathcal{U}_\partial$ coincides with a complete control Hilbert space $\mathcal{U}$.

1.1 DISTRIBUTED CONTROL OF A SYSTEM DESCRIBED BY THE DIRICHLET PROBLEM

Assume that the elliptic equation

$$-\sum_{i,j=1}^{n} \frac{\partial}{\partial x_i}\left(k_{ij}(x)\frac{\partial y}{\partial x_j}\right) + q(x)y = f(x) \qquad (1.1)$$

is specified in bounded, continuous and strictly Lipschitz domains Ω_1 and $\Omega_2 \in R^n$; in this case,

$$\sum_{i,j=1}^{n} k_{ij}\xi_i\xi_j \geq \alpha_0 \sum_{i=1}^{n} \xi_i^2,$$

$$\forall \xi_i, \xi_j \in R^1, \ i,j = \overline{1,n}, \ \forall x \in \Omega = \Omega_1 \bigcup \Omega_2, \ \alpha_0 = \text{const} > 0, \qquad (1.1')$$

$$k_{ij}\big|_{\Omega_l} = k_{ji}\big|_{\Omega_l} \in C(\bar{\Omega}_l) \cap C^1(\Omega_l), \quad q\big|_{\Omega_l}, \ f\big|_{\Omega_l} \in C(\Omega_l), \qquad l=1,2;$$

$$0 < q_0 \le q = q(x) \le q_1 < \infty; \quad q_0, q_1 = \text{const}, \ |f| < \infty.$$

The homogeneous boundary Dirichlet condition

$$y = 0 \tag{1.2}$$

is specified, in its turn, on a boundary $\Gamma = (\partial\Omega_1 \cup \partial\Omega_2) \backslash \gamma$ $(\gamma = \partial\Omega_1 \cap$ $\cap\partial\Omega_2 \ne \varnothing)$.

On a section γ of the domain $\bar{\Omega} = \bar{\Omega}_1 \cup \bar{\Omega}_2$ $(\Omega_1 \cap \Omega_2 = \varnothing)$, the conjugation conditions for an imperfect contact are

$$\left[\sum_{i,j=1}^{n} k_{ij} \frac{\partial y}{\partial x_j} \cos(\nu, x_i) \right] = 0 \tag{1.3}$$

and

$$\left\{ \sum_{i,j=1}^{n} k_{ij} \frac{\partial y}{\partial x_j} \cos(\nu, x_i) \right\}^{\pm} = r[y], \tag{1.4}$$

where $0 \le r = r(x) \le r_1 < \infty$, $r \in C(\gamma)$, $r_1 = \text{const}$, $[\varphi] = \varphi^+ - \varphi^-$, $\varphi^+ =$ $= \{\varphi\}^+ = \varphi(x)$ under $x \in \partial\Omega_2 \cap \gamma$, $\varphi^- = \{\varphi\}^- = \varphi(x)$ under $x \in \partial\Omega_1 \cap \gamma$, ν is an ort of a normal to γ and such normal is directed into the domain Ω_2.

Let there be a control Hilbert space $\mathcal{U}$ and mapping $B \in \mathcal{L}(\mathcal{U}; V')$, where V' is a space dual with respect to a state Hilbert space V. Denote a space of continuous linear mappings of a topologic space X into a topologic space Y by $\mathcal{L}(X; Y)$ [58]. Assume the following: $\mathcal{U} = L_2(\Omega)$.

For every control $u \in \mathcal{U}$, determine a system state $y = y(u)$ as a generalized solution to the problem specified by the equation

$$-\sum_{i,j=1}^{n} \frac{\partial}{\partial x_i} \left(k_{ij}(x) \frac{\partial y}{\partial x_j} \right) + q(x) y = f(x) + Bu, \ y \in V, \tag{1.5}$$

and by conditions (1.2)–(1.4).

Specify the observation

$$Z(u) = C\, y(u), \tag{1.6}$$

where $C \in \mathscr{L}\,(V; \mathscr{H})$ and $\mathscr{H}$ is some Hilbert space. Assume the following:

$$C\, y(u) \equiv y(u), \quad \mathscr{H} = V \subset L_2(\Omega). \tag{1.7}$$

Bring a value of the cost functional

$$J(u) = \left\| Cy(u) - z_g \right\|_{\mathscr{H}}^2 + (\mathscr{N}u, u)_{\mathscr{U}} \tag{1.8}$$

in correspondence with every control $u \in \mathscr{U}$; in this case, z_g is some known element of a space $\mathscr{H}$, and

$$\mathscr{N} \in \mathscr{L}(\mathscr{U}; \mathscr{U}), \quad (\mathscr{N}u, u)_{\mathscr{U}} \ge v_0 \|u\|_{\mathscr{U}}^2, \quad v_0 = \text{const} > 0, \quad \forall u \in \mathscr{U}. \tag{1.9}$$

Assume the following: $f \in L_2(\Omega)$, $Bu \equiv u \in L_2(\Omega)$, $\mathscr{N}u = \bar{a}(x)u$; in this case, $0 < a_0 \le \bar{a}(x) \le a_1 < \infty$, $\bar{a}(x)\big|_{\Omega_l} \in C(\Omega_l)$, $l = 1,2$; $a_0, a_1 =$

$= \text{const}$, $(\varphi, \psi) = (\varphi, \psi)_{\mathscr{U}} = \int_\Omega \varphi\psi\, dx$. Then, a unique state $y(u) \in V =$

$= \left\{ v\big|_{\Omega_l} \in W_2^1(\Omega_l) : l = 1,2; \ v\big|_\Gamma = 0 \right\}$, where $W_2^1(\Omega_l)$ is a set of the Sobolev

functions that are specified on the domain Ω_l, corresponds to every control $u \in \mathscr{U}$. The function y is specified, in its turn, on the domain $\bar{\Omega} = \bar{\Omega}_1 \cup \bar{\Omega}_2$, minimizes the energy functional [21]

$$\Phi(v) = \int_\Omega \left(\sum_{i,j=1}^n k_{ij} \frac{\partial v}{\partial x_j} \frac{\partial v}{\partial x_i} + qv^2 \right) dx +$$

$$+ \int_\gamma r[v]^2 d\gamma - 2(f, v) - 2(u, v) \tag{1.10}$$

on V, and it is the unique solution in V to the weakly stated problem: Find an element $y \in V$ that meets the equation

$$\int_\Omega \left(\sum_{i,j=1}^n k_{ij} \frac{\partial y}{\partial x_j} \frac{\partial v}{\partial x_i} + qyv \right) dx + \int_\gamma r[y][v] d\gamma =$$

$$= (f,v) + (u,v), \quad y \in V, \quad \forall v \in V, \tag{1.11}$$

where $(f,v) = \displaystyle\int_{\Omega} f v \, dx$.

Therefore, there exists such an operator A acting from V into L_2, that

$$y(u) = A^{-1}(f + Bu), \quad \forall u \in L_2, \tag{1.12}$$

where $L_2 = L_2(\Omega)$.

It is easy to see that $y(u_1) \neq y(u_2)$ under $u_1 \neq u_2$ since the bilinear form is expressed by the left-hand side of equality (1.11) and such form is coercive on V. The solution to linear boundary-value problem (1.5), (1.2)–(1.4) is zero only when the right-hand side of equation (1.5) is zero.

Remark. When a state $y(u)$ is determined as a solution to one of equivalent problems (1.10) and (1.11) with respect to the coefficients k_{ij} in equation (1.1), it is enough to follow ellipticity condition (1.1') and the constraint $k_{ij} \in L_{\infty}(\Omega)$.

Take the aforesaid assumptions into consideration, and the cost functional may be rewritten as

$$J(u) = \left\| y(u) - z_g \right\|^2 + (\bar{a} u, u), \tag{1.13}$$

where $\|v\| = (v,v)^{1/2}$, z_g may be, in its turn, an arbitrary fixed element of the Hilbert space $L_2(\Omega)$, and

$$J(u) = \left\| \left(y(u) - y(0) \right) + \left(y(0) - z_g \right) \right\|^2 + (\bar{a} u, u) =$$

$$= \pi(u,u) - 2L(u) + \left\| z_g - y(0) \right\|^2 \tag{1.14}$$

follows from expression (1.13); in this case, the bilinear form $\pi(\cdot,\cdot)$ and linear functional $L(\cdot)$ are expressed as

$$\pi(u,v) = \left(y(u) - y(0), y(v) - y(0) \right) + (\bar{a} u, v),$$

$$L(v) = \left(z_g - y(0), y(v) - y(0) \right). \tag{1.15}$$

The linearity of the functional $L(v)$ follows from the fact that the difference $y(v) - y(0)$ is the unique solution $\tilde{y}(v)$ to one of equivalent problems (1.10) and (1.11). It is necessary to assume $f \equiv 0$ for them, and the arbitrary element $z \in V$ must be additionally substituted for the arbitrary function v in problem (1.11). Then:

$$\tilde{y}(\alpha_1 u_1 + \alpha_2 u_2) = \alpha_1 \tilde{y}(u_1) + \alpha_2 \tilde{y}(u_2) \ \forall \alpha_1, \alpha_2 \in R^1, \ \forall u_1, u_2 \in L_2. \quad (1.16)$$

Pursuant to equality (1.16), it can also be stated that $\pi(u, v)$ is the bilinear form that is coercive on $\mathcal{U}$ and that it can be made symmetric by virtue of the following: $(\bar{a} u, v) = \left(\sqrt{\bar{a}} \, u, \sqrt{\bar{a}} \, v \right)$,

$$\pi(u, u) = \left(y(u) - y(0), y(u) - y(0) \right) + \left(\sqrt{\bar{a}} \, u, \sqrt{\bar{a}} \, u \right) \geq a_0(u, u).$$

Let $\tilde{y}' = \tilde{y}(u')$ and $\tilde{y}'' = \tilde{y}(u'')$ be solutions from V to problem (1.11) under $f = 0$ and under a function $u = u(x)$ that is equal, respectively, to u' and u''. Then, the inequality

$$\|\tilde{y}' - \tilde{y}''\|^2 \leq \|\tilde{y}' - \tilde{y}''\|_V^2 \leq \mu a(\tilde{y}' - \tilde{y}'', \tilde{y}' - \tilde{y}'') \leq$$
$$\leq \mu \|u' - u''\| \cdot \|\tilde{y}' - \tilde{y}''\|_V, \ \mu = \text{const} > 0,$$

is derived that provides the continuity of the linear functional $L(\cdot)$ and bilinear form $\pi(\cdot, \cdot)$ on $\mathcal{U}$; in this case,

$$\|v\|_V = \left\{ \sum_{i=1}^{2} \|v\|_{W_2^1(\Omega_i)}^2 \right\}^{1/2}$$

and

$$a(\varphi, \psi) = \int_\Omega \left(\sum_{i,j=1}^{n} k_{ij} \frac{\partial \varphi}{\partial x_j} \frac{\partial \psi}{\partial x_i} + q\varphi\psi \right) dx + \int_\gamma r[\varphi][\psi] d\gamma.$$

On the basis of [58, Chapter 1, Theorem 1.1], the validity of the following statement is proved.

Theorem 1.1. *Let conditions (1.1') be met, and a system state is determined as a solution to equivalent problems (1.10) and (1.11). Then, there exists a unique element u of a convex set $\mathcal{U}_\partial$ that is closed in $\mathcal{U}$, and*

$$J(u) = \inf_{v \in \mathcal{U}_\partial} J(v) \qquad (1.17)$$

takes place for u.

Definition 1.1. If an element $u \in \mathcal{U}_\partial$ meets condition (1.17), it is called an optimal control.

If $u \in \mathcal{U}_\partial$ is the optimal control, then

$$J(u) \leq J((1-\theta)u + \theta v) \quad \forall v \in \mathcal{U}_\partial, \quad \theta \in (0,1),$$

or

$$\frac{J(u+\theta(v-u)) - J(u)}{\theta} \geq 0. \tag{1.18}$$

Pass to the limit $\theta \to 0$, and

$$\lim_{\theta \to 0} \frac{J(u+\theta(v-u)) - J(u)}{\theta} \geq 0.$$

Therefore,

$$\langle J'(u), v-u \rangle \geq 0. \tag{1.19}$$

Take expressions (1.14) and (1.18) into consideration, and inequality (1.19) has the form

$$\lim_{\theta \to 0} \frac{\pi(u+\theta(v-u), u+\theta(v-u)) - 2L(u+\theta(v-u)) - \pi(u,u) + 2L(u)}{\theta} =$$

$$= \lim_{\theta \to 0} \frac{\pi(u,u) + 2\theta\,\pi(u, v-u) + \theta^2\pi(v-u, v-u) - \pi(u,u) - 2\theta L(v-u)}{\theta} =$$

$$= 2\{\pi(u, v-u) - L(v-u)\} \geq 0, \tag{1.20}$$

from which the inequality

$$\pi(u, v-u) \geq L(v-u), \quad \forall v \in \mathcal{U}_\partial, \tag{1.21}$$

is derived.

On the basis of expressions (1.15), the equality

$$\pi(u, v-u) - L(v-u) = \left(y(u) - y(0), y(v-u) - y(0) \right) +$$

$$+ (\bar{a}\,u, v-u) - \left(z_g - y(0), y(v-u) - y(0) \right) =$$

$$= \left(y(u) - z_g, y(v-u) - y(0) \right) + (\bar{a}\,u, v-u) \tag{1.22}$$

is obtained from condition (1.21), and the equality

$$\pi(u, v-u) - L(v-u) = \left(y(u) - z_g, y(v) - y(u) \right) + (\bar{a}\,u, v-u)$$

follows from equality (1.22) when the linearity of problem (1.11) is taken into account.

Then, inequality (1.21) has the form

$$\left(y(u) - z_g, y(v) - y(u) \right) + (\bar{a}\,u, v-u) \ge 0, \quad \forall v \in \mathcal{U}_\partial, \tag{1.23}$$

and it is the necessary and sufficient condition under which $u \in \mathcal{U}_\partial$ is the optimal control for the considered problem.

As for the control $v \in \mathcal{U}$, the conjugate state $p(v) \in V^*$ is specified by the equation

$$A^* p(v) = y(v) - z_g; \tag{1.24}$$

in this case, the operators A and $A^* \in \mathscr{L}(V^*; V')$ (conjugate to A) are interrelated by the bilinear form

$$(A^*\varphi, \psi) = (\varphi, A\psi) = a(\varphi, \psi), \quad \varphi \in V^*, \ \psi \in V, \tag{1.25}$$

where

$$a(\varphi, \psi) = \int_\Omega \left\{ \sum_{i,j=1}^{n} k_{ij} \frac{\partial \varphi}{\partial x_j} \frac{\partial \psi}{\partial x_i} + q\varphi\psi \right\} dx + \int_\gamma r[\varphi][\psi]\,d\gamma. \tag{1.26}$$

Consider equation (1.24), obtain the equality

$$(A^* p(u), y(v) - y(u)) = (y(u) - z_g, y(v) - y(u)) =$$

$$= (p(u), A(y(v) - y(u))) = (p(u), Ay(v) - Ay(u)) = (p(u), v-u),$$

and it is stated that inequality (1.23) is equivalent to the inequality

$$(p(u) + \bar{a}\,u, v-u) \ge 0, \quad \forall v \in \mathcal{U}_\partial.$$

Therefore, the necessary and sufficient condition for the existence of the optimal control $u \in \mathcal{U}_\partial$ is the one under which the relations

$$A\,y(u) = f + u,\tag{1.27}$$

$$A^* p(u) = y(u) - z_g\tag{1.28}$$

and

$$(p(u) + \bar{a}u, v - u) \geq 0, \quad \forall v \in \mathcal{U}_\partial,\tag{1.29}$$

are met.

If the constraints are absent, i.e. when $\mathcal{U}_\partial = \mathcal{U}$, then the equality

$$p(u) + \bar{a}\,u = 0\tag{1.30}$$

follows from condition (1.29). Therefore, when the constraints are absent, the control $u(x)$ can be excluded from equality (1.27) by means of equality (1.30). On the basis of equalities (1.27) and (1.28), the problem

$$Ay + p/\bar{a} = f, \; y \in V,\tag{1.31}$$

$$A^* p - y = -z_g, \; p \in V^*,\tag{1.32}$$

is derived, where $V^* = \left\{ v|_{\Omega_l} \in W_2^1(\Omega_l) : l = 1,2; \; v|_\Gamma = 0 \right\}$, i.e. $V^* = V$, and the vector solution $(y,p)^{\mathrm{T}}$ is found from this problem along with the optimal control

$$u = -p/\bar{a}.\tag{1.33}$$

If the vector solution $(y,p)^{\mathrm{T}}$ to problem (1.31), (1.32) is smooth enough on $\bar{\Omega}_l$, viz., $y|_{\bar{\Omega}_l}, \; p|_{\bar{\Omega}_l} \in C^1(\bar{\Omega}_l) \cap C^2(\Omega_l)$, $l = 1,2$, then the differential problem of finding the vector-function $(y,p)^{\mathrm{T}}$, that satisfies the relations

$$-\sum_{i,j=1}^{n} \frac{\partial}{\partial x_i}\left(k_{ij} \frac{\partial y}{\partial x_j} \right) + qy + p/\bar{a} = f, \quad x \in \Omega_1 \cup \Omega_2,\tag{1.34}$$

$$-\sum_{i,j=1}^{n} \frac{\partial}{\partial x_i}\left(k_{ij}\frac{\partial p}{\partial x_j}\right) + q\,p - y = -z_g, \ x \in \Omega_1 \cup \Omega_2, \tag{1.35}$$

$$y\big|_\Gamma = 0, \tag{1.36}$$

$$p\big|_\Gamma = 0, \tag{1.37}$$

$$\left[\sum_{i,j=1}^{n} k_{ij}\frac{\partial y}{\partial x_j}\cos(\nu, x_i)\right] = 0, \ x \in \gamma, \tag{1.38}$$

$$\left\{\sum_{i,j=1}^{n} k_{ij}\frac{\partial y}{\partial x_j}\cos(\nu, x_i)\right\}^{\pm} = r[y], \ x \in \gamma, \tag{1.39}$$

$$\left[\sum_{i,j=1}^{n} k_{ij}\frac{\partial p}{\partial x_j}\cos(\nu, x_i)\right] = 0, \ x \in \gamma, \tag{1.40}$$

and

$$\left\{\sum_{i,j=1}^{n} k_{ij}\frac{\partial p}{\partial x_j}\cos(\nu, x_i)\right\}^{\pm} = r[p], \ x \in \gamma, \tag{1.41}$$

corresponds to problem (1.31), (1.32).

Definition 1.2. A generalized (weak) solution to boundary-value problem (1.34)–(1.41) is called a vector-function $(y, p)^{\mathrm{T}} \in H = \left\{v = (v_1, v_2)^{\mathrm{T}} : v_i\big|_{\Omega_l} \in W_2^1(\Omega_l), i, l = 1, 2; v\big|_\Gamma = 0\right\}$ that satisfies the following integral equation $\forall z \in H$:

$$\int_\Omega\left\{\sum_{i,j=1}^{n} k_{ij}\frac{\partial y}{\partial x_j}\frac{\partial z_1}{\partial x_i} + qyz_1 + pz_1/\bar{a} + \sum_{i,j=1}^{n} k_{ij}\frac{\partial p}{\partial x_j}\frac{\partial z_2}{\partial x_i} + qpz_2 - yz_2\right\}dx +$$

$$+ \int_\gamma r[y][z_1]d\gamma + \int_\gamma r[p][z_2]d\gamma = \int_\Omega(f\,z_1 - z_g z_2)dx. \tag{1.42}$$

Let $u = (u_1, u_2)^{\mathrm{T}}$ and $v = (v_1, v_2)^{\mathrm{T}}$ be arbitrary elements of the complete

Hilbert space H with the norm $\|v\|_H = \left\{ \sum\limits_{i=1}^{2} \|v\|_{W_2^1(\Omega_i)}^2 \right\}^{1/2}$, where $\|v\|_{W_2^1(\Omega_i)}$

is the norm of the Sobolev space $W_2^1(\Omega_i)$. Specify the bilinear form

$$a(u,v) = \int\limits_{\Omega} \left\{ \sum_{i,j=1}^{n} k_{ij} \frac{\partial u_1}{\partial x_j} \frac{\partial v_1}{\partial x_i} + \sum_{i,j=1}^{n} k_{ij} \frac{\partial u_2}{\partial x_j} \frac{\partial v_2}{\partial x_i} + \right.$$

$$+ q(u_1 v_1 + u_2 v_2) + u_2 v_1 / \bar{a} - u_1 v_2 \} dx +$$

$$+ \int\limits_{\gamma} r[u_1][v_1] d\gamma + \int\limits_{\gamma} r[u_2][v_2] d\gamma \qquad (1.43)$$

on H.

Let the constraint

$$\alpha_1 = (\alpha_0 - \varepsilon)\mu + q_0 - \frac{1}{2}\left(\frac{1}{a_0} + 1\right) > 0 \qquad (1.44)$$

be met, where ε is a sufficiently small positive real number and $\mu = \mathrm{const} > 0$ is the constant in the Friedrichs inequality

$$\int\limits_{\Omega} \sum_{i=1}^{n} \left(\frac{\partial v}{\partial x_i}\right)^2 dx \geq \mu \int\limits_{\Omega} v^2 dx, \ \forall v \in V.$$

Proceed from constraints (1.1′) and (1.44), the Cauchy-Bunyakovsky and Friedrichs inequalities and embedding theorems [55], and the inequalities

$$a(v,v) \geq \bar{\alpha}_1 \|v\|_H^2, \ \forall v \in H, \ \bar{\alpha}_1 = \mathrm{const} > 0,$$

and

$$|a(u,v)| \leq c_1 \|u\|_H \|v\|_H, \ \forall u,v \in H, \ c_1 = \mathrm{const} > 0,$$

are true for the bilinear form $a(\cdot, \cdot)$, i.e. this form is H-elliptic and continuous [49] on H.

Consider the Cauchy-Bunyakovsky inequality, and $\forall v \in H$

$$|l(v)| = \left| \int_{\Omega} (f v_1 - z_g v_2) dx \right| \le c_2 \|v\|_H, \quad c_2 = const,$$

i.e. the linear functional $l(v) = \int_{\Omega} (f v_1 - z_g v_2) dx$ is continuous on H.

Use the Lax-Milgramm lemma [16], and it is concluded that the unique solution $(y, p)^T$ to problem (1.42) exists in H. It is easy to see that p is the unique solution to equation (1.32) when y is fixed.

Problem (1.42) can be solved approximately by means of the finite-element method. For this purpose, divide each domain Ω_i into N_i finite elements $\bar{e}_i^j$ $(j = \overline{1, N_i}, i = 1, 2)$ of the regular family [16]. Specify the subspace $H_k^N \subset H$ $(N = N_1 + N_2)$ of the vector-functions $V_k^N(x) = \left(v_{1k}^N(x), v_{2k}^N(x) \right)^T$. The components of $V_k^N(x) = \left(v_{1k}^N(x), v_{2k}^N(x) \right)^T$ are continuous on $\bar{\Omega}_i$, $i = 1, 2$, and they are the complete polynomials of the power k that contain the variables $x_1, x_2, ..., x_n$ at every $\bar{e}_i^j$, and $V_k^N\big|_{\Gamma} = 0$. Then, the linear algebraic equation system

$$A\bar{U} = B \tag{1.45}$$

follows from equation (1.42), and the solution $\bar{U}$ to system (1.45) exists and such solution is unique. The vector $\bar{U}$ specifies the unique approximate solution $U_k^N \in H_k^N$ to problem (1.42) as the unique one to the equation

$$a\left(U_k^N, V_k^N \right) = l\left(V_k^N \right), \quad \forall V_k^N \in H_k^N. \tag{1.46}$$

Let $U = U(x) \in H$ be the solution to problem (1.42). Then:

$$a\left(U - U_k^N, V_k^N \right) = 0, \quad \forall V_k^N \in H_k^N.$$

Therefore,

$$\bar{\alpha}_1 \left\| U - U_k^N \right\|_H^2 \le a\left(U - U_k^N, U - U_k^N \right) =$$

$$= a\left(U - U_k^N, U - \tilde{U} + \tilde{U} - U_k^N \right) = a\left(U - U_k^N, U - \tilde{U} \right), \quad \forall \tilde{U} \in H_k^N,$$

and the inequality

$$\left\| U - U_k^N \right\|_H \le \frac{c_1}{\bar{\alpha}_1} \left\| U - \tilde{U} \right\|_H \tag{1.47}$$

is thus derived since the bilinear form $a(\cdot,\cdot)$ is continuous on H.

Suppose that $\tilde{U} \in H_k^N$ is a complete interpolation polynomial for the solution U at every $\bar{e}_i^j$. Take the interpolation estimates [16] into account, assume that every component U_1 and U_2 of the solution U on Ω_l belongs to the Sobolev space $W_2^{k+1}(\Omega_l)$, $l = 1,2$, and the estimate

$$\left\| U - U_k^N \right\|_H \le ch^k, \tag{1.48}$$

where h is a maximum diameter for all the finite elements $\bar{e}_i^j$, $c = \text{const}$, follows from inequality (1.47).

Take estimate (1.48) into consideration, and the estimate

$$\left\| u - u_k^N \right\|_{W_2^1} \le c_2 \left\| p - p_k^N \right\|_{W_2^1} \le c_3 h^k,$$

where $\left\| \cdot \right\|_{W_2^1} = \left\{ \sum_{i=1}^{2} \left\| \cdot \right\|_{W_2^1(\Omega_i)} \right\}^{1/2}$, takes place for the approximation

$u_k^N(x) = -p_k^N(x)/\bar{a}(x)$ of the control $u = u(x)$.

1.2   CONTROL UNDER CONJUGATION CONDITION. THE DIRICHLET PROBLEM

Assume that elliptic equation (1.1), where the coefficients and right-hand side meet conditions (1.1'), is specified in the bounded, continuous and strictly Lipschitz domains Ω_1 and $\Omega_2 \in R^n$.

The homogeneous boundary Dirichlet condition

$$y = 0, \ x \in \Gamma, \tag{2.1}$$

is specified, in its turn, on the boundary Γ of the domain $\bar{\Omega}$.

For every control $u \in \mathcal{U} = L_2(\gamma)$, determine a state $y = y(u)$ as a generalized solution to the boundary-value problem specified by equation (1.1), boundary condition (2.1) and the conjugation conditions

$$[y] = 0, \ x \in \gamma, \tag{2.2}$$

and

$$\left[\sum_{i,j=1}^{n} k_{ij} \frac{\partial y}{\partial x_j} \cos(v, x_i) \right] = \omega + u, \ x \in \gamma, \tag{2.3}$$

where $\omega = \omega(x)$ is some known function from $L_2(\gamma)$.

Since there exists a generalized solution $y(u) \in V = \left\{ v \big|_{\Omega_i} \in W_2^1(\Omega_i) : [v]_\gamma = 0, v \big|_\Gamma = 0 \right\}$ to boundary-value problem (1.1), (2.1)–(2.3), then such solution is reasonable on $\bar{\Omega}_1$ and $\bar{\Omega}_2$. Specify the observation in the form of expression (1.7), where $C \in \mathcal{L}(V;V)$, namely:

$$C \, y(u) \equiv y(u). $$

Bring a value of the cost functional

$$J(u) = \int_\Omega (y(u) - z_g)^2 dx + (\mathcal{N}u, u)_{\mathcal{U}} \tag{2.4}$$

in correspondence with every control $u \in \mathcal{U}$; in this case, z_g is a known element from $L_2(\Omega)$; $\mathcal{N}u = \bar{a}(x)u$, $0 < a_0 \leq \bar{a}(x) \leq a_1 < \infty$, $(\varphi, \psi)_\mathcal{U} = \int_\gamma \varphi\psi\,d\gamma = (\varphi, \psi)_{L_2(\gamma)}$.

It can be shown [21] that a unique state $y(u) \in V$ corresponds to every control $u \in \mathcal{U}$. The function y is specified on the domain $\bar{\Omega}_1 \cup \bar{\Omega}_2$, minimizes the energy functional

$$\Phi(v) = \int_\Omega \left(\sum_{i,j=1}^n k_{ij} \frac{\partial v}{\partial x_j} \frac{\partial v}{\partial x_i} + qv^2 \right) dx - 2 \int_\Omega fv\,dx + 2 \int_\gamma (\omega + u)v\,d\gamma \quad (2.5)$$

on V, and it is the unique solution in V to the weakly stated problem: Find an element $y \in V$ that meets the integral equation

$$\int_\Omega \left(\sum_{i,j=1}^n k_{ij} \frac{\partial y}{\partial x_j} \frac{\partial v}{\partial x_i} + qyv \right) dx =$$

$$= \int_\Omega fv\,dx - \int_\gamma \omega v\,d\gamma - \int_\gamma uv\,d\gamma, \quad \forall v \in V. \quad (2.6)$$

The state $y(u_1) \neq y(u_2)$ is easily seen under $u_1 \neq u_2$, and

$$a_0(z, z) \geq \bar{\alpha}_0 \|z\|_V^2, \quad \bar{\alpha}_0 > 0, \quad \forall z \in V, \quad (2.6')$$

where the bilinear form $a_0(\cdot, \cdot)$ is generated by the left-hand side of equality (2.6).

Take the assumptions as for the operator $\mathcal{N}$ into account, and

$$J(u) = \left\| \left(y(u) - y(0) \right) + \left(y(0) - z_g \right) \right\|^2 + (\bar{a}\,u, u)_{L_2(\gamma)} =$$

$$= \pi(u, u) - 2L(u) + \left\| z_g - y(0) \right\|^2 \quad (2.7)$$

follows from expression (2.4); in this case, the bilinear form $\pi(\cdot, \cdot)$ and linear functional $L(\cdot)$ are expressed as

$$\pi(u,v) = \big(y(u) - y(0), y(v) - y(0)\big) + (\bar{a}\,u, v)_{L_2(\gamma)},$$

$$L(v) = \big(z_g - y(0), y(v) - y(0)\big). \tag{2.8}$$

The linearity of the functional $L(v)$ is easily seen from the fact that the difference $y(v) - y(0)$ is the unique solution $\tilde{y}(v)$ to one of equivalent problems (2.5) and (2.6). It is necessary to assume $f = 0$ and $\omega = 0$ for them, and the arbitrary element $z \in V$ must be additionally substituted for the arbitrary function v in problem (2.6). Then, equality like (1.16) takes place that allows to state the linearity of the functional $L(v)$ and the bilinearity of the form $\pi(u,v)$. The form $\pi(\cdot,\cdot)$ is coercive on $L_2(\gamma)$, i.e.:
$\pi(u,u) \geq a_0 (u,u)_{L_2(\gamma)}$.

Let $\tilde{y}' = \tilde{y}(u')$ and $\tilde{y}'' = \tilde{y}(u'')$ be solutions from V to problem (2.6) under $f = 0$ and $\omega = 0$ and under a function $u = u(x)$ that is equal, respectively, to u' and u''. Then, the inequality

$$\left\| \tilde{y}' - \tilde{y}'' \right\|_V^2 \leq \mu\, a\big(\tilde{y}' - \tilde{y}'', \tilde{y}' - \tilde{y}'' \big) \leq$$

$$\leq \mu \left\| u' - u'' \right\|_{L_2(\gamma)} \left\| \tilde{y}' - \tilde{y}'' \right\|_{L_2(\gamma)} \tag{2.9}$$

is derived, where $\mu = \mathrm{const} > 0$, the norm $\|\cdot\|_V$ is specified in point 1.1 and the bilinear form $a(\cdot,\cdot)$ is specified, in its turn, by the expression

$$a(u,v) = \int\limits_{\Omega} \left(\sum_{i,j=1}^{n} k_{ij} \frac{\partial u}{\partial x_j} \frac{\partial v}{\partial x_i} + quv \right) dx. \tag{2.10}$$

Since the inequalities

$$\left\| v \right\|_{L_2(\partial\Omega_i)} \leq c_i \left\| v \right\|_{W_2^1(\Omega_i)}, \quad c_i > 0, \ i = 1,2, \tag{2.10'}$$

are true $\forall v \in V$, then:

$$\left\| v \right\|_{L_2(\gamma)} \leq c_3 \left\| v \right\|_V, \quad c_3 = \max\{c_1, c_2\} > 0. \tag{2.11}$$

Take inequality (2.11) into account, and the inequality

$$\|\tilde{y}' - \tilde{y}''\|_{L_2(\gamma)} \le c_4 \|u' - u''\|_{L_2(\gamma)} \tag{2.12}$$

follows from inequality (2.9), i.e. the trace of the function $\tilde{y}(u)$ on γ is continuously dependent upon u. The inequality

$$\|\tilde{y}' - \tilde{y}''\| \le \sqrt{\mu\, c_4} \|u' - u''\|_{L_2(\gamma)} \tag{2.12'}$$

also follows from inequality (2.9).

Inequalities (2.12) and (2.12′) provide the continuity of the bilinear form $\pi(\cdot,\cdot)$ and linear functional $L(\cdot)$ on $\mathcal{U}$.

On the basis of [58, Chapter 1, Theorem 1.1], the validity of the following statement is proved.

Theorem 2.1. *Let conditions (1.1′) be met, and a system state is determined as a solution to equivalent problems (2.5) and (2.6). Then, there exists a unique element u of a convex set $\mathcal{U}_\partial$ that is closed in $\mathcal{U}$, and relation like (1.17) takes place for u.*

If $u \in \mathcal{U}_\partial$ is the optimal control, then

$$\pi(u, v - u) \ge L(v - u), \quad \forall v \in \mathcal{U}_\partial. \tag{2.13}$$

Proceed from expression (2.8), and it is easy to see that the equality

$$\pi(u, v - u) - L(v - u) =$$

$$= \left(y(u) - z_g, y(v - u) - y(0) \right) + (\bar{a}\, u, v - u)_{L_2(\gamma)} \tag{2.14}$$

is true. Take the linearity of problem (2.6) into account, consider equality (2.14), and the inequality

$$\left(y(u) - z_g, y(v) - y(u) \right) + (\bar{a}\, u, v - u)_{L_2(\gamma)} \ge 0, \quad \forall v \in \mathcal{U}_\partial, \tag{2.15}$$

is obtained that is the necessary and sufficient condition under which $u \in \mathcal{U}_\partial$ is the optimal control for the considered problem.

As for the control $v \in \mathcal{U}$, the conjugate state $p(v) \in V^*$ is specified by the relations

$$A^* p(v) = -y(v) + z_g, \tag{2.16}$$

$$p = 0, \ x \in \Gamma, \tag{2.17}$$

$$[p] = 0, \ x \in \gamma, \tag{2.18}$$

and

$$\left[\frac{\partial p}{\partial \nu_{A^*}} \right] = 0, \ x \in \gamma, \tag{2.19}$$

where V^* is a space conjugate to V, $V^* = V$, and

$$A^* p = -\sum_{i,j=1}^{n} \frac{\partial}{\partial x_i} \left(k_{ij} \frac{\partial p}{\partial x_j} \right) + q\, p,$$

$$\frac{\partial p}{\partial \nu_{A^*}} = \sum_{i,j=1}^{n} k_{ij} \frac{\partial p}{\partial x_j} \cos(\nu, x_i). \tag{2.19'}$$

Further on, use the Green formula [58], and the equality

$$\left(A^* p(u), y(v) - y(u) \right) = -\left(y(u) - z_g, y(v) - y(u) \right) =$$

$$= -\sum_{l=1}^{2} \int_{\partial \Omega_l} \frac{\partial p}{\partial \nu_{A^*}} (y(v) - y(u)) d\partial\Omega_l +$$

$$+ \int_{\Omega} \left(\sum_{i,j=1}^{n} k_{ij} \frac{\partial p}{\partial x_j} \frac{\partial (y(v) - y(u))}{\partial x_i} + q\, p (y(v) - y(u)) \right) dx =$$

$$= a\left(p, y(v) - y(u) \right) = \sum_{l=1}^{2} \int_{\partial \Omega_l} p \sum_{i,j=1}^{n} k_{ij} \frac{\partial (y(v) - y(u))}{\partial x_i} \cos(\nu, x_j) d\partial\Omega_l +$$

$$+ \int_{\Omega} p \left(-\sum_{i,j=1}^{n} \frac{\partial}{\partial x_j} \left(k_{ij} \frac{\partial (y(v) - y(u))}{\partial x_i} \right) + q (y(v) - y(u)) \right) dx =$$

$$= -\int_{\gamma} p(v - u) d\gamma,$$

i.e.

$$\left(y(u)-z_g, y(v)-y(u)\right) = \int_\gamma p(v-u)\,d\gamma \qquad (2.20)$$

is obtained. Take it and equality (2.14) into account, and the inequality

$$\left(p+\bar{a}\,u, v-u\right)_{L_2(\gamma)} \geq 0, \quad \forall v \in \mathcal{U}_\partial, \qquad (2.21)$$

is derived from inequality (2.15).

Therefore, if the constraints are absent, i.e. when $\mathcal{U}_\partial = \mathcal{U}$, then the equality

$$p(u)+\bar{a}\,u = 0, \ x \in \gamma, \qquad (2.22)$$

follows from condition (2.21), and, to find the optimal control $u(x)$, solve the differential problem

$$-\sum_{i,j=1}^{n} \frac{\partial}{\partial x_i}\left(k_{ij}\frac{\partial y}{\partial x_j}\right) + q\,y = f, \ x \in \Omega_1 \bigcup \Omega_2, \qquad (2.23)$$

$$-\sum_{i,j=1}^{n} \frac{\partial}{\partial x_i}\left(k_{ij}\frac{\partial p}{\partial x_j}\right) + q\,p = z_g - y, \ x \in \Omega_1 \bigcup \Omega_2, \qquad (2.24)$$

$$y = 0, \ x \in \Gamma, \qquad (2.25)$$

$$p = 0, \ x \in \Gamma, \qquad (2.26)$$

$$[y] = 0, \ x \in \gamma, \qquad (2.27)$$

$$\left[\sum_{i,j=1}^{n} k_{ij}\frac{\partial y}{\partial x_j}\cos(v, x_i)\right] = \omega - p/\bar{a}, \ x \in \gamma, \qquad (2.28)$$

$$[p] = 0, \ x \in \gamma, \qquad (2.29)$$

$$\left[\sum_{i,j=1}^{n} k_{ij}\frac{\partial p}{\partial x_j}\cos(v, x_i)\right] = 0, \ x \in \gamma, \qquad (2.30)$$

and the optimal control is

$$u = -p/\bar{a}, \ x \in \gamma. \tag{2.31}$$

Definition 2.1. A generalized (weak) solution to boundary-value problem (2.23)–(2.30) is called a vector-function $(y, p)^{\mathrm{T}} \in H = \left\{ v = (v_1, v_2)^{\mathrm{T}} : v_i \big|_{\Omega_l} \in W_2^1(\Omega_l); \ i, l = 1, 2; \ v \big|_{\Gamma} = 0, \ [v_i] \big|_{\gamma} = 0, \ i = 1, 2 \right\}$ that satisfies the following integral equation $\forall z \in H$:

$$\int_{\Omega} \left\{ \sum_{i,j=1}^{n} k_{ij} \frac{\partial y}{\partial x_j} \frac{\partial z_1}{\partial x_i} + qyz_1 + \sum_{i,j=1}^{n} k_{ij} \frac{\partial p}{\partial x_j} \frac{\partial z_2}{\partial x_i} + qpz_2 \right\} dx =$$

$$= \int_{\Omega} fz_1 dx + \int_{\Omega} (z_g - y) z_2 dx - \int_{\gamma} \omega z_1 d\gamma + \int_{\gamma} p z_1/\bar{a} \, d\gamma. \tag{2.32}$$

Let $u = (u_1, u_2)^{\mathrm{T}}$ and $v = (v_1, v_2)^{\mathrm{T}}$ be arbitrary elements of the complete Hilbert space H with the previously introduced norm $\|\cdot\|_H$. Specify the bilinear form $a(\cdot, \cdot)$ and linear functional $l(\cdot)$ on H by the expressions

$$a(u, v) =$$

$$= \int_{\Omega} \left(\sum_{i,j=1}^{n} k_{ij} \frac{\partial u_1}{\partial x_j} \frac{\partial v_1}{\partial x_i} + \sum_{i,j=1}^{n} k_{ij} \frac{\partial u_2}{\partial x_j} \frac{\partial v_2}{\partial x_i} + q(u_1 v_1 + u_2 v_2) + u_1 v_2 \right) dx -$$

$$- \int_{\gamma} u_2 v_1/\bar{a} \, d\gamma,$$

$$l(v) = \int_{\Omega} fv_1 dx + \int_{\Omega} z_g v_2 dx - \int_{\gamma} \omega v_1 d\gamma. \tag{2.33}$$

If the constraint

$$\alpha_1 = \bar{\alpha}_0 - \frac{\max\{c_1^2, c_2^2\}}{2a_0} - \frac{1}{2} > 0$$

takes place, where c_1 and c_2 are the constants from inequalities (2.10′) and μ and $\bar{\alpha}_0$ are, respectively, the constants in the Friedrichs inequality and inequality (2.6′), then:

$$a(v,v) \geq \alpha_1 \|v\|_H^2, \quad \forall v \in H,$$

and

$$|a(u,v)| \leq c_3 \|u\|_H \|v\|_H, \quad \forall u,v \in H, \quad c_3 = \text{const} > 0;$$

i.e. the bilinear form $a(\cdot,\cdot)$ is H-elliptic and continuous on H.

Consider the Cauchy-Bunyakovsky inequality, and $\forall v \in H$

$$|l(v)| = \left| \int_\Omega f v_1 dx + \int_\Omega z_g v_2 dx - \int_\gamma \omega v_1 d\gamma \right| \leq c_4 \|v\|_H, \quad c_4 = \text{const},$$

i.e. the linear functional $l(v)$ is continuous on H.

Use the Lax-Milgramm lemma, and it is concluded that the unique solution $(y,p)^T$ to problem (2.32) exists in H. Problem (2.32) can be solved approximately by means of the finite-element method. Specify the subspace $H_k^N \subset H$ of the vector-functions $V_k^N(x) = \left(v_{1k}^N(x), v_{2k}^N(x) \right)^T$. The components $v_{1k}^N(x), v_{2k}^N(x) \big|_{\Omega_i} \in C(\bar{\Omega}_i)$, $i = 1,2$, of $V_k^N(x) = \left(v_{1k}^N(x), v_{2k}^N(x) \right)^T$ are the complete polynomials of the power k that contain the variables $x_1, x_2, \ldots, x_n$ at every finite element $\bar{e}_i^j$ of the regular family [16], and $V_k^N \big|_\Gamma = 0$, $[v_{ik}^N] \big|_\gamma = 0$, $i = 1,2$. Then, the linear algebraic equation system like (1.45) is derived from equation (2.32). The solution to this system exists and such solution is unique.

Take the interpolation estimates [16] into account, assume that every component U_1 and U_2 of the solution U to problem (2.32) on Ω_l belongs to the Sobolev space $W_2^{k+1}(\Omega_l)$, $l = 1, 2$, and estimate like (1.48) follows from inequality like (1.47). Take this estimate and the embedding theorems into consideration, and the estimate

$$\left\| u - u_k^N \right\|_{L_2(\gamma)} \le c_1 h^k, \quad c_1 = \text{const},$$

takes place for the approximation $u_k^N(x) = -p_k^N(x)/\overline{a}(x)$ of the control $u = u(x)$.

1.3 BOUNDARY CONTROL OF A CORRECT SYSTEM DESCRIBED BY THE NEUMANN PROBLEM

Assume that elliptic equation (1.1), where the coefficients and right-hand side meet conditions (1.1'), is specified in the bounded, continuous and strictly Lipschitz domains Ω_1 and $\Omega_2 \in R^n$.

The heterogeneous boundary Neumann condition

$$\frac{\partial y}{\partial v_A} \equiv \sum_{i,j=1}^{n} k_{ij} \frac{\partial y}{\partial x_j} \cos(v, x_i) = g \tag{3.1}$$

is specified, in its turn, on the boundary Γ; in this case, v is an ort of an outer normal to Γ, $g \in L_2(\Gamma)$.

On the section γ of the domain Ω, the conjugation conditions have the form of expressions (1.3) and (1.4).

When $q \equiv 0$, the equality

$$\int_{\Omega} f \, d\Omega + \int_{\Gamma} g \, d\Gamma = 0 \tag{3.2}$$

is the necessary condition under which there exists the classical solution to boundary-value problem (1.1), (3.1), (1.3), (1.4).

Assume the following: $q_0 > 0$ (see point 1.1) and $\mathcal{U} = L_2(\Gamma)$. For every control $u \in \mathcal{U}$, determine a state $y = y(u)$ as a generalized solution to the boundary-value problem specified by equation (1.1), conjugation conditions (1.3) and (1.4) and the boundary condition

$$\frac{\partial y}{\partial v_A} = g + u. \tag{3.3}$$

Since there exists a generalized solution $y(u) \in V = \left\{ v|_{\Omega_i} \in W_2^1(\Omega_i) : \right.$ $\left. i = 1,2 \right\}$ to boundary-value problem (1.1), (1.3), (1.4), (3.3), then such solution is reasonable on Γ of $\bar{\Omega}$, and $\|y(u)\|_{L_2(\Gamma)} < \infty$.

Specify the observation in the form of expression (1.6), where $C \in \mathscr{L}(L_2(\Gamma); L_2(\Gamma))$, namely:

$$Cy(u) \equiv y(u), \quad x \in \Gamma.$$

Bring a value of the cost functional

$$J(u) = \int_\Gamma \left(y(u) - z_g \right)^2 d\Gamma + (\mathscr{N}u, u)_{\mathscr{U}} \tag{3.4}$$

in correspondence with every control $u \in \mathscr{U}$; in this case, z_g is a known element from $L_2(\Gamma)$; $\mathscr{N}u = \bar{a}(x)u$, $0 < a_0 \le \bar{a}(x) \le a_1 < \infty$, $(\varphi, \psi)_{\mathscr{U}} = \int_\Gamma \varphi\psi \, d\Gamma$.

It can be shown [21] that a unique state $y(u) \in V$ corresponds to every control $u \in \mathscr{U}$. The function y is specified on the domain $\bar{\Omega}_1 \cup \bar{\Omega}_2$, minimizes the energy functional

$$\Phi(v) = \int_\Omega \left(\sum_{i,j=1}^n k_{ij} \frac{\partial v}{\partial x_j} \frac{\partial v}{\partial x_i} + qv^2 \right) dx + \int_\gamma r[v]^2 d\gamma -$$

$$-2 \int_\Omega fv \, dx - 2 \int_\Gamma gv \, d\Gamma - 2 \int_\Gamma uv \, d\Gamma \tag{3.5}$$

on V, and it is the unique solution in V to the weakly stated problem: Find an element $y \in V$ that meets the following integral equation $\forall v \in V$:

$$\int_\Omega \left(\sum_{i,j=1}^n k_{ij} \frac{\partial y}{\partial x_j} \frac{\partial v}{\partial x_i} + qyv \right) dx + \int_\gamma r[y][v] d\gamma =$$

$$= \int_\Omega fv \, dx + \int_\Gamma gv \, d\Gamma + \int_\Gamma uv \, d\Gamma. \tag{3.6}$$

The state $y(u_1) \ne y(u_2)$ is easily seen under $u_1 \ne u_2$.

Take the assumptions as for the operator $\mathcal{N}$ into account, and

$$J(u) = \left\| \left(y(u) - y(0) \right) + \left(y(0) - z_g \right) \right\|_{L_2(\Gamma)}^2 + (\bar{a}u, u)_{L_2(\Gamma)} =$$

$$= \pi(u, u) - 2L(u) + \left\| z_g - y(0) \right\|_{L_2(\Gamma)}^2 \tag{3.7}$$

follows from expression (3.4); in this case, the bilinear form $\pi(\cdot, \cdot)$ and linear functional $L(\cdot)$ are expressed as

$$\pi(u, v) = \left(y(u) - y(0), \ y(v) - y(0) \right)_{L_2(\Gamma)} + (\bar{a}u, v)_{L_2(\Gamma)},$$

$$L(v) = \left(z_g - y(0), \ y(v) - y(0) \right)_{L_2(\Gamma)}, \tag{3.8}$$

$$(\varphi, \psi)_{L_2(\Gamma)} = \int_\Gamma \varphi\psi \, d\Gamma.$$

The linearity of the functional $L(v)$ is easily seen since the difference $y(v) - y(0)$ is the unique solution $\tilde{y}(v)$ to one of equivalent problems (3.5) and (3.6). It is necessary to assume $f \equiv 0$ and $g \equiv 0$ for them, and the arbitrary element $z \in V$ must be additionally substituted for the arbitrary function v in problem (3.6). Then, equality like (1.16) takes place that allows to state the linearity of the functional $L(v)$ and the bilinearity of the form $\pi(u, v)$. The form $\pi(\cdot, \cdot)$ is coercive on $L_2(\Gamma)$, i.e.:
$\pi(u, u) \ge a_0(u, u)_{L_2(\Gamma)}$.

Let $\tilde{y}' = \tilde{y}(u')$ and $\tilde{y}'' = \tilde{y}(u'')$ be solutions from V to problem (3.6) under $f = 0$ and $g = 0$ and under a function $u = u(x)$ that is equal, respectively, to u' and u''. Then, the inequality

$$\|\tilde{y}' - \tilde{y}''\|_V^2 \leq \mu\, a\left(\tilde{y}' - \tilde{y}'', \tilde{y}' - \tilde{y}''\right) \leq$$

$$\leq \mu\, \|u' - u''\|_{L_2(\Gamma)} \|\tilde{y}' - \tilde{y}''\|_{L_2(\Gamma)} \tag{3.9}$$

is derived, where $\|v\|_V$ is the norm introduced in point 1.1 and the bilinear form $a(\cdot, \cdot)$ is specified by expression (1.26).

Since the inequalities

$$\|v\|_{L_2(\partial\Omega_i)} \leq c_i \|v\|_{W_2^1(\Omega_i)}, \quad c_i > 0, \ i = 1,2, \tag{3.9'}$$

are true $\forall v \in V$ [55], then:

$$\|v\|_{L_2(\Gamma)} \leq c_3 \|v\|_V, \quad c_3 = \max_{i=1,2} c_i. \tag{3.10}$$

Take inequality (3.10) into account, and the inequality

$$\|\tilde{y}' - \tilde{y}''\|_{L_2(\Gamma)} \leq c_4 \|u' - u''\|_{L_2(\Gamma)} \tag{3.11}$$

follows from inequality (3.9), i.e. the function $\tilde{y}(u)$ is continuously dependent on u.

Inequality (3.11) provides the continuity of the linear functional $L(\cdot)$ and bilinear form $\pi(\cdot, \cdot)$ on $\mathcal{U}$.

On the basis of [58, Chapter 1, Theorem 1.1], the validity of the following statement is proved.

Theorem 3.1. *Let conditions (1.1') be met, and a system state is determined as a solution to equivalent problems (3.5) and (3.6). Then, there exists a unique element u of a convex set $\mathcal{U}_\partial$ that is closed in $\mathcal{U}$, and relation like (1.17) takes place for u.*

If $u \in \mathcal{U}_\partial$ is the optimal control, then

$$\pi(u, v - u) \geq L(v - u), \quad \forall v \in \mathcal{U}_\partial. \tag{3.12}$$

Proceed from expressions (3.8), and it is easy to see that the equality

$$\pi(u,v-u)-L(v-u)=$$

$$=\left(y(u)-z_g,\ y(v-u)-y(0)\right)_{L_2(\Gamma)}+(\bar{a}u,v-u)_{L_2(\Gamma)} \qquad (3.13)$$

is true. Consider the linearity of problem (3.6), and the equality

$$\pi(u,v-u)-L(v-u)=$$

$$=\left(y(u)-z_g,\ y(v)-y(u)\right)_{L_2(\Gamma)}+(\bar{a}u,v-u)_{L_2(\Gamma)}$$

follows from equality (3.13). Then, inequality (3.12) has the form

$$\left(y(u)-z_g,\ y(v)-y(u)\right)_{L_2(\Gamma)}+(\bar{a}u,v-u)_{L_2(\Gamma)}\geq 0,\ \forall v\in\mathscr{U}_{\partial},\quad (3.14)$$

and it is the necessary and sufficient condition under which $u\in\mathscr{U}_{\partial}$ is the optimal control for the considered problem.

Since the solution $y\in V$ to equivalent problems (3.5) and (3.6) exists and such solution is unique under arbitrary fixed $f\in L_2(\Omega)$ and $g\in L_2(\Gamma)$, then there is the operator $A:V\to L_2(\Omega)$ specified by relations (1.1), (1.3), (1.4) and (3.3) on the solutions y $\left(y|_{\Omega_l}\in C^1(\bar{\Omega}_l)\cap C^2(\Omega_l),\ l=1,2\right)$. Therefore, $\partial y/\partial v_A$ can be uniquely calculated on $\partial\Omega_l$ [58] for the solution y, where

$$\frac{\partial y}{\partial v_A}\equiv\sum_{i,j=1}^{n}k_{ij}\frac{\partial y}{\partial x_j}\cos(v,x_i) \qquad (3.14')$$

and v is an ort of an outer normal to $\partial\Omega_l$, $l=1,2$.

As for the control $v\in\mathscr{U}$, the conjugate state $p(v)\in V^*$ is specified by the relations

$$A^*p(v)=0,\ x\in\Omega, \qquad (3.15)$$

$$\frac{\partial p}{\partial v_{A^*}}=y(v)-z_g,\ x\in\Gamma, \qquad (3.16)$$

$$\left[\frac{\partial p}{\partial v_{A^*}}\right] = 0, \ x \in \gamma, \tag{3.17}$$

and

$$\left\{\frac{\partial p}{\partial v_{A^*}}\right\}^{\pm} = r[p], \ x \in \gamma, \tag{3.18}$$

where V^* is a space conjugate to V, $V^* = V$, and the operators A^* and $\dfrac{\partial}{\partial v_{A^*}}$ are specified, in their turn, by expressions (2.19′). Further on, use the Green formula [58], and the equality

$$0 = \left(A^* p(u), y(v) - y(u)\right) = -\sum_{l=1}^{2} \int_{\partial \Omega_l} \frac{\partial p}{\partial v_{A^*}}(y(v) - y(u)) d\partial \Omega_l +$$

$$+ \int_{\Omega} \left(\sum_{i,j=1}^{n} k_{ij} \frac{\partial p}{\partial x_j} \frac{\partial(y(v) - y(u))}{\partial x_i} + q\, p(y(v) - y(u))\right) dx =$$

$$= a(p, y(v) - y(u)) - \int_{\Gamma}\left(y(u) - z_g\right)(y(v) - y(u)) d\Gamma =$$

$$= -\int_{\Gamma}\left(y(u) - z_g\right)(y(v) - y(u)) d\Gamma + \int_{\Gamma} p \frac{\partial(y(v) - y(u))}{\partial v_A} d\Gamma +$$

$$+ \int_{\gamma} r[p][y(v) - y(u)] d\gamma - \int_{\gamma} r[p][y(v) - y(u)] d\gamma +$$

$$+ (p, A(y(v) - y(u))) = -\int_{\Gamma}\left(y(u) - z_g\right)(y(v) - y(u)) d\Gamma + \int_{\Gamma} p(v - u) d\Gamma$$

is obtained. Therefore:

$$\left(y(u) - z_g, \ y(v) - y(u)\right)_{L_2(\Gamma)} = (p, v - u)_{L_2(\Gamma)}. \tag{3.19}$$

Take equalities (3.13) and (3.19) into account, and the inequality

$$\left(p(u)+\bar{a}u,v-u\right)_{L_2(\Gamma)}\geq 0 \tag{3.20}$$

is derived from inequality (3.12).

Therefore, if the constraints are absent, i.e. when $\mathcal{U}_\partial=\mathcal{U}$, then the equality

$$p(u)+\bar{a}u=0,\ x\in\Gamma, \tag{3.21}$$

follows from condition (3.20).

To find the optimal control $u(x)$, solve the differential problem

$$-\sum_{i,j=1}^{n}\frac{\partial}{\partial x_i}\left(k_{ij}\frac{\partial y}{\partial x_j}\right)+qy=f,\ x\in\Omega_1\bigcup\Omega_2, \tag{3.22}$$

$$-\sum_{i,j=1}^{n}\frac{\partial}{\partial x_i}\left(k_{ij}\frac{\partial p}{\partial x_j}\right)+qp=0,\ x\in\Omega_1\bigcup\Omega_2, \tag{3.23}$$

$$\frac{\partial y}{\partial\nu_A}=g-\frac{p}{a},\ x\in\Gamma, \tag{3.24}$$

$$\frac{\partial p}{\partial\nu_{A^*}}=y-z_g,\ x\in\Gamma, \tag{3.25}$$

$$\left[\sum_{i,j=1}^{n}k_{ij}\frac{\partial y}{\partial x_j}\cos(\nu,x_i)\right]=0,\ x\in\gamma, \tag{3.26}$$

$$\left\{\sum_{i,j=1}^{n}k_{ij}\frac{\partial y}{\partial x_j}\cos(\nu,x_i)\right\}^{\pm}=r[y],\ x\in\gamma, \tag{3.27}$$

$$\left[\sum_{i,j=1}^{n}k_{ij}\frac{\partial p}{\partial x_j}\cos(\nu,x_i)\right]=0,\ x\in\gamma, \tag{3.28}$$

$$\left\{\sum_{i,j=1}^{n}k_{ij}\frac{\partial p}{\partial x_j}\cos(\nu,x_i)\right\}^{\pm}=r[p],\ x\in\gamma, \tag{3.29}$$

and the optimal control is

$$u = -p/\bar{a}, \ x \in \Gamma. \tag{3.30}$$

Definition 3.1. A generalized (weak) solution to boundary-value problem (3.22)–(3.29) is called a vector-function $(y, p)^{\mathrm{T}} \in H = \left\{ v = (v_1, v_2)^{\mathrm{T}} : \ v_i \big|_{\Omega_l} \in W_2^1(\Omega_l), \ i, l = 1, 2 \right\}$ that satisfies the following integral equation $\forall z \in H$:

$$\int_{\Omega} \left\{ \sum_{i,j=1}^{n} k_{ij} \frac{\partial y}{\partial x_j} \frac{\partial z_1}{\partial x_i} + q y z_1 + \sum_{i,j=1}^{n} k_{ij} \frac{\partial p}{\partial x_j} \frac{\partial z_2}{\partial x_i} + q p z_2 \right\} dx +$$

$$+ \int_{\gamma} r[y][z_1] d\gamma + \int_{\gamma} r[p][z_2] d\gamma + \int_{\Gamma} p z_1 / \bar{a} \, d\Gamma -$$

$$- \int_{\Gamma} y z_2 \, d\Gamma = \int_{\Omega} f z_1 \, dx + \int_{\Gamma} g z_1 \, d\Gamma - \int_{\Gamma} z_g z_2 \, d\Gamma . \tag{3.31}$$

Let $u = (u_1, u_2)^{\mathrm{T}}$ and $v = (v_1, v_2)^{\mathrm{T}}$ be arbitrary elements of the complete Hilbert space H with the previously introduced norm $\|\cdot\|_H$. Specify the bilinear form

$$a(u, v) = \int_{\Omega} \left\{ \sum_{i,j=1}^{n} k_{ij} \frac{\partial u_1}{\partial x_j} \frac{\partial v_1}{\partial x_i} + \sum_{i,j=1}^{n} k_{ij} \frac{\partial u_2}{\partial x_j} \frac{\partial v_2}{\partial x_i} + q \left(u_1 v_1 + u_2 v_2 \right) \right\} dx +$$

$$+ \int_{\gamma} r \left([u_1][v_1] + [u_2][v_2] \right) d\gamma + \int_{\Gamma} \left(u_2 v_1 / \bar{a} - u_1 v_2 \right) d\Gamma \tag{3.32}$$

and linear functional

$$l(v) = \int_{\Omega} f v_1 \, dx + \int_{\Gamma} g v_1 \, d\Gamma - \int_{\Gamma} z_g v_2 \, d\Gamma \tag{3.32'}$$

on H.

If the constraint

$$\alpha_1 \equiv \min\{\alpha_0, q_0\} - c_3^2 \max(1/a_0, 1) > 0,$$

where α_0 is the ellipticity condition constant, $c_3 = \max(c_1, c_2)$ and c_i is the constant from the inequalities

$$\|\varphi\|_{L_2(\partial\Omega_i)} \le c_i \|\varphi\|_{W_2^1(\Omega_i)}, \quad i = 1, 2, \tag{3.33}$$

are met, then $\forall u \in H$ the inequalities

$$a(u, u) \ge \bar{\alpha}_1 \|u\|_H^2 \text{ and } |a(u, v)| \le c_4 \|u\|_H \|v\|_H, \quad \bar{\alpha}_1, c_4 = \text{const} > 0,$$

follow from expression (3.32), i.e. the bilinear form $a(\cdot, \cdot)$ is coercive and continuous on H.

It is easy to see the following:

$$|l(v)| = \left| \int_\Omega f v_1 \, dx + \int_\Gamma g v_1 \, d\Gamma - \int_\Gamma z_g v_2 \, d\Gamma \right| \le c_5 \|v\|_H;$$

i.e. the linear functional $l(v)$ is continuous on H.

Use the Lax-Milgramm lemma, and it is concluded that the unique solution $U = (u_1, u_2)^T$ to problem (3.31) exists in H. Problem (3.31) can be solved approximately by means of the finite-element method. Then, linear algebraic equation system like (1.45) is derived from equation (3.31). The solution $\bar{U}$ to this system exists and such solution is unique. The vector $\bar{U}$ specifies the unique approximate solution $U_k^N \in H_k^N$ to problem (3.31) as the unique one to equation like (1.46), where the bilinear form $a(\cdot, \cdot)$ and the linear functional $l(\cdot)$ are specified, in their turn, respectively, by expressions (3.32) and (3.32'). In this case, H_k^N is the space of the vector-functions $V_k^N(x) = \left(v_{1k}^N(x), v_{2k}^N(x)\right)^T$. The components $v_{1k}^N, v_{2k}^N\big|_{\bar{\Omega}_i} \in C(\bar{\Omega}_i), \quad i = 1, 2, \text{ of } V_k^N(x) = \left(v_{1k}^N(x), v_{2k}^N(x)\right)^T$ are the complete polynomials of the power k that contain the variables $x_1, x_2, ..., x_n$ at every finite element $\bar{e}_i^j$ of partitioning of the domain $\bar{\Omega}$.

The control $u = u(x)$ is specified on the boundary Γ and such control is equal to $-p/\bar{a}$, where p is the trace of the function $p = p(x) \in H$ on Γ. Therefore, the function $u(x)$ has the extension $-p/\bar{a}$ to the domain $\bar{\Omega}_1 \bigcup \bar{\Omega}_2$. Then, the estimate

$$\left\| u - \tilde{u}_k^N \right\|_{L_2(\Gamma)}^2 \leq \frac{\bar{c}_i}{a_0^2} \left\| p - p_k^N \right\|_{W_2^1}^2 \leq c_3 \, h^{2k}$$

can be written, where $\tilde{u}_k^N = -p_k^N/\bar{a}$, $\bar{c}_i = \max\limits_{l=1,2}\{c_l\}$, $c_3 = \text{const} > 0$, and h is the largest diameter for all the finite elements $\bar{e}_i^j$ of the regular family.

1.4 DISTRIBUTED CONTROL OF A SYSTEM: A COMPLICATED THIN INCLUSION CASE

Assume that the elliptic equation

$$-\sum_{i,j=1}^{n} \frac{\partial}{\partial x_i}\left(k_{ij}(x) \frac{\partial y}{\partial x_j} \right) + q(x)y = f(x), \tag{4.1}$$

where the coefficients and right-hand side meet conditions (1.1'), is specified in the bounded, continuous and strictly Lipschitz domains Ω_1 and $\Omega_2 \in R^n$.

The homogeneous boundary Dirichlet condition

$$y = 0 \tag{4.2}$$

is specified, in its turn, on the boundary $\Gamma = (\partial\Omega_1 \bigcup \partial\Omega_2) \setminus \gamma$ $(\gamma = \partial\Omega_1 \bigcap \partial\Omega_2 \neq \varnothing)$.

On the section γ of the domain $\bar{\Omega} = \bar{\Omega}_1 \cup \bar{\Omega}_2$ $(\Omega_1 \cap \Omega_2 = \varnothing)$, the conjugation conditions for an imperfect contact are

$$R_1 \left\{ \sum_{i,j=1}^n k_{ij} \frac{\partial y}{\partial x_j} \cos(v, x_i) \right\}^- +$$

$$+ R_2 \left\{ \sum_{i,j=1}^n k_{ij} \frac{\partial y}{\partial x_j} \cos(v, x_i) \right\}^+ = [y] + \delta \qquad (4.3)$$

and

$$\left[\sum_{i,j=1}^n k_{ij} \frac{\partial y}{\partial x_j} \cos(v, x_i) \right] = \omega, \qquad (4.4)$$

where $R_1, R_2, \omega, \delta \in C(\gamma)$, $R_1, R_2 \geq 0$, $R_1 + R_2 \geq R_0 > 0$, $R_0 = \text{const}$, v is an ort of a normal to γ and such normal is directed into the domain Ω_2.

Let there be the control Hilbert space $\mathcal{U}$ and mapping $B \in \mathcal{L}(\mathcal{U}; V')$, where V' is the space dual with respect to the state Hilbert space V. Assume the following: $\mathcal{U} = L_2(\Omega)$.

For every control $u \in \mathcal{U}$, determine a system state y as a generalized solution to the boundary-value problem specified by the equation

$$-\sum_{i,j=1}^n \frac{\partial}{\partial x_i} \left(k_{ij}(x) \frac{\partial y}{\partial x_j} \right) + q(x)y = f(x) + Bu, \quad y \in V, \qquad (4.5)$$

and by conditions (4.2)–(4.4).

Specify the observation

$$Z(u) = Cy(u), \qquad (4.6)$$

where $C \in \mathcal{L}(V; \mathcal{H})$ and $\mathcal{H}$ is some Hilbert space. Assume the following:

$$Cy(u) \equiv y(u), \quad \mathcal{H} = V \subset L_2(\Omega). \qquad (4.7)$$

Bring a value of the cost functional

$$J(u) = \left\| Cy(u) - z_g \right\|_{\mathcal{H}}^2 + (\mathcal{N} u, u)_{\mathcal{U}} \qquad (4.8)$$

in correspondence with every control $u \in \mathcal{U}$; in this case, z_g is a known element of the space $\mathcal{H}$, and

$$\mathcal{N} \in \mathcal{L}(\mathcal{U}; \mathcal{U}), \ (\mathcal{N}u, u)_{\mathcal{U}} \geq v_0 \|u\|_{\mathcal{U}}^2, \ v_0 = \text{const} > 0, \ \forall u \in \mathcal{U}. \quad (4.9)$$

Assume the following: $f \in L_2(\Omega)$, $Bu \equiv u \in L_2(\Omega)$, $\mathcal{N}u = \bar{a}(x)u$, $0 < a_0 \leq \bar{a}(x) \leq a_1 < \infty$, $\bar{a}(x)|_{\Omega_l} \in C(\Omega_l)$, $l = 1, 2$, $a_0, a_1 = \text{const}$, $(\varphi, \psi)_{\mathcal{U}} =$

$= (\varphi, \psi) = \int_{\Omega} \varphi \psi dx$. Then, a unique state, namely, a function $y(u) \in V =$

$= \{ v|_{\Omega_l} \in W_2^1(\Omega_l) : l = 1, 2; \ v|_\Gamma = 0 \}$ corresponds to every control $u \in \mathcal{U}$, delivers the minimum to the energy functional [21]

$$\Phi(v) = \int_{\Omega} \left(\sum_{i,j=1}^{n} k_{ij} \frac{\partial v}{\partial x_j} \frac{\partial v}{\partial x_i} + q v^2 \right) dx + \int_{\gamma} \frac{[v]^2}{R_1 + R_2} d\gamma - 2(f, v) -$$

$$-2(u, v) - 2 \int_{\gamma} \frac{R_2 \omega - \delta}{R_1 + R_2} [v] d\gamma + 2 \int_{\gamma} \omega v^+ d\gamma \quad (4.10)$$

on V, and it is the unique solution in V to the weakly stated problem: Find an element $y = V$ that meets the equation

$$\int_{\Omega} \left(\sum_{i,j=1}^{n} k_{ij} \frac{\partial y}{\partial x_j} \frac{\partial v}{\partial x_i} + q y v \right) dx + \int_{\gamma} \frac{[y][v]}{R_1 + R_2} d\gamma =$$

$$= (f, v) + (u, v) + \int_{\gamma} \frac{R_2 \omega - \delta}{R_1 + R_2} [v] d\gamma - \int_{\gamma} \omega v^+ d\gamma, \ \forall v \in V. \quad (4.11)$$

Therefore, there exists such an operator A acting from V into L_2, that

$$y(u) = A^{-1}(f + Bu), \ \forall u \in L_2, \quad (4.12)$$

where $L_2 = L_2(\Omega)$.

It is easy to see that $y(u_1) \neq y(u_2)$ under $u_1 \neq u_2$ ($Bu_1 \neq Bu_2$) because the operator A is linear, and the non-zero solution $\tilde{y}$ corresponds to problem (4.1)–(4.4) with the right-hand side $\tilde{f} = u_2 - u_1 \neq 0$ under $\omega = 0$, and $y(u_2) = y(u_1) + \tilde{y}$.

Remark. When a state $y(u)$ is determined as a solution to one of equivalent problems (4.10) and (4.11) with respect to the coefficients k_{ij} in equation (4.1), it is enough to follow ellipticity condition (1.1') and the constraint $k_{ij} \in L_\infty(\Omega)$.

Take the aforesaid assumptions into consideration, and the cost functional may be rewritten as

$$J(u) = \|y(u) - z_g\|^2 + (\bar{a}u, u), \qquad (4.13)$$

where $\|v\| = (v, v)^{1/2}$, z_g may be, in its turn, an arbitrary fixed element of the Hilbert space $L_2(\Omega)$, and

$$J(u) = \pi(u, u) - 2L(u) + \|z_g - y(0)\|^2 \qquad (4.14)$$

follows from expression (4.13); in this case, the bilinear form $\pi(\cdot, \cdot)$ and linear functional $L(\cdot)$ are expressed as

$$\pi(u, v) = \big(y(u) - y(0), y(v) - y(0)\big) + (\bar{a}u, v),$$
$$L(v) = \big(z_g - y(0), y(v) - y(0)\big). \qquad (4.15)$$

The linearity of the functional $L(\cdot)$ is easily seen from the fact that the difference $y(v) - y(0)$ is the unique solution $\tilde{y}(v)$ to one of equivalent problems (4.10) and (4.11). It is necessary to assume $f = 0$, $\delta = 0$ and $\omega = 0$ for them, and the arbitrary element $z \in V$ must be additionally substituted for the arbitrary function v in problem (4.11). Then:

$$\tilde{y}\big(\alpha_1 u_1 + \alpha_2 u_2\big) = \alpha_1 \tilde{y}(u_1) + \alpha_2 \tilde{y}(u_2), \quad \forall \alpha_1, \alpha_2 \in R^1, \forall u_1, u_2 \in \mathcal{U}. \quad (4.16)$$

Pursuant to equality (4.16), the linearity of the functional $L(v)$ and the bilinearity of the form $\pi(u,v)$ are stated. The form $\pi(\cdot,\cdot)$ is coercive on $\mathcal{U}$, i.e.: $\pi(u,u) = \left(y(u) - y(0), y(u) - y(0)\right) + (\bar{a}u, u) \geq a_0(u,u)$.

Let $\tilde{y}' = \tilde{y}(u')$ and $\tilde{y}'' = \tilde{y}(u'')$ be solutions from V to problem (4.11) under $f = 0$, $\delta = 0$ and $\omega = 0$ and under a function $u = u(x)$ that is equal, respectively, to u' and u''. Then, the inequality

$$\left\|\tilde{y}' - \tilde{y}''\right\|^2 \leq \left\|\tilde{y}' - \tilde{y}''\right\|_V^2 \leq \mu a\left(\tilde{y}' - \tilde{y}'', \tilde{y}' - \tilde{y}''\right) \leq$$

$$\leq \mu \left\|u' - u''\right\| \cdot \left\|\tilde{y}' - \tilde{y}''\right\|, \quad \mu = \text{const} > 0, \tag{4.16'}$$

is derived that provides the continuity of the linear functional $L(\cdot)$ and bilinear form $\pi(\cdot,\cdot)$ on $\mathcal{U}$; in this case,

$$\|v\|_V = \left\{ \sum_{i=1}^{2} \|v\|_{W_2^1(\Omega_i)}^2 \right\}^{1/2}$$

and

$$a(\varphi, \psi) = \int_{\Omega} \left(\sum_{i,j=1}^{n} k_{ij} \frac{\partial \varphi}{\partial x_j} \frac{\partial \psi}{\partial x_i} + q\varphi\psi \right) dx + \int_{\gamma} \frac{[\varphi][\psi]}{R_1 + R_2} d\gamma.$$

On the basis of [58, Chapter 1, Theorem 1.1], the validity of the following statement is proved.

Theorem 4.1. *Let conditions (1.1') be met, and a system state is determined as a solution to equivalent problems (4.10) and (4.11). Then, there exists a unique element u of a convex set $\mathcal{U}_\partial$ that is closed in $\mathcal{U}$, and*

$$J(u) = \inf_{v \in \mathcal{U}_\partial} J(v) \tag{4.17}$$

takes place for u.

If $u \in \mathcal{U}_\partial$ is the optimal control, then

$$\langle J'(u), v - u \rangle \geq 0. \tag{4.18}$$

Take expression (4.14) and the inequality

$$\frac{J\left(u+\theta(v-u)\right)-J(u)}{\theta}\geq 0$$

into consideration, and relation (4.18) takes the form

$$\pi(u,v-u)\geq L(v-u),\ \forall v\in \mathcal{U}_{\partial}. \tag{4.19}$$

Proceed from expressions (4.15), and

$$\pi(u,v-u)-L(v-u)=\left(y(u)-z_g,y(v)-y(u)\right)+(\bar{a}u,v-u). \tag{4.20}$$

Then, inequality (4.19) has the form

$$\left(y(u)-z_g,y(v)-y(u)\right)+(\bar{a}u,v-u)\geq 0,\ \forall v\in \mathcal{U}_{\partial}, \tag{4.21}$$

and it is the necessary and sufficient condition under which $u\in \mathcal{U}_{\partial}$ is the optimal control for the considered problem.

As for the control $v\in \mathcal{U}$, the conjugate state $p(v)\in V^*$ is specified by the relations

$$A^*p(v)=y(v)-z_g, \tag{4.22}$$

$$p=0,\ x\in \Gamma, \tag{4.23}$$

$$\left[\frac{\partial p}{\partial v_{A^*}}\right]=0,\ x\in \gamma, \tag{4.24}$$

and

$$\left\{\frac{\partial p}{\partial v_{A^*}}\right\}^{\pm}=\frac{1}{R_1+R_2}[p],\ x\in \gamma, \tag{4.25}$$

where V^* is a space conjugate to V, $V^*=V$, and

$$A^*p=-\sum_{i,j=1}^{n}\frac{\partial}{\partial x_i}\left(k_{ij}\frac{\partial p}{\partial x_j}\right)+qp,\quad \frac{\partial p}{\partial v_{A^*}}=\sum_{i,j=1}^{n}k_{ij}\frac{\partial p}{\partial x_j}\cos(v,x_i). \tag{4.26}$$

Further on, use the Green formula [58], and the equality

$$\left(A^*p(u), y(v) - y(u)\right) = \left(y(u) - z_g, y(v) - y(u)\right) =$$

$$= a\left(p, y(v) - y(u)\right) = \left(p(u), A\left(y(v) - y(u)\right)\right) = \left(p(u), v - u\right) \quad (4.27)$$

is obtained, where

$$a(u,v) = \int_{\Omega}\left(\sum_{i,j=1}^{n} k_{ij}\frac{\partial u}{\partial x_j}\frac{\partial v}{\partial x_i} + quv\right)dx + \int_{\gamma}\frac{[u][v]}{R_1 + R_2}d\gamma. \quad (4.28)$$

Consider equality (4.27), and it is stated that inequality (4.21) is equivalent to the inequality

$$\left(p(u) + \bar{a}u, v - u\right) \geq 0, \quad \forall v \in \mathcal{U}_\partial. \quad (4.29)$$

Therefore, the necessary and sufficient condition for the existence of the optimal control $u \in \mathcal{U}_\partial$ is the one under which the relations

$$Ay(u) = f + u, \quad (4.30)$$

$$A^*p(u) = y(u) - z_g \quad (4.31)$$

and

$$\left(p(u) + \bar{a}u, v - u\right) \geq 0, \quad \forall v \in \mathcal{U}_\partial, \quad (4.32)$$

are met.

If the constraints are absent, i.e. when $\mathcal{U}_\partial = \mathcal{U}$, then the equality

$$p(u) + \bar{a}u = 0 \quad (4.33)$$

follows from condition (4.32). Therefore, when the constraints are absent, the control $u(x)$ can be excluded from equality (4.30) by means of equality (4.33). On the basis of equalities (4.30) and (4.31), the problem

$$Ay + p/\bar{a} = f, \quad y \in V, \quad (4.34)$$

$$A^*p - y = -z_g, \quad p \in V^*, \quad (4.35)$$

is derived, where $V^* = \left\{ v|_{\Omega_l} \in W_2^1(\Omega_l) : l = 1,2; \ v|_\Gamma = 0 \right\}$, and the vector

solution $(y, p)^T$ is found from this problem along with the optimal control

$$u = -p/\bar{a} . \tag{4.36}$$

If the vector solution $(y, p)^T$ to problem (4.34), (4.35) is smooth enough

on $\bar{\Omega}_l$, viz. $y|_{\bar{\Omega}_l}, \ p|_{\bar{\Omega}_l} \in C^1(\bar{\Omega}_l) \cap C^2(\Omega_l)$, $l = 1,2$, then the differential

problem of finding the vector-function $(y, p)^T$, that satisfies the relations

$$-\sum_{i,j=1}^{n} \frac{\partial}{\partial x_i} \left(k_{ij} \frac{\partial y}{\partial x_j} \right) + qy + p/\bar{a} = f, \ x \in \Omega_1 \cup \Omega_2, \tag{4.37}$$

$$-\sum_{i,j=1}^{n} \frac{\partial}{\partial x_i} \left(k_{ij} \frac{\partial p}{\partial x_j} \right) + qp - y = -z_g, \ x \in \Omega_1 \cup \Omega_2, \tag{4.38}$$

$$y|_\Gamma = 0, \tag{4.39}$$

$$p|_\Gamma = 0, \tag{4.40}$$

$$R_1 \left\{ \frac{\partial y}{\partial v_A} \right\}^- + R_2 \left\{ \frac{\partial y}{\partial v_A} \right\}^+ = [y] + \delta, \ x \in \gamma, \tag{4.41}$$

$$\left[\frac{\partial y}{\partial v_A} \right] = \omega, \ x \in \gamma, \tag{4.42}$$

$$\left[\frac{\partial p}{\partial v_{A^*}} \right] = 0, \ x \in \gamma, \tag{4.43}$$

and

$$\left\{ \frac{\partial p}{\partial v_{A^*}} \right\}^\pm = \frac{[p]}{R_1 + R_2}, \ x \in \gamma, \tag{4.44}$$

where

$$\frac{\partial y}{\partial \nu_A} = \sum_{i,j=1}^{n} k_{ij} \frac{\partial y}{\partial x_j} \cos(\nu, x_i),$$

corresponds to problem (4.34), (4.35).

Definition 4.1. A generalized (weak) solution to boundary-value problem (4.37)–(4.44) is called a vector-function $(y, p)^T \in H = \left\{ v = (v_1, v_2)^T : v_i\big|_{\Omega_l} \in W_2^1(\Omega_l),\ i, l = 1, 2;\ v\big|_\Gamma = 0 \right\}$ that satisfies the following integral equation $\forall z \in H$:

$$\int_\Omega \left\{ \sum_{i,j=1}^{n} k_{ij} \frac{\partial y}{\partial x_j} \frac{\partial z_1}{\partial x_i} + qyz_1 + pz_1/\bar{a} + \sum_{i,j=1}^{n} k_{ij} \frac{\partial p}{\partial x_j} \frac{\partial z_2}{\partial x_i} + qpz_2 - yz_2 \right\} dx +$$

$$+ \int_\gamma \frac{[y][z_1]}{R_1 + R_2} d\gamma + \int_\gamma \frac{[p][z_2]}{R_1 + R_2} d\gamma = \int_\Omega \left(fz_1 - z_g z_2 \right) dx +$$

$$+ \int_\gamma \frac{R_2 \omega - \delta}{R_1 + R_2} [z_1] d\gamma - \int_\gamma \omega z_1^+ d\gamma. \tag{4.45}$$

Let $u = (u_1, u_2)^T$ and $v = (v_1, v_2)^T$ be arbitrary elements of the complete Hilbert space H with the norm $\|v\|_H = \left\{ \sum_{i=1}^{2} \|v\|_{W_2^1(\Omega_i)}^2 \right\}^{1/2}$. Specify the bilinear form

$$a(u, v) = \int_\Omega \left\{ \sum_{i,j=1}^{n} k_{ij} \frac{\partial u_1}{\partial x_j} \frac{\partial v_1}{\partial x_i} + qu_1 v_1 + u_2 v_1/\bar{a} + \right.$$

$$+ \left. \sum_{i,j=1}^{n} k_{ij} \frac{\partial u_2}{\partial x_j} \frac{\partial v_2}{\partial x_i} + qu_2 v_2 - u_1 v_2 \right\} dx +$$

$$+ \int\limits_{\gamma} \frac{[u_1][v_1]}{R_1 + R_2} d\gamma + \int\limits_{\gamma} \frac{[u_2][v_2]}{R_1 + R_2} d\gamma \qquad (4.46)$$

and linear functional

$$l(v) = \int\limits_{\Omega} \left(f v_1 - z_g v_2 \right) dx + \int\limits_{\gamma} \frac{R_2 \omega - \delta}{R_1 + R_2} [v_1] d\gamma - \int\limits_{\gamma} \omega v_1^+ d\gamma \qquad (4.47)$$

on H.

Let the constraint

$$\alpha_1 = \alpha_0 \mu + 2 q_0 - \left(\frac{1}{a_0} + 1 \right) > 0 \qquad (4.48)$$

be met, where $\mu = \mathrm{const} > 0$ is the constant in Friedrichs inequality

$$\int\limits_{\Omega} \sum_{i=1}^{n} \left(\frac{\partial v}{\partial x_i} \right)^2 dx \geq \mu \int\limits_{\Omega} v^2 dx, \ \forall v \in V.$$

Proceed from constraints $(1.1')$ and (4.48), the Cauchy-Bunyakovsky and Friedrichs inequalities and embedding theorems [55], and the inequalities

$$a(v,v) \geq \bar{\alpha}_1 \|v\|_H^2, \ \forall v \in H, \ \bar{\alpha}_1 = \mathrm{const} > 0,$$

and

$$|a(u,v)| \leq c_1 \|u\|_H \|v\|_H, \ \forall u, v \in H, \ c_1 = \mathrm{const} > 0,$$

are true for the bilinear form $a(\cdot,\cdot)$, i.e. this form is H-elliptic and continuous [49] on H.

Consider the Cauchy-Bunyakovsky inequality and embedding theorems, and the following inequality is obtained $\forall v \in H$:

$$|l(v)| \leq c_2 \|v\|_H, \ c_2 = \mathrm{const}.$$

Use the Lax-Milgramm lemma [16], and it is concluded that the unique solution (y, p) to problem (4.45) exists in H.

Problem (4.45) can be solved approximately by means of the finite-element method. For this purpose, divide the domains $\bar{\Omega}_i$ into N_i finite elements $\bar{e}_i^j$ $(j = \overline{1, N_i},\ i = 1, 2)$ of the regular family [16]. Specify the subspace $H_k^N \subset H$ $(N = N_1 + N_2)$ of the vector-functions $V_k^N(x)$. The components $v_{1k}^N, v_{2k}^N\big|_{\bar{\Omega}_i} \in C(\bar{\Omega}_i)$ $(i = 1, 2)$ of $V_k^N(x)$ are the complete polynomials of the power k that contain the variables $x_1, x_2, \ldots, x_n$ at every $\bar{e}_i^j$, and $V_k^N\big|_\Gamma = 0$. Then, the linear algebraic equation system

$$A\bar{U} = B \qquad (4.49)$$

follows from equation (4.45), and the solution $\bar{U}$ to system (4.49) exists and such solution is unique. The vector $\bar{U}$ specifies the unique approximate solution $U_k^N \in H_k^N$ to problem (4.45) as the unique one to the equation

$$a\left(U_k^N, V_k^N\right) = l\left(V_k^N\right), \ \forall V_k^N \in H_k^N. \qquad (4.50)$$

Let $U = U(x) \in H$ be the solution to problem (4.45). Then:

$$a\left(U - U_k^N, V_k^N\right) = 0, \ \forall V_k^N \in H_k^N.$$

Therefore,

$$\bar{\alpha}_1 \left\| U - U_k^N \right\|_H^2 \le a\left(U - U_k^N, U - U_k^N\right) = a\left(U - U_k^N, U - \tilde{U} + \tilde{U} - U_k^N\right) =$$

$$= a\left(U - U_k^N, U - \tilde{U}\right), \ \forall \tilde{U} \in H_k^N, \qquad (4.51)$$

and the inequality

$$\left\| U - U_k^N \right\|_H \le \frac{c_1}{\alpha_1} \left\| U - \tilde{U} \right\|_H \qquad (4.52)$$

is thus derived since the bilinear form $a(\cdot, \cdot)$ is continuous on H.

Suppose that $\tilde{U} \in H_k^N$ is a complete interpolation polynomial for the solution U at every $\bar{e}_i^j$. Take the interpolation estimates [16] into account, assume that every component U_1 and U_2 of the solution U on Ω_l belongs to the Sobolev space $W_2^{k+1}(\Omega_l)$ $(l = 1, 2)$, and the estimate

$$\left\| U - U_k^N \right\|_H \le c h^k, \qquad (4.53)$$

where h is a maximum diameter of all the finite elements $\bar{e}_i^j$, $c = \text{const}$, follows from inequality (4.52).

Take estimate (4.53) into consideration, and the estimate

$$\left\| u - u_k^N \right\|_{W_2^1} \le c_2 \left\| p - p_k^N \right\|_{W_2^1} \le c_3 h^k,$$

where $\|\cdot\|_{W_2^1} = \left\{ \sum_{i=1}^{2} \|\cdot\|_{W_2^1(\Omega_i)}^2 \right\}^{1/2}$, takes place for the approximation $u_k^N(x) = -p_k^N(x) / \bar{a}(x)$ of the control $u = u(x)$.

1.5 CONTROL UNDER CONJUGATION CONDITION: A COMPLICATED THIN INCLUSION CASE

Assume that elliptic equation (4.1), where the coefficients and right-hand side meet the conditions of point 1.1, is specified in the bounded, continuous and strictly Lipschitz domains Ω_1 and $\Omega_2 \in R^n$.

Homogeneous Dirichlet condition (4.2) is specified, in its turn, on the boundary, and the conjugation conditions have the form of expressions (4.3) and (4.4).

For every control $u \in \mathcal{U} = L_2(\gamma)$, determine a system state as a generalized solution to the boundary-value problem specified by equation (4.1), condition (4.2) and the heterogeneous conjugation conditions

$$R_1 \left\{ \frac{\partial y}{\partial v_A} \right\}^- + R_2 \left\{ \frac{\partial y}{\partial v_A} \right\}^+ = [y] + \delta, \quad x \in \gamma, \tag{5.1}$$

and

$$\left[\frac{\partial y}{\partial v_A} \right] = \omega + u, \quad x \in \gamma, \tag{5.2}$$

where

$$\frac{\partial y}{\partial v_A} = \sum_{i,j=1}^{n} k_{ij} \frac{\partial y}{\partial x_j} \cos(v, x_i).$$

Specify the observation

$$Cy(u) \equiv y(u), \quad x \in \Omega.$$

Bring a value of the cost functional

$$J(u) = \int_{\Omega} (y(u) - z_g)^2 dx + (\mathcal{N} u, u)_{\mathcal{U}} \tag{5.3}$$

in correspondence with every control $u \in \mathcal{U}$; in this case, z_g is a known element from $L_2(\Omega)$, $\mathcal{N} u = \bar{a}(x)u$, $0 < a_0 \le \bar{a}(x) \le a_1 < \infty$, $\bar{a}(x) \in C(\gamma)$, $a_0, a_1 = \text{const}$, $(\varphi, \psi)_{\mathcal{U}} = (\varphi, \psi)_{L_2(\gamma)}$.

It can be shown [21] that a unique state, namely, a function $y(u) \in V$ corresponds to every control $u \in \mathcal{U}$, minimizes the energy functional

$$\Phi(v) = \int_{\Omega} \left(\sum_{i,j=1}^{n} k_{ij} \frac{\partial v}{\partial x_j} \frac{\partial v}{\partial x_i} + qv^2 \right) dx + \int_{\gamma} \frac{[v]^2}{R_1 + R_2} d\gamma - 2(f, v) -$$

$$-2 \int_{\gamma} \frac{R_2(\omega + u) - \delta}{R_1 + R_2} [v] d\gamma + 2 \int_{\gamma} (\omega + u) v^+ d\gamma \tag{5.4}$$

on V, and it is the unique solution in V to the weakly stated problem: Find an element $y(u) \in V$ that meets the equation

$$\int_{\Omega} \left(\sum_{i,j=1}^{n} k_{ij} \frac{\partial y}{\partial x_j} \frac{\partial v}{\partial x_i} + qyv \right) dx + \int_{\gamma} \frac{[y][v]}{R_1 + R_2} d\gamma =$$

$$= (f, v) + \int_{\gamma} \frac{R_2(\omega + u) - \delta}{R_1 + R_2} [v] d\gamma - \int_{\gamma} (\omega + u) v^+ d\gamma, \quad \forall v \in V. \quad (5.5)$$

The space V is specified in point 1.4.

Take the assumptions as for the operator $\mathcal{N}$ into account, and expression like (4.14) is obtained from expression (5.3), where the bilinear form $\pi(\cdot, \cdot)$ is expressed as

$$\pi(u, v) = \left(y(u) - y(0), y(v) - y(0) \right) + (\bar{a}u, v)_{L_2(\gamma)}$$

and

$$\pi(u, u) \geq a_0(u, u)_{L_2(\gamma)}$$

and the linear functional $L(\cdot)$ is specified by expression (4.15).

The linearity of the functional $L(v)$ follows from the fact that the difference $y(v) - y(0)$ is the unique solution $\tilde{y}(v)$ to one of equivalent problems (5.4) and (5.5). It is necessary to assume $f \equiv 0$, $\delta \equiv 0$ and $\omega \equiv 0$ for them, and the arbitrary element $z \in V$ must be additionally substituted for the arbitrary function v. Then, equality like (4.16) takes place that allows to state the bilinearity of the form $\pi(\cdot, \cdot)$. This form is coercive on $\mathcal{U}$.

Let $\tilde{y}' = \tilde{y}(u')$ and $\tilde{y}'' = \tilde{y}(u'')$ be solutions from V to problem (5.5) under $f = 0$, $\delta = 0$ and $\omega = 0$ and under a function $u = u(x)$ that is equal, respectively, to u' and u''. Then:

$$\| \tilde{y}' - \tilde{y}'' \|^2 \leq \| \tilde{y}' - \tilde{y}'' \|_V^2 \leq \mu a (\tilde{y}' - \tilde{y}'', \tilde{y}' - \tilde{y}'') \leq$$

$$\leq \mu \left(\left| \int_{\gamma} \frac{R_2(u' - u'')}{R_1 + R_2} [\tilde{y}' - \tilde{y}''] d\gamma \right| + \int_{\gamma} |u' - u''| \cdot |(\tilde{y}' - \tilde{y}'')^+| d\gamma \right) \leq$$

$$\leq c_1 \|u' - u''\|_{L_2(\gamma)} \left(\sum_{i=1}^{2} \|\tilde{y}' - \tilde{y}''\|_{L_2(\partial\Omega_i)} \right) \leq c_2 \|u' - u''\|_{L_2(\gamma)} \|\tilde{y}' - \tilde{y}''\|_V.$$

Therefore, the inequality

$$\|\tilde{y}' - \tilde{y}''\| \leq c_2 \|u' - u''\|_{L_2(\gamma)} \tag{5.6}$$

is derived that provides the continuity of the linear functional $L(\cdot)$ and bilinear form $\pi(\cdot, \cdot)$ on $\mathcal{U}$.

On the basis of [58, Chapter 1, Theorem 1.1], the validity of the following statement is proved.

Theorem 5.1. *Let conditions (1.1′) be met, and a system state is determined as a solution to equivalent problems (5.4) and (5.5). Then, there exists a unique element u of a convex set $\mathcal{U}_\partial$ that is closed in $\mathcal{U}$, and relation like (4.17) takes place for u.*

If $u \in \mathcal{U}_\partial$ is the optimal control, then the inequality

$$\langle J'(u), v - u \rangle \geq 0$$

is true $\forall v \in \mathcal{U}_\partial$ and it is transformed into

$$\left(y(u) - z_g, \, y(v) - y(u) \right) + (\bar{a}u, v - u)_{L_2(\gamma)} \geq 0, \quad \forall v \in \mathcal{U}_\partial. \tag{5.7}$$

As for the control $v \in \mathcal{U}$, the conjugate state $p(v) \in V$ is specified by the relations

$$A^* p(v) = y(v) - z_g,$$
$$p = 0, \quad x \in \Gamma,$$

$$R_1 \left\{ \frac{\partial p}{\partial \nu_{A^*}} \right\}^{-} + R_2 \left\{ \frac{\partial p}{\partial \nu_{A^*}} \right\}^{+} = [p], \quad x \in \gamma, \tag{5.8}$$

$$\left[\frac{\partial p}{\partial \nu_{A^*}} \right] = 0, \quad x \in \gamma;$$

in this case, the function $\dfrac{\partial p}{\partial v_{A^*}}$ is specified, in its turn, by one of expressions (2.19').

Further on, use Green formula [58], and the equality

$$\left(A^* p(u),\ y(v) - y(u) \right) = \left(y(u) - z_g,\ y(v) - y(u) \right) =$$

$$= a\left(p, y(v) - y(u) \right) = \sum_{l=1}^{2} \int_{\partial\Omega_l} \sum_{i,j=1}^{n} k_{ij} \frac{\partial(y(v) - y(u))}{\partial x_j} \cos(v, x_i) p(u) d\partial\Omega_l +$$

$$+ \int_{\gamma} \frac{[p][y(v) - y(u)]}{R_1 + R_2} d\gamma = \int_{\gamma} \frac{[p][y(v) - y(u)]}{R_1 + R_2} d\gamma +$$

$$+ \int_{\gamma} \left(\left\{ \frac{\partial(y(v) - y(u))}{\partial v_A} \right\}^{-} p^{-} - \left\{ \frac{\partial(y(v) - y(u))}{\partial v_A} \right\}^{+} p^{+} \right) d\gamma =$$

$$= \int_{\gamma} \frac{R_2(v - u)}{R_1 + R_2} [p] d\gamma - \int_{\gamma} (v - u) p^{+} d\gamma \qquad (5.9)$$

is obtained, where

$$a(\varphi, \psi) = \int_{\Omega} \left(\sum_{i,j=1}^{n} k_{ij} \frac{\partial\varphi}{\partial x_j} \frac{\partial\psi}{\partial x_i} + q\varphi\psi \right) dx + \int_{\gamma} \frac{[\varphi][\psi]}{R_1 + R_2} d\gamma . \qquad (5.10)$$

Take equality (5.9) into account, and it is stated that inequality (5.7) is equivalent to the inequality

$$\int_{\gamma} \left(-\frac{R_1 p^{+} + R_2 p^{-}}{R_1 + R_2} + \bar{a}\, u \right) (v - u) d\gamma \geq 0, \quad \forall v \in \mathcal{U}_\partial .$$

The necessary and sufficient condition for $u \in \mathcal{U}_\partial$ to be the optimal control is the one under which the relations

$$a(y, v) = l_1(u, v), \quad y \in V, \ \forall v \in V , \qquad (5.11)$$

$$a(p,v) = l_2(y,v), \ p \in V, \ \forall v \in V, \tag{5.12}$$

and

$$\int_\gamma \left(-\frac{R_1 p^+ + R_2 p^-}{R_1 + R_2} + \bar{a}\, u \right)(v - u)\, d\gamma \geq 0, \ \forall v \in \mathcal{U}_\partial, \tag{5.13}$$

are met.

In this case, the bilinear form $a(\cdot, \cdot)$ is specified by expression (5.10) and the functionals $l_1(\cdot, \cdot)$ and $l_2(\cdot, \cdot)$ are expressed as

$$l_1(u,v) = (f,v) + \int_\gamma \frac{R_2(\omega + u) - \delta}{R_1 + R_2}[v]\, d\gamma - \int_\gamma (\omega + u) v^+ d\gamma \tag{5.14}$$

and

$$l_2(y,v) = (y - z_g, v). \tag{5.15}$$

If the constraints are absent, i.e. when $\mathcal{U}_\partial = \mathcal{U}$, then the equality

$$-\frac{R_1 p^+ + R_2 p^-}{R_1 + R_2} + \bar{a}\, u = 0, \ x \in \gamma, \tag{5.16}$$

follows from condition (5.13).

Therefore, when the constraints are absent, the control $u(x)$ can be excluded from expression (5.14) by means of equality (5.16), and problem (5.11), (5.12) can be obtained, where $l_1(u,v) = l_1\big(u(p),v\big)$. The solution to it is $(y,p)^{\mathrm{T}}$ and the optimal control is

$$u = \frac{R_1 p^+ + R_2 p^-}{(R_1 + R_2)\bar{a}}, \ x \in \gamma. \tag{5.17}$$

In this case: $V^* = V$.

If the vector solution $(y,p)^{\mathrm{T}}$ to problem (5.11), (5.12), (5.17) is smooth enough on $\bar{\Omega}_l$, viz. $y, p|_{\bar{\Omega}_l} \in C^1(\bar{\Omega}_l) \cap C^2(\Omega_l)$, $l = 1, 2$, then the differential problem of finding the vector-function $(y,p)^{\mathrm{T}}$, that satisfies the relations

$$-\sum_{i,j=1}^{n}\frac{\partial}{\partial x_i}\left(k_{ij}\frac{\partial y}{\partial x_j}\right)+qy=f,\ x\in\Omega_1\cup\Omega_2,$$

$$-\sum_{i,j=1}^{n}\frac{\partial}{\partial x_i}\left(k_{ij}\frac{\partial p}{\partial x_j}\right)+qp-y=-z_g,\ x\in\Omega_1\cup\Omega_2,$$

$$y|_{\Gamma}=0,$$

$$p|_{\Gamma}=0,$$

$$R_1\left\{\frac{\partial y}{\partial v_A}\right\}^-+R_2\left\{\frac{\partial y}{\partial v_A}\right\}^+=[y]+\delta,\ x\in\gamma,$$

$$\left[\frac{\partial y}{\partial v_A}\right]=\omega+\frac{R_1p^++R_2p^-}{(R_1+R_2)\overline{a}},\ x\in\gamma,\qquad(5.18)$$

$$\left[\frac{\partial p}{\partial v_{A^*}}\right]=0,\ x\in\gamma,$$

$$\left\{\frac{\partial p}{\partial v_{A^*}}\right\}^{\pm}=\frac{[p]}{R_1+R_2},\ x\in\gamma,$$

corresponds to problem (5.11), (5.12), (5.17).

Definition 5.1. A generalized (weak) solution to boundary-value problem (5.18) is called a vector-function $(y,p)^{\mathrm{T}}\in H=\left\{v=(v_1,v_2)^{\mathrm{T}}:\ v_i|_{\Omega_l}\in W_2^1(\Omega_l);\ i,l=1,2;\ v|_{\Gamma}=0\right\}$ that satisfies the following integral equation $\forall v\in H$:

$$\int_{\Omega}\left\{\sum_{i,j=1}^{n}k_{ij}\frac{\partial y}{\partial x_j}\frac{\partial z_1}{\partial x_i}+qyz_1+\sum_{i,j=1}^{n}k_{ij}\frac{\partial p}{\partial x_j}\frac{\partial z_2}{\partial x_i}+qpz_2-yz_2\right\}dx+$$

$$+ \int_\gamma \frac{[y][z_1]+[p][z_2]}{R_1 + R_2} d\gamma = \int_\Omega \left(fz_1 - z_g z_2 \right) dx + \tag{5.19}$$

$$+ \int_\gamma \frac{R_2 \left(\omega + \frac{R_1 p^+ + R_2 p^-}{(R_1 + R_2)\overline{a}} \right) - \delta}{R_1 + R_2} [z_1] d\gamma - \int_\gamma \left(\omega + \frac{R_1 p^+ + R_2 p^-}{(R_1 + R_2)\overline{a}} \right) z_1^+ d\gamma .$$

Let $u = (u_1, u_2)^T$ and $v = (v_1, v_2)^T$ be arbitrary elements of the complete Hilbert space H. Specify the bilinear form

$$a(u,v) = \int_\Omega \left\{ \sum_{i,j=1}^n k_{ij} \frac{\partial u_1}{\partial x_j} \frac{\partial v_1}{\partial x_i} + \sum_{i,j=1}^n k_{ij} \frac{\partial u_2}{\partial x_j} \frac{\partial v_2}{\partial x_i} + q u_1 v_1 + q u_2 v_2 - u_1 v_2 \right\} dx +$$

$$+ \int_\gamma \frac{\sum_{i=1}^2 [u_i][v_i]}{R_1 + R_2} d\gamma - \int_\gamma \frac{R_2 \left(R_1 u_2^+ + R_2 u_2^- \right)}{(R_1 + R_2)^2 \overline{a}} [v_1] d\gamma +$$

$$+ \int_\gamma \frac{\left(R_1 u_2^+ + R_2 u_2^- \right) v_1^+}{(R_1 + R_2) \overline{a}} d\gamma \tag{5.20}$$

and linear functional

$$l(v) = \int_\Omega \left(fv_1 - z_g v_2 \right) dx + \int_\gamma \frac{R_2 \omega - \delta}{R_1 + R_2} [v_1] d\gamma - \int_\gamma \omega v_1^+ d\gamma \tag{5.21}$$

on H.

If the constraint

$$\alpha_1 = \frac{\alpha_0}{2} \min\{1, \mu\} -$$

$$- 6 \sup_{x \in \gamma} \left\{ \frac{R_1 R_2}{(R_1 + R_2)^2 \overline{a}}, \frac{R_2^2}{(R_1 + R_2)^2 \overline{a}}, \frac{R_1}{(R_1 + R_2)\overline{a}}, \frac{R_2}{(R_1 + R_2)\overline{a}} \right\} c_3^2 > 0,$$

where c_3 is the constant from the embedding theorem, is met, then $\forall u, v \in H$ the inequalities

$$a(u,u) \geq \alpha_1 \|u\|_H^2 \text{ and } |a(u,v)| \leq c_4 \|u\|_H \|v\|_H$$

follow from expression (5.20). It is easy to see that

$$|l(v)| \leq c_5 \|v\|_H.$$

Use the Lax-Milgramm lemma, and it is concluded that the unique solution $U = (u_1, u_2)^T$ to problem (5.19) exists in H. Problem (5.19) can be solved approximately by means of the finite-element method. Therefore, it is possible to derive the approximate solution $U_k^N \in H_k^N$ to problem (5.19) for which estimate like (4.53) is true. The control $u = u(x)$ is specified on γ and such control is equal to $\dfrac{R_1 p^+ + R_2 p^-}{(R_1 + R_2)\bar{a}}$, where $p^\pm$ is the trace on $\gamma^\pm$ of the function $p = p(x)$. Then, the estimate

$$\left\| u - \tilde{u}_{2k}^N \right\|_{L_2(\gamma)} \leq c_6 h^k, \tag{5.22}$$

where $c_6 = \text{const} > 0$, can be written for $\tilde{u}_{2k}^N = \dfrac{R_1 p_k^{N+} + R_2 p_k^{N-}}{(R_1 + R_2)\bar{a}}$.

1.6 BOUNDARY CONTROL: THIRD BOUNDARY-VALUE PROBLEM

Assume that elliptic equation (4.1), where the coefficients and right-hand side meet the conditions of point 1.1, is specified in the domains Ω_1 and $\Omega_2 \in R^n$. The conjugation conditions have the form of expressions (4.3) and (4.4) and the third boundary condition

$$\sum_{i,j=1}^{n} k_{ij} \frac{\partial y}{\partial x_j} \cos(\nu, x_i) = -\alpha\, y + \beta, \ \ x \in \Gamma, \tag{6.1}$$

where $\alpha, \beta \in R^1$ and $\alpha > 0$, is specified, in its turn, on the boundary Γ.

For every control $u \in \mathcal{U} = L_2(\Gamma)$, determine a system state as a generalized solution to the boundary-value problem specified by equation (4.1), conjugation conditions (4.3) and (4.4) and the boundary condition

$$\sum_{i,j=1}^{n} k_{ij} \frac{\partial y}{\partial x_j} \cos(\nu, x_i) = -\alpha\, y + \beta + u, \ \ x \in \Gamma. \tag{6.2}$$

Specify the observation
$$Cy(u) \equiv y(u).$$

Bring a value of the cost functional

$$J(u) = \int_{\Omega} \left(y(u) - z_g \right)^2 dx + (\mathcal{N}\, u, u)_{\mathcal{U}} \tag{6.3}$$

in correspondence with every control $u \in \mathcal{U}$; in this case, z_g is a known element from $L_2(\Omega)$, $\mathcal{N}u = \bar{a}(x)u$, $0 < a_0 \le \bar{a}(x) \le a_1 < \infty$, $\bar{a}(x) \in C(\Gamma)$, $a_0, a_1 = \text{const}$, $(\varphi, \psi)_{\mathcal{U}} = (\varphi, \psi)_{L_2(\Gamma)}$.

It can be shown [21] that a unique state, namely, a function $y(u) \in V = \left\{ v|_{\Omega_i} \in W_2^1(\Omega_i) : i = 1, 2 \right\}$ corresponds to every control $u = L_2(\Gamma)$, minimizes the energy functional

$$\Phi(v) = \int_{\Omega} \left(\sum_{i,j=1}^{n} k_{ij} \frac{\partial v}{\partial x_j} \frac{\partial v}{\partial x_i} + q v^2 \right) dx +$$

$$+ \int_{\gamma} \frac{[v]^2}{R_1 + R_2} d\gamma - 2(f, v) - 2 \int_{\gamma} \frac{R_2 \omega - \delta}{R_1 + R_2} [v] d\gamma +$$

$$+ \alpha \int_{\Gamma} v^2 d\Gamma - 2\beta \int_{\Gamma} v d\Gamma - 2 \int_{\Gamma} uv d\Gamma + 2 \int_{\gamma} \omega v^+ d\gamma \qquad (6.4)$$

on V, and it is the unique solution in V to the weakly stated problem: Find an element $y(u) \in V$ that meets the equation

$$\int_{\Omega} \left(\sum_{i,j=1}^{n} k_{ij} \frac{\partial y}{\partial x_j} \frac{\partial v}{\partial x_i} + qyv \right) dx + \int_{\gamma} \frac{[y][v]}{R_1 + R_2} d\gamma + \int_{\Gamma} \alpha yv \, d\Gamma =$$

$$= (f, v) + \int_{\gamma} \frac{R_2 \omega - \delta}{R_1 + R_2} [v] d\gamma - \int_{\gamma} \omega v^+ d\gamma + \int_{\Gamma} (\beta + u) v d\Gamma, \ \forall v \in V. \qquad (6.5)$$

Take the assumptions as for the operator $\mathcal{N}$ into account, and expression like (4.14) follows from representation (6.3), where the bilinear form $\pi(\cdot, \cdot)$ and linear functional $L(\cdot)$ are expressed as

$$\pi(u, v) = \left(y(u) - y(0), y(v) - y(0) \right) + (\bar{a}u, v)_{L_2(\Gamma)}$$

and

$$L(v) = \left(z_g - y(0), y(v) - y(0) \right).$$

The form $\pi(\cdot, \cdot)$ is coercive on $\mathcal{U}$, i.e.: $\pi(u, u) \geq a_0(u, u)_{L_2(\Gamma)}$, $\forall u \in L_2(\Gamma)$.

Let $\tilde{y}' = \tilde{y}(u')$ and $\tilde{y}'' = \tilde{y}(u'')$ be solutions from V to problem (6.5) under $f = 0$, $\delta = 0$, $\omega = 0$ and $\beta = 0$ and under a function u that is equal, respectively, to u' and u''. Then, the inequality

$$\|\tilde{y}' - \tilde{y}''\|_{L_2(\Gamma)}^2 \leq c_1 \|\tilde{y}' - \tilde{y}''\|_V^2 \leq \mu \, a\left(\tilde{y}' - \tilde{y}'', \tilde{y}' - \tilde{y}'' \right) \leq$$

$$\leq \mu \|u' - u''\|_{L_2(\Gamma)} \|\tilde{y}' - \tilde{y}''\|_{L_2(\Gamma)}$$

is obtained, i.e.

$$\|\tilde{y}' - \tilde{y}''\|_{L_2(\Gamma)} \leq \mu \|u' - u''\|_{L_2(\Gamma)}, \ \|\tilde{y}' - \tilde{y}''\| \leq \mu_1 \|u' - u''\|_{L_2(\Gamma)}, \qquad (6.6)$$

where

$$a(\varphi, \psi) = \int\limits_{\Omega} \left(\sum_{i,j=1}^{n} k_{ij} \frac{\partial \varphi}{\partial x_j} \frac{\partial \psi}{\partial x_i} + q\varphi\psi \right) dx +$$

$$+ \int\limits_{\gamma} \frac{[\varphi][\psi]}{R_1 + R_2} d\gamma + \alpha \int\limits_{\Gamma} \varphi\psi \, d\Gamma. \tag{6.7}$$

Inequalities (6.6) provide the continuity of the linear functional $L(\cdot)$ and bilinear form $\pi(\cdot, \cdot)$ on $\mathcal{U}$.

On the basis of [58, Chapter 1, Theorem 1.1], the validity of the following statement is proved.

Theorem 6.1. *Let conditions (1.1ʹ) be met, and a system state is determined as a solution to equivalent problems (6.4) and (6.5). Then, there exists a unique element u of a convex closed set $\mathcal{U}_\partial \subset \mathcal{U}$, and relation like (4.17) takes place for u.*

If $u \in \mathcal{U}_\partial$ is the optimal control, then inequality like (4.19) is true $\forall v \in \mathcal{U}_\partial$. Represent this inequality as

$$\left(y(u) - z_g, \, y(v) - y(u) \right) + (\bar{a}u, v - u)_{L_2(\Gamma)} \geq 0, \quad \forall v \in \mathcal{U}_\partial. \tag{6.8}$$

As for the control $v \in \mathcal{U}$, the conjugate state $p(v) \in V$ is specified by the relations

$$A^* p(v) = y(v) - z_g,$$

$$R_1 \left\{ \frac{\partial p}{\partial v_{A^*}} \right\}^- + R_2 \left\{ \frac{\partial p}{\partial v_{A^*}} \right\}^+ = [p], \quad x \in \gamma,$$

$$\left[\frac{\partial p}{\partial v_{A^*}} \right] = 0, \quad x \in \gamma, \tag{6.9}$$

$$\frac{\partial p}{\partial v_{A^*}} = -\alpha \, p, \ x \in \Gamma.$$

Further on, use the Green formula [58], and the equality

$$\left(A^* p(u), \ y(v) - y(u)\right) =$$

$$= \left(y(u) - z_g, y(v) - y(u)\right) = a\left(p, y(v) - y(u)\right) =$$

$$= \sum_{l=1}^{2} \int_{\partial \Omega_l} \sum_{i,j=1}^{n} k_{ij} \frac{\partial\left(y(v) - y(u)\right)}{\partial x_j} \cos(v, x_i) p(u) d\partial\Omega_l +$$

$$+ \int_{\gamma} \frac{[p][y(v) - y(u)]}{R_1 + R_2} d\gamma +$$

$$+ \alpha \int_{\Gamma} p\left(y(v) - y(u)\right) d\Gamma = \int_{\Gamma} (v - u) p(u) d\Gamma \qquad (6.10)$$

is obtained. Take it into account, and it is stated that inequality (6.8) is equivalent to the inequality

$$\int_{\Gamma} \left(p(u) + \bar{a}u\right)(v - u) d\Gamma \geq 0, \ \forall v \in \mathcal{U}_{\partial}.$$

Therefore, the necessary and sufficient condition for the existence of the optimal control $u \in \mathcal{U}_{\partial}$ is the one under which the relations

$$a(y, v) = l_1(u, v), \ y \in V, \ \forall v \in V, \qquad (6.11)$$

$$a(p, v) = l_2(y, v), \ p \in V, \ \forall v \in V, \qquad (6.12)$$

and

$$\int_{\Gamma}(p(u)+\bar{a}u)(v-u)d\Gamma \geq 0, \ \forall v \in \mathcal{U}_{\partial},\tag{6.13}$$

are met, where the bilinear form $a(\cdot,\cdot)$ is specified by expression (6.7) and the functionals $l_1(\cdot,\cdot)$ and $l_2(\cdot,\cdot)$ are

$$l_1(u,v)=(f,v)+\int_{\gamma}\frac{R_2\omega-\delta}{R_1+R_2}[v]d\gamma-\int_{\gamma}\omega\, v^{+}d\gamma+$$

$$+\int_{\Gamma}(\beta+u)vd\Gamma, \ u \in \mathcal{U}_{\partial}, \ \forall v \in V,\tag{6.14}$$

$$l_2(y,v)=(y-z_g,v).$$

If the constraints are absent, i.e. when $\mathcal{U}_{\partial}=\mathcal{U}$, then the equality

$$p(u)+\bar{a}u=0, \ x \in \Gamma,\tag{6.15}$$

follows from condition (6.13). Therefore, when the constraints are absent, the control $u(x)$ can be excluded from equality (6.11) by means of equality (6.15), and problem (6.11), (6.12) can be obtained, where $l_1(u,v)=l_1\big(u(p),v\big)$. The solution to it is $(y,p)^{T}$ and the optimal control is

$$u=-p/\bar{a}, \ x \in \Gamma.\tag{6.16}$$

If the vector solution $(y,p)^{T}$ to problem (6.11), (6.12), (6.15) is smooth enough on $\bar{\Omega}_l$, viz. $y|_{\bar{\Omega}_l}$, $p|_{\bar{\Omega}_l} \in C^1(\bar{\Omega}_l)\cap C^2(\Omega_l)$, $l=1,2$, then the differential problem of finding the vector-function $(y,p)^{T}$, that satisfies the relations

$$-\sum_{i,j=1}^{n}\frac{\partial}{\partial x_i}\left(k_{ij}\frac{\partial y}{\partial x_j}\right)+qy=f, \ x \in \Omega_1\bigcup\Omega_2,$$

$$-\sum_{i,j=1}^{n}\frac{\partial}{\partial x_i}\left(k_{ij}\frac{\partial p}{\partial x_j}\right)+qp-y=-z_g, \quad x\in\Omega_1\cup\Omega_2,$$

$$\sum_{i,j=1}^{n}k_{ij}\frac{\partial y}{\partial x_j}\cos(v,x_i)=-\alpha y+\beta-\frac{p}{a}, \quad x\in\Gamma,$$

$$\sum_{i,j=1}^{n}k_{ij}\frac{\partial p}{\partial x_j}\cos(v,x_i)=-\alpha p, \quad x\in\Gamma,$$

$$R_1\left\{\frac{\partial y}{\partial v_A}\right\}^{-}+R_2\left\{\frac{\partial y}{\partial v_A}\right\}^{+}=[y]+\delta, \quad x\in\gamma,$$

$$\left[\frac{\partial y}{\partial v_A}\right]=\omega, \quad x\in\gamma, \tag{6.17}$$

$$\left[\frac{\partial p}{\partial v_{A^*}}\right]=0, \quad x\in\gamma,$$

$$\left\{\frac{\partial p}{\partial v_{A^*}}\right\}^{\pm}=\frac{[p]}{R_1+R_2},$$

corresponds to problem (6.11), (6.12), (6.15).

Definition 6.1. A generalized (weak) solution to problem (6.17) is called a vector-function $(y,p)^{\mathrm{T}}\in H=\left\{v=(v_1,v_2)^{\mathrm{T}}: \left.v_i\right|_{\Omega_l}\in W_2^1(\Omega_l);\right.$ $\left.i,l=1,2\right\}$ that satisfies the following integral equation $\forall z\in H$:

$$\int_{\Omega}\left\{\sum_{i,j=1}^{n}k_{ij}\frac{\partial y}{\partial x_j}\frac{\partial z_1}{\partial x_i}+qyz_1+\sum_{i,j=1}^{n}k_{ij}\frac{\partial p}{\partial x_j}\frac{\partial z_2}{\partial x_i}+qpz_2-yz_2\right\}dx+$$

$$+ \int_\gamma \frac{[y][z_1] + [p][z_2]}{R_1 + R_2} d\gamma + \alpha \int_\Gamma (y z_1 + p z_2) d\Gamma + \int_\Gamma p\, z_1 / \bar{a}\, d\Gamma =$$

$$= \int_\Omega \left(f z_1 - z_g z_2 \right) dx + \int_\gamma \frac{R_2 \omega - \delta}{R_1 + R_2} [z_1] d\gamma - \int_\gamma \omega\, z_1^+ d\gamma. \tag{6.18}$$

Let $u = (u_1, u_2)^{\mathrm{T}}$ and $v = (v_1, v_2)^{\mathrm{T}}$ be arbitrary elements of the complete Hilbert space H. Specify the bilinear form

$$a(u,v) = \int_\Omega \left\{ \sum_{i,j=1}^{n} k_{ij} \sum_{l=1}^{2} \frac{\partial u_l}{\partial x_j} \frac{\partial v_l}{\partial x_i} + q \sum_{l=1}^{2} u_l v_l - u_1 v_2 \right\} dx +$$

$$+ \int_\gamma \frac{\sum_{i=1}^{2} [u_i][v_i]}{R_1 + R_2} d\gamma + \alpha \int_\Gamma \sum_{l=1}^{2} u_l v_l d\Gamma + \int_\Gamma u_2\, v_1 / \bar{a}\, d\Gamma \tag{6.19}$$

and linear functional

$$l(v) = \int_\Omega \left(f\, v_1 - z_g v_2 \right) dx + \int_\gamma \frac{R_2 \omega - \delta}{R_1 + R_2} [v_1] d\gamma - \int_\gamma \omega\, v_1^+ d\gamma$$

on H.

Let the constraint

$$\alpha_1 = \min \left\{ \frac{\alpha_0 \mu}{2} + q_0 - \frac{1}{2}, \; \alpha - \frac{1}{2a_0} \right\} > 0$$

be met. Then, $\forall u \in H$ the inequalities

$$a(u,u) \geq \bar{\alpha}_1 \|u\|_H^2 \quad \text{and} \quad |a(u,v)| \leq c_4 \|u\|_H \|v\|_H, \quad \bar{\alpha}_1, c_4 = \text{const} > 0, \tag{6.20}$$

follow from expression (6.19).

The following is evident:

$$|l(v)| \leq c_5 \|v\|_H. \tag{6.21}$$

Use the Lax-Milgramm lemma, and it is concluded that the unique solution $U = (u_1, u_2)^T$ to problem (6.18) exists in H. If problem (6.18) is solved by means of the finite-element method, then the estimate

$$\left\| u - \tilde{u}_{2k}^N \right\|_{L_2(\Gamma)} \leq c h^k \tag{6.22}$$

takes place for the approximation $\tilde{u}_{2k}^N = -p_k^N / \bar{a}$ of the optimal control u.

1.7 BOUNDARY CONTROL AND OBSERVATION: THIRD BOUNDARY-VALUE PROBLEM

Assume that elliptic equation (4.1), where the coefficients and right-hand side meet conditions of point 1.1, is specified in the domains Ω_1 and $\Omega_2 \in R^n$. The conjugation conditions have the form of expressions (4.3) and (4.4) and third boundary condition (6.1) is specified, in its turn, on the boundary Γ.

For every control $u \in \mathcal{U} = L_2(\Gamma)$, determine a system state as a generalized solution to problem (4.1), (4.3), (4.4), (6.2). Represent the observation by expression like (4.6), where $C \in \mathcal{L}(L_2(\Gamma); L_2(\Gamma))$ is specified by the relation $Cy(u) \equiv y(u)$ under $x \in \Gamma$. Bring the value of the cost functional

$$J(u) = \int_\Gamma \left(y(u) - z_g \right)^2 d\Gamma + (\mathcal{N} u, u)_{\mathcal{U}} \tag{7.1}$$

in correspondence with every control $u \in \mathcal{U}_\partial$; in this case, z_g is a known element from $L_2(\Gamma)$, $\mathcal{N}u = \bar{a}u$, $0 < a_0 \leq \bar{a}(x) \leq a_1 < \infty$, $\bar{a}(x) \in C(\Gamma)$, $a_0, a_1 =$ =const, $(\varphi, \psi)_{\mathcal{U}} = \int_\Gamma \varphi \psi \, d\Gamma = (\varphi, \psi)_{L_2(\Gamma)}$.

It can be shown [21] that a unique state, namely, a function $y(u) \in V$ corresponds to every control $u \in L_2(\Gamma)$, minimizes energy functional (6.4) on V, and it is the unique solution in V to weakly stated problem (6.5).

Take the assumptions as for the operator $\mathcal{N}$ into account, and

$$J(u) = \pi(u,u) - 2L(u) + \left\| z_g - y(0) \right\|^2_{L_2(\Gamma)} \tag{7.2}$$

follows from expression (7.1); in this case, the bilinear form $\pi(\cdot,\cdot)$ and linear functional $L(\cdot)$ are expressed as

$$\pi(u,v) = \big(y(u) - y(0), y(v) - y(0) \big)_{L_2(\Gamma)} + (\bar{a}u, v)_{L_2(\Gamma)}$$

and

$$L(v) = \big(z_g - y(0), y(v) - y(0) \big)_{L_2(\Gamma)}.$$

The form $\pi(\cdot,\cdot)$ is coercive on $\mathcal{U}$, i.e.: $\pi(u,u) \geq a_0(u,u)_{L_2(\Gamma)}$.

Let $\tilde{y}' = \tilde{y}(u')$ and $\tilde{y}'' = \tilde{y}(u'')$ be solutions from V to problem like (6.5) under $f = 0$, $\delta = 0$, $\omega = 0$ and $\beta = 0$ and under a function u that is equal, respectively, to u' and u''. Then, inequalities like (6.6) are true and they provide the continuity of the bilinear form $\pi(\cdot,\cdot)$ and linear functional $L(\cdot)$ on $\mathcal{U}$.

On the basis of [58, Chapter 1, Theorem 1.1], the validity of the following statement is proved.

Theorem 7.1. *Let conditions (1.1′) be met, and a system state is determined as a solution to equivalent problems (6.4) and (6.5). Then, there exists a unique element u of a convex closed set $\mathcal{U}_\partial \subset \mathcal{U}$, and relation like (4.17) takes place for u, where a cost functional is specified by expression (7.1).*

If $u \in \mathcal{U}_\partial$ is the optimal control, then inequality like (4.19) is true. Represent this inequality as

$$\big(y(u) - z_g, y(v) - y(u) \big)_{L_2(\Gamma)} + (\bar{a}u, v - u)_{L_2(\Gamma)} \geq 0, \ \forall v \in \mathcal{U}_\partial. \tag{7.3}$$

As for the control $v \in \mathcal{U}$, the conjugate state $p(v) \in V$ is specified by the relations

$$A^* p(v) = 0,$$

$$\left[\frac{\partial p}{\partial v_{A^*}} \right] = 0, \ x \in \gamma,$$

$$R_1 \left\{ \frac{\partial p}{\partial v_{A^*}} \right\}^- + R_2 \left\{ \frac{\partial p}{\partial v_{A^*}} \right\}^+ = [p], \ x \in \gamma, \tag{7.4}$$

$$\frac{\partial p}{\partial v_{A^*}} = -\alpha \, p + y(v) - z_g, \ x \in \Gamma.$$

The equality

$$0 = \left(A^* p(u), \, y(v) - y(u) \right) = a \left(p, y(v) - y(u) \right) -$$

$$- \int_\Gamma \left(y(u) - z_g \right) \left(y(v) - y(u) \right) d\Gamma =$$

$$= \int_\gamma \frac{[p][y(v) - y(u)]}{R_1 + R_2} d\gamma + \alpha \int_\Gamma p \left(y(v) - y(u) \right) d\Gamma +$$

$$+ \sum_{l=1}^{2} \int_{\partial\Omega_l} \sum_{i,j=1}^{n} k_{ij} \frac{\partial \left(y(v) - y(u) \right)}{\partial x_j} \cos(v, x_i) p \, d\partial\Omega_l -$$

$$- \int_\Gamma \left(y(u) - z_g \right) \left(y(v) - y(u) \right) d\Gamma =$$

$$= \int_\Gamma \left(p(v-u) - (y(u) - z_g)(y(v) - y(u)) \right) d\Gamma \tag{7.5}$$

is obtained. The bilinear form $a(\cdot, \cdot)$ is specified here by expression (6.7). Take equality (7.5) into account, and the inequality

$$\int_{\Gamma}(p+\bar{a}u)(v-u)\,d\Gamma \geq 0, \quad \forall v \in \mathcal{U}_{\partial}, \tag{7.6}$$

follows from inequality (7.3).

The necessary and sufficient condition for the existence of the optimal control $u \in \mathcal{U}_{\partial}$ is the one under which relations like (6.11), (6.12) and (7.6) are met, where the functional $l_1(u,v)$ has the form of expression (6.14), and

$$l_2(y,v) = \left(y(v) - z_g, v\right)_{L_2(\Gamma)}. \tag{7.7}$$

If the constraints are absent, i.e. when $\mathcal{U}_{\partial} = \mathcal{U}$, then equality (6.15) follows from inequality (7.6). Therefore, when the constraints are absent, the control $u(x)$ can be excluded from equality (6.11) by means of equality (6.15), and problem like (6.11), (6.12) can be solved, where $l_1(u,v) = = l_1\left(u(p), v\right)$. The solution $(y,p)^{\mathrm{T}}$ to it is found and the optimal control u is found, in its turn, by formula (6.16).

If the vector solution $(y,p)^{\mathrm{T}}$ to problem like (6.11), (6.12) is smooth enough on $\bar{\Omega}_l$, viz. $y|_{\bar{\Omega}_l}, \; p|_{\bar{\Omega}_l} \in C^1(\bar{\Omega}_l) \cap C^2(\Omega_l)$, $l=1,2$, then the differential problem of finding the vector-function $(y,p)^{\mathrm{T}}$, that meets relations (6.17), except the second and fourth ones, and that satisfies the equalities

$$-\sum_{i,j=1}^{n} \frac{\partial}{\partial x_i}\left(k_{ij}\frac{\partial p}{\partial x_j}\right) + qp = 0, \quad x \in \Omega_1 \bigcup \Omega_2, \tag{7.8}$$

and

$$\sum_{i,j=1}^{n} k_{ij}\frac{\partial p}{\partial x_j}\cos(\nu, x_i) = -\alpha\, p + y(u) - z_g, \quad x \in \Gamma, \tag{7.9}$$

corresponds to problem like (6.11), (6.12), where the functional $l_2(y,v)$ has the form of expression (7.7).

Definition 7.1. A generalized (weak) solution to a problem, specified by equalities (6.17), except the second and fourth ones, and by constraints

(7.8) and (7.9), is called a vector-function $(y,p)^T \in H$ that satisfies the following integral equation $\forall z \in H$:

$$\int_\Omega \left\{ \sum_{i,j=1}^n k_{ij} \frac{\partial y}{\partial x_j} \frac{\partial z_1}{\partial x_i} + qyz_1 + \sum_{i,j=1}^n k_{ij} \frac{\partial p}{\partial x_j} \frac{\partial z_2}{\partial x_i} + qpz_2 \right\} dx +$$

$$+ \int_\gamma \frac{[y][z_1] + [p][z_2]}{R_1 + R_2} d\gamma + \int_\Gamma \alpha(yz_1 + pz_2) d\Gamma +$$

$$+ \int_\Gamma p z_1 /\bar{a}\, d\Gamma - \int_\Gamma (y - z_g) z_2\, d\Gamma =$$

$$= \int_\Omega fz_1 dx + \int_\gamma \frac{R_2\omega - \delta}{R_1 + R_2} [z_1]\, d\gamma - \int_\gamma \omega z_1^+ d\gamma. \qquad (7.10)$$

Let $u = (u_1, u_2)^T$ and $v = (v_1, v_2)^T$ be arbitrary elements of the complete Hilbert space H. Specify the bilinear form

$$a(u,v) = \int_\Omega \left\{ \sum_{i,j=1}^n k_{ij} \sum_{l=1}^2 \frac{\partial u_l}{\partial x_j} \frac{\partial v_l}{\partial x_i} + q \sum_{l=1}^2 u_l v_l \right\} dx +$$

$$+ \int_\gamma \frac{\sum_{i=1}^2 [u_i][v_i]}{R_1 + R_2} d\gamma + \int_\Gamma \alpha \sum_{l=1}^2 u_l v_l d\Gamma + \int_\Gamma u_2\, v_1 /\bar{a}\, d\Gamma - \int_\Gamma u_1 v_2 d\Gamma \qquad (7.11)$$

and linear functional

$$l(v) = \int_\Omega f v_1\, dx + \int_\gamma \frac{R_2\omega - \delta}{R_1 + R_2} [v_1]\, d\gamma -$$

$$- \int_\gamma \omega v_1^+\, d\gamma - \int_\Gamma z_g v_2\, d\Gamma \qquad (7.12)$$

on H. If the constraint

$$\alpha_1 = \alpha - \frac{1}{2}\left(\frac{1}{a_0} + 1\right) > 0$$

is met, then $\forall u, v \in H$ inequalities like (6.20) follow from expression (7.11) and estimate like (6.21) is derived from expression (7.12).

Use the Lax-Milgramm lemma, and it is concluded that the unique solution $U = (u_1, u_2)^{\mathrm{T}}$ to problem (7.10) exists in H. If problem (7.10) is solved by means of the finite-element method, then estimate like (6.22) takes place for the approximation $\tilde{u}_{2k}^N = -p_k^N / \bar{a}$ of the optimal control u.

CONTROL OF A CONDITIONALLY CORRECT SYSTEM DESCRIBED BY THE NEUMANN PROBLEM FOR AN ELLIPTIC-TYPE EQUATION UNDER CONJUGATION CONDITIONS

2.1   DISTRIBUTED CONTROL WITH OBSERVATION THROUGHOUT A WHOLE DOMAIN

Assume that the elliptic equation

$$Lu \equiv -\sum_{i,j=1}^{n} \frac{\partial}{\partial x_i}\left(k_{ij}(x)\frac{\partial y}{\partial x_j}\right) = f(x) \tag{1.1}$$

is specified in a domain Ω that consists of two bounded convex domains, namely, Ω_1 and $\Omega_2 \in R^n$, where R^n is an n-dimensional real linear space. The second-type boundary Neumann condition

$$\sum_{i,j=1}^{n} k_{ij}(x)\frac{\partial y}{\partial x_j}\cos(v, x_i) = g(x) \tag{1.2}$$

is specified, in its turn, on a boundary $\Gamma = (\partial\Omega_1 \cup \partial\Omega_2)\backslash\gamma$ $(\gamma = \partial\Omega_1 \cap \cap\partial\Omega_2 \neq \varnothing)$; in this case, v is an outer normal to Γ, $k_{ij}(x)\big|_{\overline{\Omega}_l} = k_{ji}\big|_{\overline{\Omega}_l} = k_{ij}\big|_{\overline{\Omega}_l} \in C(\overline{\Omega}_l) \cap C^1(\Omega_l)$, $f(x)\big|_{\Omega_l} = f\big|_{\Omega_l} \in C(\Omega_l)$, $\left|D^1 k_{ij}\right| < \infty$, $g(x)\big|_{\Gamma\cap\partial\Omega_l} = g\big|_{\Gamma\cap\partial\Omega_l} \in C(\Gamma \cap \partial\Omega_l)$, $i,j = \overline{1,n}$; $l = 1,2$, $\left|f\right| \leq c_1 < \infty$,

$$\sum_{i,j=1}^{n} k_{ij}(x)\xi_i \xi_j \geq \alpha_0 \sum_{i=1}^{n} \xi_i^2 \quad \forall x \in \Omega,$$

$$\forall \xi_i, \xi_j \in R^1, \quad i,j = \overline{1,n}, \quad \alpha_0 = \text{const} > 0; \tag{1.2'}$$

and the conjugation conditions

$$[y] = 0 \tag{1.3}$$

and

$$\left[\sum_{i,j=1}^{n} k_{ij} \frac{\partial y}{\partial x_j} \cos(v, x_i) \right] = \omega \tag{1.4}$$

are specified, also in their turn, on a section γ of the domain $\bar{\Omega}$; in this case, $\omega \in C(\gamma)$, $[\varphi] = \varphi^+ - \varphi^-$, $\varphi^\pm = \{\varphi\}^\pm = \varphi(x)$ under $x \in \gamma^\pm$, $\gamma^+ = = \gamma \cap \partial\Omega_2$, $\gamma^- = \gamma \cap \partial\Omega_1$, v is a normal to γ and such normal is directed into the domain Ω_2.

Let $y(x) \in \bar{M} = \left\{ v(x): \ v|_{\bar{\Omega}_l} \in C^1(\bar{\Omega}_l) \cap C^2(\Omega_l), \ l = 1, 2, \ \left| D^2 v \right| < \infty \right\}$

be a classical solution to boundary-value problem (1.1)–(1.4). It is easy to see that a solution $y + c$ is also classical to it for an arbitrary constant c.

The necessary condition for the existence of the classical solution y to problem (1.1)–(1.4) is the one under which the equality

$$\int_\Omega f \, dx + \int_\Gamma g \, d\Gamma = \int_\gamma \omega \, d\gamma \tag{1.5}$$

is met. Find this solution under the constraint

$$\int_\Omega y \, dx = Q, \tag{1.6}$$

where Q is some known real number. Assume the following: $\bar{H} =$
$= \left\{ v(x): v|_{\Omega_i} \in W_2^1(\Omega_i), \ i = 1, 2 \right\}$, $V_Q = \left\{ v \in \bar{H}: [v] = 0, \ (v, 1) = Q \right\}$,
$(\varphi, \psi) = \int_\Omega \varphi \psi \, dx$.

Let there be a control Hilbert space $\mathcal{U}$ and mapping $B \in \mathcal{L}(\mathcal{U}; V')$, where V' is a space dual with respect to a state Hilbert space V. Assume the following: $\mathcal{U} = L_2(\Omega)$.

For every control $u \in \mathcal{U}$, determine system state $y = y(u)$ as a generalized solution to the boundary-value problem specified by the equation

$$-\sum_{i, j = 1}^n \frac{\partial}{\partial x_i} \left(k_{ij}(x) \frac{\partial y}{\partial x_j} \right) = f(x) + Bu, \ y \in V_Q, \tag{1.7}$$

and by conditions (1.2)–(1.4) and (1.6).

Specify the observation

$$Z(u) = C\, y(u), \tag{1.8}$$

where $C \in \mathcal{L}(V; \mathcal{H})$ and $\mathcal{H}$ is some Hilbert space. Assume the following:

$$C\, y(u) \equiv y(u), \ \mathcal{H} = V \subset L_2(\Omega). \tag{1.9}$$

Bring a value of the cost functional

$$J(u) = \left\| Cy(u) - z_g \right\|_{\mathcal{H}}^2 + (\mathcal{N}u, u)_{\mathcal{U}} \tag{1.10}$$

in correspondence with every control $u \in \mathcal{U}$; in this case, z_g is some known element of $\mathcal{H}$, $\mathcal{N} \in \mathcal{L}(\mathcal{U}; \mathcal{U})$, $(\mathcal{N}u, u)_{\mathcal{U}} \geq v_0 \|u\|_{\mathcal{U}}^2$, $v_0 = \text{const} > 0$ $\forall u \in \mathcal{U}$.

Assume the following: $f \in L_2(\Omega)$, $Bu \equiv u \in L_2(\Omega)$, $\mathcal{N}u = \bar{a}(x)u$, $0 < a_0 \leq \bar{a}(x) \leq a_1 < \infty$, $\bar{a}(x)|_{\Omega_l} \in C(\Omega_l)$, $l = 1, 2$; $a_0, a_1 = \text{const}$, $(\varphi, \psi)_{\mathcal{U}} = \int_\Omega \varphi \psi \, dx$. A unique state, namely, a function $y(u) \in V_Q$

corresponds to every control $u \in \mathscr{U}$, delivers the minimum to the energy functional [21]

$$\Phi_1(v) = a_1(v,v) - 2l_1(v) \tag{1.11}$$

on V_Q, and it is the unique solution in V_Q to the weakly stated problem: Find an element $y \in V_Q$ that meets the equation

$$a_1(y,v) = l_1(v) \quad \forall v \in V_0, \tag{1.12}$$

where $V_0 = \left\{ v \in \bar{H} : [v] = 0, (v,1) = 0 \right\}$, $a_1(u,v) = \int\limits_{\Omega} \sum\limits_{i,j=1}^{n} k_{ij} \dfrac{\partial u}{\partial x_j} \dfrac{\partial v}{\partial x_i} dx$,

$$l_1(v) = \int\limits_{\Omega} (f+u)v\,dx + \int\limits_{\Gamma} g v\,d\Gamma - \int\limits_{\gamma} \omega v\,d\gamma .$$

The following statement is valid [21].

Lemma 1.1. *Problems (1.11) and (1.12) are equivalent* $\forall f \in L_2(\Omega)$, $\forall \omega \in L_2(\gamma)$, $\forall u \in \mathscr{U}$ *and have a unique solution* $y = y(u) \in V_Q$.

Remark 1.1. If a solution $y \in V_Q$ to problems (1.11) and (1.12) belongs to a set $\bar{M}$, then y is classical to boundary-value problem (1.7), (1.2)–(1.4), (1.6) under the constraint

$$\int\limits_{\Omega} (f+u)\,dx + \int\limits_{\Gamma} g\,d\Gamma = \int\limits_{\gamma} \omega\,d\gamma . \tag{1.13}$$

Remark 1.2. If a solution y to problems (1.11) and (1.12) exists, it is not necessary to meet constraint (1.13).

Remark 1.3. If equality (1.5) takes place, then, to meet constraint (1.13), it is necessary for a control u to satisfy the condition

$$\int\limits_{\Omega} u\,dx = 0 . \tag{1.14}$$

Rewrite cost functional (1.10) as

$$J(u) = \pi(u,u) - 2L(u) + \left\| z_g - y(0) \right\|^2 ; \tag{1.15}$$

in this case, $\|\varphi\| = \|\varphi\|_{L_2(\Omega)} = (\varphi, \varphi)^{1/2}$ and the bilinear form $\pi(\cdot, \cdot)$ and linear functional $L(\cdot)$ are expressed as

$$\pi(u, v) = (y(u) - y(0), \, y(v) - y(0)) + (\bar{a}\, u, v)$$

$$L(v) = (z_g - y(0), \, y(v) - y(0)). \tag{1.16}$$

Let $\tilde{y}' = \tilde{y}(u')$ and $\tilde{y}'' = \tilde{y}(u'')$ be solutions from V_Q to problem (1.12) under $f = 0$, $g = 0$ and $\omega = 0$ and under a function $u = u(x)$ that is equal, respectively, to u' and u''. Then, take the ellipticity condition and generalized Poincare inequality into account, and the inequality

$$\bar{\alpha}_0 \|\tilde{y}' - \tilde{y}''\|^2 \leq \bar{\alpha}_0 \|\tilde{y}' - \tilde{y}''\|_V^2 \leq a_1 (\tilde{y}' - \tilde{y}'', \tilde{y}' - \tilde{y}'') \leq$$

$$\leq \|u' - u''\| \cdot \|\tilde{y}' - \tilde{y}''\|_V, \quad \bar{\alpha}_0 = \text{const} > 0,$$

is derived, where $\|v\|_V = \left\{ \sum_{i=1}^{2} \|v\|_{W_2^1(\Omega_i)}^2 \right\}^{1/2}$ and $\|\cdot\|_{W_2^1(\Omega_i)}$ is the norm of the Sobolev space $W_2^1(\Omega_i)$.

On the basis of [58, Theorem 1.1, Chapter 1], the validity of the following statement is proved.

Theorem 1.1. *Let a system state be determined as a solution to equivalent problems (1.11) and (1.12). Then, there exists a unique element u of a convex set $\mathcal{U}_\partial$ that is closed in $\mathcal{U}$, and*

$$J(u) = \inf_{v \in \mathcal{U}_\partial} J(v) \tag{1.17}$$

takes place for u.

Definition 1.1. If an element $u \in \mathcal{U}_\partial$ meets condition (1.17), it is called an optimal control.

Let the equation

$$-\sum_{i,j=1}^{n} \frac{\partial}{\partial x_i} \left(k_{ij} \frac{\partial y}{\partial x_j} \right) + \int_\Omega y \, dx = f(x) + Q + u \tag{1.18}$$

be specified on the domain Ω instead of equation (1.7). Neumann condition (1.2) and conjugation conditions (1.3)–(1.4) are specified, in their turn, respectively, on the boundary Γ and section γ.

If y is a classical solution to boundary-value problem (1.7), (1.2)–(1.4), (1.6) (Problem 1), then it is easy to see that y is classical to problem (1.18), (1.2)–(1.4) (Problem 1′). It can be shown [21] that a classical solution to Problem 1′ is also classical to Problem 1 if constraint (1.13) is satisfied.

Let observation (1.8) be specified, where the operator C is given by expression (1.9). Cost functional (1.10) is specified, in its turn, for every control $u \in \mathcal{U}$. Then, a unique state, namely, a function $y(u) \in$ $\in V = \{v \in \bar{H} : \ [v] = 0\}$, corresponds to every $u \in \mathcal{U}$, minimizes the energy functional

$$\Phi(v) = a_1'(v,v) - 2 l_1'(v) \tag{1.19}$$

on V, and it is the unique solution in V to the weakly stated problem: Find an element $y \in V$ that meets the equation

$$a_1'(y,v) = l_1'(v), \quad \forall v \in V, \tag{1.20}$$

where $a_1'(y,v) = a_1(y,v) + (y,1)(v,1)$ and $l_1'(v) = l_1(v) + Q(v,1)$.

Lemma 1.2. *Problems (1.19) and (1.20) are equivalent* $\forall f \in L_2(\Omega)$, $\forall u \in \mathcal{U}$ *and have a unique solution* $y(u) \in V$.

Remark 1.4. If a solution $y \in V$ to problems (1.19) and (1.20) belongs to a set $\bar{M}$, then y is classical to boundary-value Problem 1′, and it is also classical to Problem 1 if constraint (1.13) is met.

Therefore, there exists such an operator A generated by problems (1.19), (1.20) and acting from V into $L_2(\Omega)$, that

$$y(u) = A^{-1}(f + Q + Bu), \quad \forall u \in L_2 = L_2(\Omega).$$

Let $\tilde{y}' = \tilde{y}(u')$ and $\tilde{y}'' = \tilde{y}(u'')$ be solutions from V to Problem 1′ under $f = 0$, $g = 0$ and $\omega = 0$ and under a function $u = u(x)$ that is equal, respectively, to u' and u''.

Then, on the basis of the generalized Poincare inequality, the following one, i.e.

$$\bar{\alpha}_1 \|\tilde{y}' - \tilde{y}''\|^2 \le \bar{\alpha}_1 \|\tilde{y}' - \tilde{y}''\|_V^2 \le a \left(\tilde{y}' - \tilde{y}'', \tilde{y}' - \tilde{y}''\right) \le$$

$$\le \|u' - u''\| \cdot \|\tilde{y}' - \tilde{y}''\|, \quad \bar{\alpha}_1 = \text{const} > 0,$$

is derived that provides the continuity of the linear functional $L(\cdot)$ and bilinear form $\pi(\cdot,\cdot)$ of expressions (1.16) on $\mathcal{U}$.

On the basis of [58, Theorem 1.1, Chapter 1], the validity of the following statement is proved.

Theorem 1.2. *Let a system state be determined as a solution to equivalent problems (1.19) and (1.20). Then, there exists a unique element u of a convex set $\mathcal{U}_\partial$ that is closed in $\mathcal{U}$, and relation like (1.17) takes place for u.*

Remark 1.5. If equality (1.13) is satisfied, then problems (1.11) and (1.19) are equivalent. Therefore, optimal controls coincide when states are described by boundary-value Problems 1 and 1'.

Here is the problem of finding the control $u \in \mathcal{U}_\partial$ that satisfies relation (1.17). It is optimization Problem 1 if a system state is a generalized solution to boundary-value Problem 1, and it is optimization Problem 1' if a system state is a generalized solution to boundary-value Problem 1'.

Remark 1.6. If constraint (1.5) is met and $\mathcal{U}_\partial =$

$$= \left\{ u \in L_2(\Omega) : \int_\Omega u dx = 0 \right\},$$ then optimization Problems 1 and 1' are

equivalent.

If $u \in \mathcal{U}_\partial$ is the optimal control, then the following inequality is true $\forall v \in \mathcal{U}_\partial$:

$$\left(y(u) - z_g, \, y(v) - y(u)\right) + \left(\bar{a}\, u, v - u\right) \ge 0 . \tag{1.21}$$

As for the control $v \in \mathcal{U}$, the conjugate state $p(v) \in V^* = V$ is specified by the relations

$$A^* p(v) = y(v) - z_g, \quad x \in \Omega_1 \cup \Omega_2,$$

$$\frac{\partial p}{\partial v_{A^*}} = 0, \ x \in \Gamma,$$

$$[p] = 0, \quad \left[\frac{\partial p}{\partial v_{A^*}}\right] = 0, \ x \in \gamma, \qquad (1.22)$$

where

$$A^* p = -\sum_{i,j=1}^{n} \frac{\partial}{\partial x_i}\left(k_{ij}\frac{\partial p}{\partial x_j}\right) + \int_{\Omega} p\, dx,$$

$$\frac{\partial p}{\partial v_{A^*}} = \sum_{i,j=1}^{n} k_{ij}\frac{\partial p}{\partial x_j}\cos(v, x_i). \qquad (1.23)$$

The equality

$$\left(A^* p(u), y(v) - y(u)\right) = \left(y(u) - z_g, y(v) - y(u)\right) = a_1'\left(p, y(v) - y(u)\right) =$$

$$= \sum_{l=1}^{2}\int_{\partial\Omega_l}\sum_{i,j=1}^{n} k_{ij}\frac{\partial(y(v) - y(u))}{\partial x_j}\cos(v, x_i)\, p(u)\, d\partial\Omega_l +$$

$$+ (p, v - u) = (p, v - u),$$

i.e. $\left(y(u) - z_g, y(v) - y(u)\right) = (p, v - u)$ is obtained. Take it into account, and the inequality

$$\left(p + \bar{a}u, v - u\right) \geq 0, \ \ \forall v \in \mathcal{U}_{\partial}, \qquad (1.24)$$

is derived from inequality (1.21).

To make the element $u \in \mathcal{U}_{\partial}$ the optimal control of a state described by boundary-value Problem 1′, it is necessary and sufficient to meet inequality (1.24) and the relations

$$a_1'(y, v) = l_1(u, v), \ \ y \in V, \ \forall v \in V, \qquad (1.25)$$

and

$$a_1'(p, v) = l_2(y, v), \ \ p \in V, \ \forall v \in V, \qquad (1.26)$$

where

$$l_1(u,v) = (f+Q, v) + (u,v) + \int_\Gamma gv\,d\Gamma - \int_\gamma \omega v\,d\gamma$$

and

$$l_2(y,v) = (y,v) - (z_g,v).$$

If the constraints are absent, i.e. when $\mathcal{U}_\partial = \mathcal{U}$, then the equality

$$p + \bar{a}u = 0 \qquad (1.27)$$

follows from condition (1.24). Therefore, when the constraints are absent, the control u can be excluded from equality (1.25) by means of equality (1.27). On the basis of equalities (1.25) and (1.26), the problem

$$Ay + p/\bar{a} = f, \quad y \in V, \qquad (1.28)$$

$$A^* p - y = -z_g, \quad p \in V^*, \qquad (1.29)$$

is derived, and the vector solution $(y,p)^T$ is found from this problem along with the optimal control $u = -p/\bar{a}$ of the system specified by boundary-value Problem 1′.

If the vector solution $(y,p)^T$ to problem (1.28), (1.29) is smooth enough on $\bar{\Omega}_l$, viz., $y|_{\bar{\Omega}_l}$, $p|_{\bar{\Omega}_l} \in C^1(\bar{\Omega}_l) \cap C^2(\Omega_l)$, $l = 1,2$, then the differential problem of finding the vector-function $(y,p)^T$, that satisfies the relations

$$-\sum_{i,j=1}^{n} \frac{\partial}{\partial x_i}\left(k_{ij}\frac{\partial y}{\partial x_j}\right) + p/\bar{a} + \int_\Omega y\,dx = f + Q, \quad x \in \Omega_1 \cup \Omega_2,$$

$$-\sum_{i,j=1}^{n} \frac{\partial}{\partial x_i}\left(k_{ij}\frac{\partial p}{\partial x_j}\right) + \int_\Omega p\,dx - y = -z_g, \quad x \in \Omega_1 \cup \Omega_2,$$

$$\sum_{i,j=1}^{n} k_{ij} \frac{\partial y}{\partial x_j} \cos(\nu, x_i) = g, \quad x \in \Gamma,$$

$$\sum_{i,j=1}^{n} k_{ij} \frac{\partial p}{\partial x_j} \cos(\nu, x_i) = 0, \quad x \in \Gamma,$$

$$[y] = 0, \quad [p] = 0, \quad x \in \gamma,$$

$$\left[\sum_{i,j=1}^{n} k_{ij} \frac{\partial y}{\partial x_j} \cos(\nu, x_i)\right] = \omega, \quad \left[\sum_{i,j=1}^{n} k_{ij} \frac{\partial p}{\partial x_j} \cos(\nu, x_i)\right] = 0, \quad x \in \gamma, (1.30)$$

corresponds to problem (1.28), (1.29).

Definition 1.2. A generalized (weak) solution to boundary-value problem (1.30) is called a vector-function $(y, p)^{\mathrm{T}} \in H = $
$= \left\{ v = (v_1, v_2)^{\mathrm{T}} : v_i|_{\Omega_j} \in W_2^1(\Omega_j), \ i, j = 1, 2; \ [v] = 0 \right\}$ that satisfies the following integral equation $\forall z \in H$:

$$\int_{\Omega} \left\{ \sum_{i,j=1}^{n} k_{ij} \frac{\partial y}{\partial x_j} \frac{\partial z_1}{\partial x_i} + p z_1 / \bar{a} + \sum_{i,j=1}^{n} k_{ij} \frac{\partial p}{\partial x_j} \frac{\partial z_2}{\partial x_i} - y z_2 \right\} dx + $$

$$+ \int_{\Omega} y \, dx \int_{\Omega} z_1 dx + \int_{\Omega} p \, dx \int_{\Omega} z_2 dx = $$

$$= \int_{\Omega} \left((f + Q) z_1 - z_g z_2 \right) dx + \int_{\Gamma} g z_1 d\Gamma - \int_{\gamma} \omega z_1 d\gamma. \tag{1.31}$$

Let $u = (u_1, u_2)^{\mathrm{T}}$ and $v = (v_1, v_2)^{\mathrm{T}}$ be arbitrary elements of the complete Hilbert space H with the norm $\|\cdot\|_H = \left\{ \sum_{i=1}^{2} \|\cdot\|_{W_2^1(\Omega_i)} \right\}^{1/2}$. Specify the bilinear form

$$a(u,v) = \int_\Omega \left\{ \sum_{l=1}^{2} \sum_{i,j=1}^{n} k_{ij} \frac{\partial u_l}{\partial x_j} \frac{\partial v_l}{\partial x_i} + u_2 v_1 / \bar{a} - u_1 v_2 \right\} dx + \sum_{l=1}^{2} \int_\Omega u_l dx \int_\Omega v_l dx$$

and linear functional

$$l(v) = \int_\Omega \left((f+Q)v_1 - z_g v_2 \right) dx + \int_\Gamma g\, v_1 d\Gamma - \int_\gamma \omega\, v_1 d\gamma$$

on H.

Assume that the constraint $\alpha_1 = \min\left\{\dfrac{\alpha_0}{2}, 1\right\}\mu - \dfrac{1}{2}\left\{\dfrac{1}{a_0}+1\right\} > 0$ is met,

where μ is the constant in the generalized Poincare inequality. Take the generalized Poincare inequality [21] and Cauchy-Bunyakovsky one into account, and the relations

$$a(v,v) \geq \bar{\alpha}_1 \|v\|_H^2 \quad \forall v \in H, \ \bar{\alpha}_1 = \text{const} > 0,$$

and

$$|a(u,v)| \leq c_1 \|u\|_H \|v\|_H \quad \forall u,v \in H, \ c_1 = \text{const} > 0,$$

are true for the bilinear form $a(\cdot,\cdot)$, i.e. this form is H-elliptic and continuous [49] on H.

Consider the Cauchy-Bunyakovsky inequality and embedding theorems [55], and the following inequality is obtained $\forall v \in H$:

$$|l(v)| \leq c_2 \|v\|_H, \ c_2 = \text{const}.$$

Use the Lax-Milgramm lemma [16], and it is concluded that the unique solution (y, p) to problem (1.31) exists in H.

Problem (1.31) can be solved approximately by means of the finite-element method. For this purpose, divide the domains $\bar{\Omega}_i$ into N_i finite elements $\bar{e}_i^j$ ($j = \overline{1, N_i}$, $i = 1,2$) of the regular family [16]. Specify the subspace $H_k^N \subset H$ ($N = N_1 + N_2$) of the vector-functions $V_k^N(x)$. The components $v_{1k}^N\big|_{\bar\Omega_i}$, $v_{2k}^N\big|_{\bar\Omega_i} \in C(\bar\Omega_i)$ $(i=1,2)$ of $V_k^N(x)$ are the complete polynomials of the power k that contain the variables $x_1, x_2, ..., x_n$ at every $\bar e_i^j$, and $\left[V_k^N\right]=0$. Then, the linear algebraic equation system

$$A\bar U = B \tag{1.32}$$

follows from equation (1.31), and the solution $\bar U$ to system (1.32) exists and such solution is unique. The vector $\bar U$ specifies the unique approximate solution $U_k^N \in H_k^N$ to problem (1.31) as the unique one to the equation

$$a\left(U_k^N, V_k^N\right)=l\left(V_k^N\right), \quad \forall V_k^N \in H_k^N. \tag{1.33}$$

Let $U = U(x) \in H$ be the solution to problem (1.31). Then:

$$a\left(U - U_k^N, V_k^N\right)=0, \quad \forall V_k^N \in H_k^N.$$

Therefore,

$$\bar\alpha_1 \left\|U - U_k^N\right\|_H^2 \le a\left(U - U_k^N, U - U_k^N\right)=$$

$$= a\left(U - U_k^N, U - \tilde U\right), \quad \forall \tilde U \in H_k^N,$$

and the inequality

$$\left\|U - U_k^N\right\|_H \le c_0\left\|U - \tilde U\right\|_H, \quad c_0 = \text{const}, \tag{1.34}$$

is thus derived since the bilinear form $a(\cdot,\cdot)$ is continuous on H.

Suppose that $\tilde U \in H_k^N$ is a complete interpolation polynomial for the solution U at every $\bar e_i^j$. Take the interpolation estimates [16] into account, assume that every component U_1 and U_2 of the solution U on Ω_l belongs to the Sobolev space $W_2^{k+1}(\Omega_l)$ $(l=1,2)$, and the estimate

$$\left\| U - U_k^N \right\|_H \leq ch^k, \tag{1.35}$$

where h is a maximum diameter of all the finite elements $\bar{e}_i^j$, $c = \text{const}$, follows from inequality (1.34).

Take estimate (1.35) into consideration, and the estimate

$$\left\| u - u_k^N \right\|_{W_2^1} \leq c_2 \left\| p - p_k^N \right\|_{W_2^1} \leq c_3 h^k, \tag{1.35'}$$

where $\|\cdot\|_{W_2^1} = \left\{ \sum_{i=1}^{2} \|\cdot\|_{W_2^1(\Omega_i)}^2 \right\}^{1/2}$, takes place for the approximation $u_k^N(x) = -p_k^N / \bar{a}(x)$ of the control $u = u(x)$ of a state described by Problem 1' and $p_k^N = u_{2k}^N$ is the second component of the vector U_k^N.

Remark 1.7. If constraint (1.13) is met, then the first component of a classical solution to problem (1.30) is a classical solution to boundary-value Problem 1.

2.2 DISTRIBUTED CONTROL WITH OBSERVATION ON A THIN INCLUSION

Assume that equation (1.1), where the coefficients and right-hand side meet conditions (1.2'), is specified in the bounded, continuous and strictly Lipschitz domains Ω_1 and Ω_2. Condition (1.2) is specified, in its turn, on the boundary Γ and the conjugation conditions have the form of expressions (1.3) and (1.4).

For every control $u \in \mathcal{U} = L_2(\Omega)$, determine a system state as a generalized solution to the boundary-value problem specified by equation (1.7) and by conditions (1.2)–(1.4), where $Bu \equiv u$ and $u \in L_2(\Omega)$.

Equality (1.13) is the necessary condition under which there exists a classical solution $y = y(u)$ to boundary-value problem (1.7), (1.2)–(1.4) (Problem 2). Find this solution under constraint (1.6).

Bring a value of the cost functional

$$J(u) = \int_\gamma (y(u) - z_g)^2 d\gamma + (\mathcal{N}u, u)_{\mathcal{U}} \qquad (2.1)$$

in correspondence with every control $u \in \mathcal{U} = L_2(\Omega)$; in this case, z_g is a known element from the space $L_2(\gamma)$; $\mathcal{N}u = \bar{a}u$, $0 < a_0 \le \bar{a}(x) \le \le a_1^0 < \infty$, $\bar{a}(x)\big|_{\Omega_i} \in C(\Omega_i)$, $i = 1, 2$; a_0, $a_1^0 = \text{const}$.

A unique state, namely, a function $y(u) \in V_Q$ corresponds to every control $u \in \mathcal{U}$, minimizes energy functional (1.11) on V_Q, and it is the unique solution in V_Q to weakly stated problem (1.12). Lemma 1.1 and Remarks 1.1 and 1.2 hold here.

Rewrite cost functional (2.1) as

$$J(u) = \pi(u, u) - 2L(u) + \left\| z_g - y(0) \right\|_{L_2(\gamma)}^2, \qquad (2.2)$$

where

$$\|v\|_{L_2(\gamma)} = (v, v)_{L_2(\gamma)}^{1/2}, \quad (\varphi, \psi)_{L_2(\gamma)} = \int_\gamma \varphi\psi d\gamma,$$

$$\pi(u, v) = \left(y(u) - y(0),\ y(v) - y(0) \right)_{L_2(\gamma)} + (\bar{a}u, v)$$

and

$$L(v) = \left(z_g - y(0),\ y(v) - y(0) \right)_{L_2(\gamma)}.$$

Take the embedding theorems, ellipticity condition and generalized Poincare inequality into account, and the inequality

$$\bar{\alpha}_2 \left\| \tilde{y}' - \tilde{y}'' \right\|_{L_2(\gamma)}^2 \le \bar{\alpha}_1 \left\| \tilde{y}' - \tilde{y}'' \right\|_V^2 \le$$

$$\le a_1(\tilde{y}' - \tilde{y}'',\ \tilde{y}' - \tilde{y}'') \le \left\| u' - u'' \right\| \left\| \tilde{y}' - \tilde{y}'' \right\|_V,$$

i.e.

$$\|\tilde{y}' - \tilde{y}''\|_{L_2(\gamma)} \le \frac{1}{\sqrt{\bar{\alpha}_1 \bar{\alpha}_2}} \|u' - u''\|, \quad \alpha_1, \alpha_2 = \text{const} > 0,$$

is derived, where $\tilde{y}' = \tilde{y}(u')$ and $\tilde{y}'' = \tilde{y}(u'')$ are the generalized solutions from V_Q to boundary-value Problem 2 under $f = 0$, $g = 0$ and $\omega = 0$ and under a function $u = u(x)$ that is equal, respectively, to u' and u''.

On the basis of the derived inequality and [58, Chapter 1, Theorem 1.1], the validity of the following statement is proved.

Theorem 2.1. *Let a system state be determined as a solution to equivalent problems (1.11) and (1.12). Then, there exists a unique element u of a convex set $\mathcal{U}_\partial$ that is closed in $\mathcal{U} = L_2(\Omega)$, and relation like (1.17) takes place for u, where the cost functional $J(u)$ is specified by expression (2.1).*

Let equation (1.18) be specified on Ω instead of equation (1.7). Neumann condition (1.2) and conjugation conditions (1.3) and (1.4) are specified, in their turn, respectively, on Γ and γ. I.e., boundary-value problem (1.18), (1.2)–(1.4) (Problem 2′) is obtained.

Remark 2.1. Boundary-value Problems 1, 2 and 1′, 2′ coincide pairwise. Optimization Problems do not coincide because their cost functionals $J(u)$ are different.

Consider optimization Problem 2′: Find a control $u \in \mathcal{U}_\partial \subset \mathcal{U} = L_2(\Omega)$, for which relation like (1.17) is satisfied, where the cost functional $J(u)$ is specified by expression (2.1), and a state $y(u)$ is a generalized solution to boundary-value Problem 2′.

If $u \in \mathcal{U}_\partial$ is the optimal control for optimization Problem 2′, then the following inequality is true:

$$\left(y(u) - z_g, \ y(v) - y(u) \right)_{L_2(\gamma)} + (\bar{a}u, v - u) \ge 0, \ \forall v \in \mathcal{U}_\partial. \tag{2.3}$$

As for the control $v \in \mathcal{U}$, the conjugate state $p(v) \in V^* = V$ is specified by the relations

$$A^* p(v) = 0, \ x \in \Omega_1 \cup \Omega_2,$$

$$\frac{\partial p}{\partial \nu_{A^*}} = 0, \ x \in \Gamma,$$

$$[p] = 0, \quad \left[\frac{\partial p}{\partial \nu_{A^*}}\right] = -y(v) + z_g, \ x \in \gamma, \tag{2.4}$$

where the operators A^* and $\dfrac{\partial}{\partial \nu_{A^*}}$ are specified, in their turn, by expressions (1.23).

The equality

$$0 = \left(A^* p(u), \ y(v) - y(u)\right) = a_1'\left(p, \ y(v) - y(u)\right) -$$

$$-\left(y(u) - z_g, \ y(v) - y(u)\right)_{L_2(\gamma)} =$$

$$= -\left(y(u) - z_g, y(v) - y(u)\right)_{L_2(\gamma)} + (p, \ v - u),$$

i.e. $\left(y(u) - z_g, y(v) - y(u)\right)_{L_2(\gamma)} = (p, \ v - u) \ \forall v \in \mathcal{U}_\partial$ is obtained. Take it into account, and the inequality

$$\left(\bar{a} u + p, v - u\right) \geq 0, \ \ \forall v \in \mathcal{U}_\partial, \tag{2.5}$$

is derived from inequality (2.3).

An element $u \in \mathcal{U}_\partial$ is an optimal control for optimization Problem 2′ if and only if inequality (2.5) and the equalities

$$a_1'(y, v) = l_1(u, v), \ y \in V, \ \forall v \in V, \tag{2.6}$$

and

$$a_1'(p, v) = l_2(y, v), \ p \in V, \ \forall v \in V, \tag{2.7}$$

are met; the bilinear form $a_1'(\cdot, \cdot)$ and functional $l_1(u, v)$ are specified in point 2.1, and

$$l_2(y,v) = -\int_\gamma (z_g - y)v\,d\gamma.$$

If the constraints are absent, i.e. when $\mathcal{U}_\partial = \mathcal{U}$, then the equality

$$p + \bar{a}u = 0 \tag{2.8}$$

follows from condition (2.5). Therefore, when the constraints are absent, the control u can be excluded from equality (2.6) by means of equality (2.8). Let the solution $(y,p)^{\mathrm{T}}$ to problem (2.6), (2.7), where $l_1(u,y) = l_1(u(p),y)$, be sufficiently smooth on $\bar{\Omega}_1$ and $\bar{\Omega}_2$. Then, such solution satisfies the relations

$$-\sum_{i,j=1}^n \frac{\partial}{\partial x_i}\left(k_{ij}\frac{\partial y}{\partial x_j}\right) + p/\bar{a} + \int_\Omega y\,dx = f + Q, \quad x \in \Omega_1 \cup \Omega_2,$$

$$-\sum_{i,j=1}^n \frac{\partial}{\partial x_i}\left(k_{ij}\frac{\partial p}{\partial x_j}\right) + \int_\Omega p\,dx = 0, \quad x \in \Omega_1 \cup \Omega_2,$$

$$\sum_{i,j=1}^n k_{ij}\frac{\partial y}{\partial x_j}\cos(v,x_i) = g, \quad x \in \Gamma,$$

$$\sum_{i,j=1}^n k_{ij}\frac{\partial p}{\partial x_j}\cos(v,x_i) = 0, \quad x \in \Gamma,$$

$$[y] = 0, \quad [p] = 0, \quad x \in \gamma,$$

$$\left[\sum_{i,j=1}^n k_{ij}\frac{\partial y}{\partial x_j}\cos(v,x_i)\right] = \omega, \quad x \in \gamma,$$

$$\left[\sum_{i,j=1}^n k_{ij}\frac{\partial p}{\partial x_j}\cos(v,x_i)\right] = -y + z_g, \quad x \in \gamma. \tag{2.9}$$

Definition 2.1. A generalized (weak) solution to boundary-value problem (2.9) is called a vector-function $(y, p)^T \in H$ that satisfies the following integral equation $\forall z \in H$:

$$\int_\Omega \left\{ \sum_{i,j=1}^n k_{ij} \frac{\partial y}{\partial x_j} \frac{\partial z_1}{\partial x_i} + p \, z_1/\bar{a} + \sum_{i,j=1}^n k_{ij} \frac{\partial p}{\partial x_j} \frac{\partial z_2}{\partial x_i} \right\} dx +$$

$$+ \int_\Omega y \, dx \int_\Omega z_1 dx + \int_\Omega p \, dx \int_\Omega z_2 dx =$$

$$= \int_\Omega (f + Q) z_1 \, dx + \int_\Gamma g \, z_1 \, d\Gamma - \int_\gamma \omega z_1 \, d\gamma + \int_\gamma \left(y - z_g \right) z_2 \, d\gamma. \quad (2.10)$$

Let $u = (u_1, u_2)^T$ and $v = (v_1, v_2)^T$ be arbitrary elements of the complete Hilbert space H. Specify the bilinear form

$$a(u, v) = \int_\Omega \left\{ \sum_{l=1}^2 \sum_{i,j=1}^n k_{ij} \frac{\partial u_l}{\partial x_j} \frac{\partial v_l}{\partial x_i} + u_2 v_1/\bar{a} \right\} dx +$$

$$+ \sum_{l=1}^2 \int_\Omega u_l \, dx \int_\Omega v_l \, dx - \int_\gamma u_1 v_2 \, d\gamma$$

and linear functional

$$l(v) = \int_\Omega (f + Q) v_1 \, dx + \int_\Gamma g \, v_1 \, d\Gamma - \int_\gamma \omega v_1 \, d\gamma - \int_\gamma z_g v_2 \, d\gamma$$

on H. If the constraint

$$\min \left\{ \frac{\alpha_0}{2}, \min \left\{ \frac{\alpha_0}{2}, 1 \right\} \mu \right\} - \frac{1}{2a_0} - \frac{c_0^2}{2} > 0, \quad (2.11)$$

where μ and c_0 are the positive constants, respectively, in the generalized Poincare inequality and embedding theorem, is met, then the unique solution $(y, p)^T$ to problem (2.10) exists in H. Problem (2.10) can be solved by means of the finite-element method. Estimates like (1.35) and

(1.35′) are true, respectively, for its approximate solution $U_k^N \in H_k^N \subset H$ and for the approximation $u_k^N(x)$ of the control u.

2.3   DISTRIBUTED CONTROL WITH BOUNDARY OBSERVATION

Assume that equation (1.1), where the coefficients and right-hand side meet conditions (1.2′), is specified in the bounded, continuous and strictly Lipschitz domains Ω_1 and Ω_2. The conjugation conditions have the form of expressions (1.3) and (1.4) and the boundary condition has the form of expression (1.2).

For every control $u \in \mathcal{U} = L_2(\Omega)$, determine a system state as a generalized solution to the boundary-value problem specified by equation (1.7) and by conditions (1.2)–(1.4). Equality (1.13) is the necessary condition under which there exists a classical solution y to boundary-value problem (1.7), (1.2)–(1.4) (Problem 3): Find a solution y that meets constraint (1.6).

Bring a value of the cost functional

$$J(u) = \int_\Gamma (y(u) - z_g)^2 d\Gamma + (\mathcal{N}u, u)_{\mathcal{U}} \tag{3.1}$$

in correspondence with every control $u \in \mathcal{U} = L_2(\Omega)$; in this case, z_g is a known element from the space $L_2(\Gamma)$, $\mathcal{N}u = \bar{a}u$, $0 < a_0 \le \bar{a}(x) \le$ $\le a_1^0 < \infty$, $\bar{a}(x)\big|_{\Omega_i} \in C(\Omega_i)$, $i = 1, 2$; a_0, $a_1^0 = \text{const}$.

A unique state, namely, a function $y(u) \in V_Q$ corresponds to every control $u \in \mathcal{U}$, delivers the minimum to energy functional (1.11) on V_Q, and it is the unique solution in V_Q to weakly stated problem (1.12). Lemma 1.1 and Remarks 1.1 and 1.2 take place here.

Rewrite cost functional (3.1) as

$$J(u) = \pi(u,u) - 2L(u) + \left\| z_g - y(0) \right\|^2_{L_2(\Gamma)}, \qquad (3.2)$$

where

$$\|v\|_{L_2(\Gamma)} = (v,v)^{1/2}_{L_2(\Gamma)}, \quad (\varphi,\psi)_{L_2(\Gamma)} = \int_\Gamma \varphi\psi\, d\Gamma,$$

$$\pi(u,v) = \left(y(u) - y(0),\; y(v) - y(0) \right)_{L_2(\Gamma)} + (\bar{a}\, u, v)$$

and

$$L(v) = \left(z_g - y(0),\; y(v) - y(0) \right)_{L_2(\Gamma)}.$$

Take the embedding theorems, ellipticity condition and generalized Poincare inequality into account, and the inequality

$$\bar{\alpha}_2 \left\| \tilde{y}' - \tilde{y}'' \right\|^2_{L_2(\Gamma)} \le \bar{\alpha}_1 \left\| \tilde{y}' - \tilde{y}'' \right\|^2_V \le$$

$$\le a_1(\tilde{y}' - \tilde{y}'', \tilde{y}' - \tilde{y}'') \le \left\| u' - u'' \right\| \left\| \tilde{y}' - \tilde{y}'' \right\|_V,$$

i.e.

$$\left\| \tilde{y}' - \tilde{y}'' \right\|_{L_2(\Gamma)} \le \frac{1}{\sqrt{\bar{\alpha}_1 \bar{\alpha}_2}} \left\| u' - u'' \right\|, \quad \bar{\alpha}_1, \bar{\alpha}_2 = \text{const} > 0,$$

is derived, where $\tilde{y}' = \tilde{y}(u')$ and $\tilde{y}'' = \tilde{y}(u'')$ are the generalized solutions from V_Q to boundary-value Problem 3 under $f = 0$, $g = 0$ and $\omega = 0$ and under a function $u = u(x)$ that is equal, respectively, to u' and u''.

On the basis of the derived inequality and [58, Chapter 1, Theorem 1.1], the validity of the following statement is proved.

Theorem 3.1. *Let a system state be determined as a solution to equivalent problems (1.11) and (1.12). Then, there exists a unique element u of a convex set $\mathcal{U}_\partial$ that is closed in $\mathcal{U}$, and relation like (1.17) takes place for u, where the cost functional J(u) is specified by expression (3.1).*

Let equation (1.18) be specified on the domain Ω instead of equation (1.7). Neumann condition (1.2) and conjugation conditions (1.3) and (1.4) are specified, in their turn, respectively, on Γ and γ. I.e., boundary-value problem (1.18), (1.2)–(1.4) (Problem 3') is obtained.

Remark 3.1. Boundary-value Problem 3 coincides with Problems 1 and 2. Problem 3′ coincides with Problems 1′ and 2′. Optimization Problems do not coincide because their cost functionals $J(u)$ are different.

Consider optimization Problem 3′: Find a control $u \in \mathcal{U}_\partial \subset \mathcal{U} = L_2(\Omega)$, for which relation like (1.17) is satisfied and where the cost functional $J(u)$ is specified by expression (3.1). A state $y = y(u)$ is a generalized solution to boundary-value Problem 3′ and such solution is unique for one of equivalent problems (1.19) and (1.20).

If $u \in \mathcal{U}_\partial$ is the optimal control for optimization Problem 3′, then the following inequality is true:

$$\left(y(u) - z_g,\, y(v) - y(u)\right)_{L_2(\Gamma)} + (\bar{a}\, u, v - u) \geq 0, \quad \forall v \in \mathcal{U}_\partial. \tag{3.3}$$

As for the control $v \in \mathcal{U}$, the conjugate state $p(v) \in V^* = V$ is specified by the relations

$$A^* p(v) = 0, \quad x \in \Omega_1 \cup \Omega_2,$$

$$\frac{\partial p}{\partial v_{A^*}} = -z_g + y, \quad x \in \Gamma,$$

$$[p] = 0, \quad \left[\frac{\partial p}{\partial v_{A^*}}\right] = 0, \quad x \in \gamma, \tag{3.4}$$

where the operators A^* and $\dfrac{\partial}{\partial v_{A^*}}$ are specified, in their turn, by expressions (1.23).

The equality

$$0 = \left(A^* p(u),\, y(v) - y(u)\right) = a_1'\left(p,\, y(v) - y(u)\right) -$$

$$- \left(y(u) - z_g,\, y(v) - y(u)\right)_{L_2(\Gamma)} =$$

$$= -\left(y(u) - z_g,\, y(v) - y(u)\right)_{L_2(\Gamma)} + (p,\, v - u),$$

i.e. $\left(y(u)-z_g,y(v)-y(u)\right)_{L_2(\Gamma)}=(p,v-u)$ $\forall v\in\mathcal{U}_\partial$ is obtained. Take it into account, and the inequality

$$(p+\bar{a}u,v-u)\geq 0, \quad \forall v\in\mathcal{U}_\partial, \tag{3.5}$$

is derived from inequality (3.3).

To make the element $u\in\mathcal{U}_\partial$ the optimal control for optimization Problem 3′, it is necessary and sufficient for inequality (3.5) and the equalities

$$a_1'(y,v)=l_1(u,v), \quad y\in V, \quad \forall v\in V, \tag{3.6}$$

and

$$a_1'(p,v)=l_2(y,v), \quad p\in V, \quad \forall v\in V, \tag{3.7}$$

to be met; the bilinear form $a_1'(\cdot,\cdot)$ and linear functional $l_1(\cdot,\cdot)$ are specified in point 2.1, and

$$l_2(y,v)=-\int_\Gamma(z_g-y)v\,d\Gamma .$$

If the constraints are absent, i.e. when $\mathcal{U}_\partial=\mathcal{U}$, then the equality

$$p+\bar{a}u=0 \tag{3.8}$$

follows from condition (3.5).

Therefore, when the constraints are absent, the control u can be excluded from equality (3.6) by means of equality (3.8), and the following may be written: $l_1(u,y)=l_1(u(p),y)$. If the solution $(y,p)^T$ to problem (3.6), (3.7), where $l_1(u,y)=l_1(u(p),y)$, is smooth enough on $\bar{\Omega}_1$ and $\bar{\Omega}_2$, then such solution satisfies the relations

$$-\sum_{i,j=1}^{n}\frac{\partial}{\partial x_i}\left(k_{ij}\frac{\partial y}{\partial x_j}\right)+p/\bar{a}+\int_\Omega y\,dx=f+Q, \quad x\in\Omega_1\cup\Omega_2,$$

$$-\sum_{i,j=1}^{n}\frac{\partial}{\partial x_i}\left(k_{ij}\frac{\partial p}{\partial x_j}\right)+\int_{\Omega}p\,dx=0,\quad x\in\Omega_1\bigcup\Omega_2,$$

$$\sum_{i,j=1}^{n}k_{ij}\frac{\partial y}{\partial x_j}\cos(\nu,x_i)=g,\quad x\in\Gamma,$$

$$\sum_{i,j=1}^{n}k_{ij}\frac{\partial p}{\partial x_j}\cos(\nu,x_i)=y-z_g,\quad x\in\Gamma,$$

$$[y]=0,\quad [p]=0,\quad x\in\gamma,$$

$$\left[\sum_{i,j=1}^{n}k_{ij}\frac{\partial y}{\partial x_j}\cos(\nu,x_i)\right]=\omega,\quad\left[\sum_{i,j=1}^{n}k_{ij}\frac{\partial p}{\partial x_j}\cos(\nu,x_i)\right]=0,\quad x\in\gamma.\quad(3.9)$$

Definition 3.1. A generalized (weak) solution to boundary-value problem (3.9) is called a vector-function $(y,p)^{\mathrm{T}}\in H$ that satisfies the following integral equation $\forall z\in H$:

$$\int_{\Omega}\left\{\sum_{i,j=1}^{n}k_{ij}\frac{\partial y}{\partial x_j}\frac{\partial z_1}{\partial x_i}+p\,z_1/\bar{a}+\sum_{i,j=1}^{n}k_{ij}\frac{\partial p}{\partial x_j}\frac{\partial z_2}{\partial x_i}\right\}dx+$$

$$+\int_{\Omega}y\,dx\int_{\Omega}z_1\,dx+\int_{\Omega}p\,dx\int_{\Omega}z_2\,dx=$$

$$=\int_{\Omega}(f+Q)\,z_1\,dx+\int_{\Gamma}g\,z_1\,d\Gamma+$$

$$+\int_{\Gamma}(y-z_g)z_2\,d\Gamma-\int_{\gamma}\omega\,z_1\,d\gamma.\quad(3.10)$$

Let $u = (u_1, u_2)^T$ and $v = (v_1, v_2)^T$ be arbitrary elements of the complete Hilbert space H. Specify the bilinear form

$$a(u,v) = \int_\Omega \left\{ \sum_{l=1}^{2} \sum_{i,j=1}^{n} k_{ij} \frac{\partial u_l}{\partial x_j} \frac{\partial v_l}{\partial x_i} + u_2 v_1 / \overline{a} \right\} dx +$$

$$+ \sum_{l=1}^{2} \int_\Omega u_l \, dx \int_\Omega v_l \, dx - \int_\Gamma u_1 v_2 \, d\Gamma$$

and linear functional

$$l(v) = \int_\Omega (f + Q) v_1 \, dx + \int_\Gamma g v_1 \, d\Gamma - \int_\gamma \omega v_1 \, d\gamma - \int_\Gamma z_g v_2 \, d\Gamma$$

on H.

Let constraint like (2.11) be met. Then, the unique solution $(y, p)^T$ to problem (3.10) exists in H. Problem (3.10) can be solved by means of the finite-element method. Estimates like (1.35) and (1.35′) take place, respectively, for the approximate solution $U_k^N \in H_k^N$ to problem like (3.10) and for the approximation $u_k^N(x)$ of the control u.

2.4   CONTROL UNDER CONJUGATION CONDITION WITH BOUNDARY OBSERVATION

Assume that equation (1.1), where the coefficients and right-hand side meet conditions (1.2′), is specified in the domains Ω_1 and Ω_2. Neumann condition (1.2), constraint (1.3) and the condition

$$\left[\sum_{i,j=1}^{n} k_{ij} \frac{\partial y}{\partial x_j} \cos(\nu, x_i) \right] = \omega + u, \quad x \in \gamma, \tag{4.1}$$

where ω is a fixed function from $L_2(\gamma)$ and the control is $u \in \mathcal{U} = L_2(\gamma)$, are specified, in their turn, on the boundary Γ.

For every control $u \in \mathcal{U}$, determine a system state as a generalized solution to the boundary-value problem specified by equation (1.1) and by conditions (1.2), (1.3) and (4.1) (Problem 4). The equality

$$\int_\Omega f \, dx + \int_\Gamma g \, d\Gamma = \int_\gamma (\omega + u) \, d\gamma \qquad (4.2)$$

is the necessary condition under which there exists a classical solution to the latter problem. Find this solution under constraint (1.6).

Bring a value of the cost functional

$$J(u) = \int_\Gamma \left(y(u) - z_g \right)^2 d\Gamma + (\mathcal{N}u, u)_{L_2(\gamma)} \qquad (4.3)$$

in correspondence with every control $u \in \mathcal{U} = L_2(\gamma)$; in this case, z_g is a known element from the space $L_2(\Gamma)$, $\mathcal{N}u = \bar{a}\,u$, $0 < a_0 \le \bar{a}(x) \le \le a_1 < \infty$, $\bar{a} \in L_2(\gamma)$, $a_0, a_1 = \text{const}$.

A unique state, namely, a function $y(u) \in V_Q$ corresponds to every control $u \in \mathcal{U}$, minimizes the functional

$$\Phi(v) = a_4(v, v) - 2l_4(v) \qquad (4.4)$$

on V_Q, and it is the unique solution in V_Q to the weakly stated problem: Find a function $y \in V_Q$ that meets the equation

$$a_4(y, v) = l_4(v) \quad \forall v \in V_0, \qquad (4.5)$$

where

$$a_4(y, v) = \int_\Omega \sum_{i,j=1}^n k_{ij} \frac{\partial y}{\partial x_j} \frac{\partial v}{\partial x_i} \, dx,$$

$$l_4(v) = \int_\Omega fv \, dx + \int_\Gamma gv \, d\Gamma - \int_\gamma (\omega + u)v \, d\gamma. \qquad (4.6)$$

Lemma 4.1. Problems *(4.4) and (4.5) are equivalent* $\forall f \in L_2(\Omega)$, $\forall \omega \in L_2(\gamma)$, $\forall u \in \mathcal{U}$ *and have a unique solution* $y \in V_Q$.

Rewrite cost functional (4.3) in the form of expression (3.2), where

$$\pi(u,v) = \left(y(u) - y(0),\ y(v) - y(0)\right)_{L_2(\Gamma)} + \left(\bar{a}\,u, v\right)_{L_2(\gamma)}$$

and

$$L(v) = \left(z_g - y(0),\ y(v) - y(0)\right)_{L_2(\Gamma)}.$$

Take the embedding theorems, ellipticity condition and generalized Poincare inequality into account, and the inequality

$$\bar{\alpha}_2 \left\| \tilde{y}' - \tilde{y}'' \right\|_{L_2(\Gamma)}^2 \le \bar{\alpha}_1 \left\| \tilde{y}' - \tilde{y}'' \right\|_V^2 \le a_4 \left(\tilde{y}' - \tilde{y}'', \tilde{y}' - \tilde{y}''\right) \le$$

$$\le c_1 \left\| u' - u'' \right\|_{L_2(\gamma)} \left\| \tilde{y}' - \tilde{y}'' \right\|_V,\quad \bar{\alpha}_1, \bar{\alpha}_2, c_1 = \mathrm{const} > 0,$$

i.e.

$$\left\| \tilde{y}' - \tilde{y}'' \right\|_{L_2(\Gamma)} \le \frac{c_1}{\sqrt{\bar{\alpha}_1 \bar{\alpha}_2,}} \left\| u' - u'' \right\|_{L_2(\gamma)},$$

is derived, where $\tilde{y}' = \tilde{y}(u')$ and $\tilde{y}'' = \tilde{y}(u'')$ are the generalized solutions to boundary-value Problem 4 under $f = 0$, $g = 0$ and $\omega = 0$ and under a function $u = u(x)$ that is equal, respectively, to u' and u''.

On the basis of the derived inequality and [58, Chapter 1, Theorem 1.1], the validity of the following statement is proved.

Theorem 4.1. *Let a system state y be determined as a solution to equivalent problems (4.4) and (4.5). Then, there exists a unique element* $u = u(x)$ *of a convex set* $\mathcal{U}_\partial$ *that is closed in* $\mathcal{U}$, *and relation like (1.17) takes place for* $u = u(x)$, *where the cost functional J(u) is specified by expression (4.3).*

Let the equation

$$-\sum_{i,j=1}^{n} \frac{\partial}{\partial x_i} \left(k_{ij} \frac{\partial y}{\partial x_j} \right) + \int_\Omega y\, dx = f(x) + Q \tag{4.7}$$

be specified on Ω instead of equation (1.1). Neumann condition (1.2) and conjugation conditions (1.3) and (4.1) are specified, in their turn, respectively, on Γ and γ. I.e., boundary-value problem (4.7), (1.2), (1.3), (4.1) (Problem 4$'$) is obtained.

Consider optimization Problem 4$'$: Find a control $u \in \mathcal{U}_\partial \subset \mathcal{U} = L_2(\gamma)$, for which relation like (1.17) is satisfied and where the cost functional $J(u)$ is specified by expression (4.3). A state $y = y(u)$ is a generalized solution to boundary-value Problem 4$'$, where the energy functional is

$$\Phi(v) = a'_4(v, v) - 2\, l'_4(v), \quad \forall v \in V, \tag{4.8}$$

and the weakly stated problem is to find a function $y \in V$ that meets the following equation $\forall z \in V$:

$$a'_4(y, z) = l'_4(z); \tag{4.8$'$}$$

in this case:

$$a'_4(y,v) = \int_\Omega \sum_{i,j=1}^n k_{ij} \frac{\partial y}{\partial x_j} \frac{\partial v}{\partial x_i} dx + \int_\Omega y\, dx \int_\Omega v\, dx,$$

$$l'_4(v) = \int_\Omega (f + Q)v\, dx + \int_\Gamma gv\, d\Gamma - \int_\gamma (\omega + u)v\, d\gamma. \tag{4.9}$$

If $u \in \mathcal{U}_\partial$ is the optimal control for optimization Problem 4$'$, then the following inequality is true $\forall v \in \mathcal{U}_\partial$:

$$\left(y(u) - z_g,\ y(v) - y(u)\right)_{L_2(\Gamma)} + \left(\bar{a}\, u, v - u\right)_{L_2(\gamma)} \geq 0. \tag{4.10}$$

As for the control $v \in \mathcal{U}$, the conjugate state $p(v) \in V^* = V$ is specified by relations like (3.4). The equality

$$0 = \left(A^* p(u),\ y(v) - y(u)\right) = a'_4\left(p,\ y(v) - y(u)\right) +$$

$$+ \left(z_g - y,\ y(v) - y(u)\right)_{L_2(\Gamma)} =$$

$$= -(p, v-u)_{L_2(\gamma)} + (z_g - y, \, y(v) - y(u))_{L_2(\Gamma)},$$

i.e.

$$(y(u) - z_g, y(v) - y(u))_{L_2(\Gamma)} = -(p, \, v-u)_{L_2(\gamma)}$$

is obtained. Take it into account, and the inequality

$$(-p + \bar{a} u, v - u)_{L_2(\gamma)} \geq 0, \quad \forall v \in \mathcal{U}_{\partial}, \tag{4.11}$$

is derived from inequality (4.10).

An element $u \in \mathcal{U}_{\partial}$ is an optimal control for optimization Problem 4′ if and only if inequality (4.11) and the equalities

$$a_4'(y, v) = l_1(u, v), \quad y \in V, \ \forall v \in V, \tag{4.12}$$

and

$$a_4'(p, v) = l_2(y, v), \quad p \in V, \ \forall v \in V, \tag{4.13}$$

are met; in this case, the bilinear form $a_4'(\cdot, \cdot)$ is specified by the first formula of expressions (4.9) and the functionals $l_1(\cdot, \cdot)$ and $l_2(\cdot, \cdot)$ are

$$l_1(u, v) = \int_{\Omega} (f + Q) v \, dx + \int_{\Gamma} g v \, d\Gamma - \int_{\gamma} (\omega + u) v \, d\gamma$$

and

$$l_2(y, v) = -\int_{\Gamma} (z_g - y) v \, d\Gamma.$$

If the constraints are absent, i.e. when $\mathcal{U}_{\partial} = \mathcal{U}$, then the equality

$$-p + \bar{a} u = 0, \quad x \in \gamma, \tag{4.14}$$

follows from condition (4.11).

Therefore, when the constraints are absent and if the solution $(y, p)^{\mathrm{T}}$ to problem (4.12)–(4.14) is smooth enough on $\bar{\Omega}_i \, (i = 1, 2)$, then the boundary-value problem is obtained:

$$-\sum_{i,j=1}^{n} \frac{\partial}{\partial x_i} \left(k_{ij} \frac{\partial y}{\partial x_j} \right) + \int_{\Omega} y \, dx = f + Q, \quad x \in \Omega_1 \cup \Omega_2,$$

$$-\sum_{i,j=1}^{n}\frac{\partial}{\partial x_i}\left(k_{ij}\frac{\partial p}{\partial x_j}\right)+\int_{\Omega}p\,dx=0,\quad x\in\Omega_1\cup\Omega_2,$$

$$\sum_{i,j=1}^{n}k_{ij}\frac{\partial y}{\partial x_j}\cos(\nu,x_i)=g,\quad x\in\Gamma,$$

$$\sum_{i,j=1}^{n}k_{ij}\frac{\partial p}{\partial x_j}\cos(\nu,x_i)=y-z_g,\quad x\in\Gamma,$$

$$[y]=0,\quad[p]=0,\quad x\in\gamma,$$

$$\left[\sum_{i,j=1}^{n}k_{ij}\frac{\partial y}{\partial x_j}\cos(\nu,x_i)\right]=\omega+p/\bar{a},\quad x\in\gamma,$$

$$\left[\sum_{i,j=1}^{n}k_{ij}\frac{\partial p}{\partial x_j}\cos(\nu,x_i)\right]=0,\quad x\in\gamma.\qquad(4.15)$$

Definition 4.1. A generalized (weak) solution to boundary-value problem (4.15) is called a vector-function $(y,p)^{\mathrm{T}}\in H$ that satisfies the following integral equation $\forall z\in H$:

$$\int_{\Omega}\left\{\sum_{i,j=1}^{n}k_{ij}\frac{\partial y}{\partial x_j}\frac{\partial z_1}{\partial x_i}+\sum_{i,j=1}^{n}k_{ij}\frac{\partial p}{\partial x_j}\frac{\partial z_2}{\partial x_i}\right\}dx+$$

$$+\int_{\Omega}y\,dx\int_{\Omega}z_1\,dx+\int_{\Omega}p\,dx\int_{\Omega}z_2\,dx=\int_{\Omega}(f+Q)z_1\,dx+$$

$$+\int_{\Gamma}g\,z_1\,d\Gamma+\int_{\Gamma}\left(y-z_g\right)z_2\,d\Gamma-\int_{\gamma}(\omega+p/\bar{a})z_1\,d\gamma.\qquad(4.16)$$

Let $u=(u_1,u_2)^{\mathrm{T}}$ and $v=(v_1,v_2)^{\mathrm{T}}$ be arbitrary elements of the complete Hilbert space H. Specify the bilinear form

$$a(u,v) = \int_{\Omega}\left\{\sum_{l=1}^{2}\sum_{i,j=1}^{n}k_{ij}\frac{\partial u_l}{\partial x_j}\frac{\partial v_l}{\partial x_i}\right\}dx +$$

$$+\sum_{l=1}^{2}\int_{\Omega}u_l\,dx\int_{\Omega}v_l\,dx - \int_{\Gamma}u_1v_2\,d\Gamma + \int_{\gamma}u_2v_1/\bar{a}\,d\gamma$$

and linear functional

$$l(v) = \int_{\Omega}(f+Q)v_1\,dx + \int_{\Gamma}gv_1\,d\Gamma - \int_{\Gamma}z_gv_2\,d\Gamma - \int_{\gamma}\omega v_1\,d\gamma$$

on H.

If the constraint

$$\min\left\{\frac{\alpha_0}{2}, \min\left\{\frac{\alpha_0}{2}, 1\right\}\mu\right\} - \frac{c_0'^2}{2a_0} - \frac{c_0^2}{2} > 0, \qquad (4.16')$$

where μ is the constant in the Poincare inequality and c_0' and c_0 are the positive constants derived on the basis of the inequalities proved within the framework of the embedding theorems, is met, then the unique solution $(y,p)^{\mathrm{T}}$ to problem (4.16) exists in H. Estimate like (1.35) is true for its approximate solution $U_k^N \in H_k^N$ and the estimate

$$\left\|u - u_k^N\right\|_{L_2(\gamma)} \le ch^k \qquad (4.17)$$

takes place for the approximation $u_k^N(x)$ of the control u.

2.5 BOUNDARY CONTROL WITH OBSERVATION ON A THIN INCLUSION

Assume that equation (1.1), where the coefficients and right-hand side meet conditions (1.2'), is specified in the domains Ω_1 and Ω_2. The condition

$$\sum_{i,j=1}^{n} k_{ij} \frac{\partial y}{\partial x_j} \cos(\nu, x_i) = g + u \tag{5.1}$$

is specified, in its turn, on the boundary Γ and the conjugation conditions have the form of expressions (1.3) and (1.4) on γ, where g is a fixed function from $L_2(\Gamma)$ and the control is $u \in \mathcal{U} = L_2(\Gamma)$.

For every control $u \in \mathcal{U}$, determine a system state as a generalized solution to the boundary-value problem specified by equation (1.1) and by constraints (1.3), (1.4) and (5.1) (Problem 5). The equality

$$\int_{\Omega} f \, dx + \int_{\Gamma} g \, d\Gamma + \int_{\Gamma} u \, d\Gamma = \int_{\gamma} \omega \, d\gamma \tag{5.2}$$

is the necessary condition under which there exists a classical solution y to Problem 5: Find this solution under constraint (1.6).

Bring a value of the cost functional

$$J(u) = \int_{\gamma} \left(y(u) - z_g \right)^2 d\gamma + (\mathcal{N} u, u)_{L_2(\Gamma)} \tag{5.3}$$

in correspondence with every control $u \in \mathcal{U}$; in this case, z_g is a known element from the space $L_2(\gamma)$, $\mathcal{N} u = \bar{a} u$, $0 < a_0 \leq \bar{a}(x) \leq a_1 < \infty$, $\bar{a} \in L_2(\Gamma)$; $a_0, a_1 = \mathrm{const}$.

A unique state, namely, a function $y(u) \in V_Q$ corresponds to every control $u \in \mathcal{U}$, delivers the minimum to functional (4.4) on V_Q, and it is the unique solution in V_Q to the weakly stated problem specified by equation like (4.5), where

$$l_4(v) = \int_{\Omega} f \, v \, dx + \int_{\Gamma} (g + u) v \, d\Gamma - \int_{\gamma} \omega v \, d\gamma . \tag{5.4}$$

Lemma 5.1. *Problems like (4.4) and (4.5), where the bilinear form $a_4(\cdot, \cdot)$ is specified by the first formula of expressions (4.6) and the linear*

functional $l_4(\cdot)$ is specified by formula (5.4), are equivalent $\forall f \in L_2(\Omega)$,
$\forall \omega \in L_2(\gamma)$, $\forall u \in \mathcal{U}$ and have a unique solution $y(u) \in V_Q$.

Rewrite cost functional (5.3) as

$$J(u) = \pi(u, u) - 2L(u) + \left\| z_g - y(0) \right\|^2_{L_2(\gamma)},$$

where

$$\pi(u, v) = \left(y(u) - y(0), \; y(v) - y(0) \right)_{L_2(\gamma)} + \left(\bar{a} u, v \right)_{L_2(\Gamma)}$$

and

$$L(v) = \left(z_g - y(0), \; y(v) - y(0) \right)_{L_2(\gamma)}.$$

Take the embedding theorems, ellipticity condition and generalized Poincare inequality into account, and the inequality

$$\bar{\alpha}_2 \left\| \tilde{y}' - \tilde{y}'' \right\|^2_{L_2(\gamma)} \leq \bar{\alpha}_1 \left\| \tilde{y}' - \tilde{y}'' \right\|^2_V \leq a_4 \left(\tilde{y}' - \tilde{y}'', \tilde{y}' - \tilde{y}'' \right) \leq$$

$$\leq c_0 \left\| u' - u'' \right\|_{L_2(\Gamma)} \left\| \tilde{y}' - \tilde{y}'' \right\|_V, \quad \bar{\alpha}_1, \bar{\alpha}_2, c_0 = \text{const} > 0,$$

i.e. $\left\| \tilde{y}' - \tilde{y}'' \right\|_{L_2(\gamma)} \leq c_1 \left\| u' - u'' \right\|_{L_2(\Gamma)}$ is derived, where $\tilde{y}' = \tilde{y}(u')$ and $\tilde{y}'' = \tilde{y}(u'')$.

On the basis of the derived inequality and [58, Chapter 1, Theorem 1.1], the validity of the following statement is proved.

Theorem 5.1. *If a system state y is determined as a solution to equivalent problems (4.4) and (4.5) that correspond to boundary-value Problem 5, then there exists a unique element $u = u(x)$ of a convex set $\mathcal{U}_\partial$ that is closed in $\mathcal{U}$, and relation like (1.17) takes place for $u = u(x)$, where the cost functional $J(u)$ is specified by expression (5.3).*

Let equation (4.7) be specified on Ω instead of equation (1.1). Condition (5.1) and constraints (1.3) and (1.4) are specified, in their turn, respectively, on Γ and γ. I.e., boundary-value problem (4.7), (1.3), (1.4), (5.1) (Problem 5$'$) is obtained.

Consider the following problem: Find a control $u \in \mathcal{U}_\partial \subset \mathcal{U} = L_2(\Gamma)$ for which relation like (1.17) is satisfied and where the cost functional $J(u)$ is specified by expression (5.3); a state $y = y(u)$ is a generalized solution to boundary-value Problem 5′, where the energy functional and weakly stated problem are given, respectively, by expression (4.8) and equality (4.8′). The form $a'_4(\cdot\,,\,\cdot)$ is specified by expression (4.9) and the linear functional is

$$l'_4(v) = \int_\Omega (f + Q)v dx + \int_\Gamma (g + u)v d\Gamma - \int_\gamma \omega v d\gamma.$$

If $u \in \mathcal{U}_\partial$ is the optimal control for optimization Problem 5′, then the following inequality is true $\forall v \in \mathcal{U}_\partial$:

$$\left(y(u) - z_g, \, y(v) - y(u) \right)_{L_2(\gamma)} + \left(\bar{a}\,u, v - u \right)_{L_2(\Gamma)} \geq 0. \tag{5.5}$$

As for the control $v \in \mathcal{U}$, the conjugate state $p(v) \in V^* = V$ is specified by the relations

$$A^* p(v) = 0, \quad x \in \Omega_1 \cup \Omega_2,$$

$$\frac{\partial p}{\partial v}\Big|_{A^*} = 0, \quad x \in \Gamma,$$

$$[p] = 0, \quad \left[\frac{\partial p}{\partial v}_{A^*} \right] = z_g - y, \quad x \in \gamma, \tag{5.6}$$

where the operators A^* and $\dfrac{\partial}{\partial v}_{A^*}$ are specified, in their turn, by expressions (1.23).

The equality

$$0 = \left(A^* p(u), \, y(v) - y(u) \right) = a'_4\left(p, \, y(v) - y(u) \right) +$$

$$+ \left(z_g - y(u), \, y(v) - y(u) \right)_{L_2(\gamma)} =$$

$$= \left(p, v - u \right)_{L_2(\Gamma)} + \left(z_g - y(u), y(v) - y(u) \right)_{L_2(\gamma)},$$

i.e. $\left(y(u)-z_g, y(v)-y(u)\right)_{L_2(\gamma)} = \left(p, v-u\right)_{L_2(\Gamma)}$ is obtained. Take it into account, and the inequality

$$\left(p+\bar{a}u, v-u\right)_{L_2(\Gamma)} \geq 0 \tag{5.7}$$

is derived from inequality (5.5).

To make the element $u \in \mathcal{U}_\partial$ an optimal control for optimization Problem 5', it is necessary and sufficient to meet inequality (5.7) and equalities like (4.12) and (4.13), where the bilinear form $a'_4(\cdot, \cdot)$ is specified by expression (4.9) and, besides this, the linear functionals are

$$l_1(u,v) = \int_\Omega (f+Q)v\,dx + \int_\Gamma (g+u)v\,d\Gamma - \int_\gamma \omega v\,d\gamma$$

and

$$l_2(y,v) = -\int_\gamma \left(z_g - y\right)v\,d\gamma.$$

If the constraints are absent, i.e. when $\mathcal{U}_\partial = \mathcal{U}$, then the equality

$$p+\bar{a}u = 0, \ x \in \Gamma, \tag{5.8}$$

follows from condition (5.7).

Therefore, when the constraints are absent and if the solution $(y, p)^T$ to problem (4.12), (4.13), (5.8) is smooth enough on $\bar{\Omega}_i$ $(i = 1, 2)$, then, take equality (5.8) into account, and the boundary-value problem is obtained:

$$-\sum_{i,j=1}^n \frac{\partial}{\partial x_i}\left(k_{ij}\frac{\partial y}{\partial x_j}\right) + \int_\Omega y\,dx = f+Q, \ x \in \Omega_1 \bigcup \Omega_2,$$

$$-\sum_{i,j=1}^n \frac{\partial}{\partial x_i}\left(k_{ij}\frac{\partial p}{\partial x_j}\right) + \int_\Omega p\,dx = 0, \ x \in \Omega_1 \bigcup \Omega_2,$$

$$\sum_{i,j=1}^n k_{ij}\frac{\partial y}{\partial x_j}\cos(v,x_i) = g - p/\bar{a}, \ x \in \Gamma,$$

$$\sum_{i,j=1}^{n} k_{ij} \frac{\partial p}{\partial x_j} \cos(v, x_i) = 0, \quad x \in \Gamma, \tag{5.9}$$

$$[y] = 0, \quad [p] = 0, \quad x \in \gamma,$$

$$\left[\sum_{i,j=1}^{n} k_{ij} \frac{\partial y}{\partial x_j} \cos(v, x_i) \right] = \omega, \quad x \in \gamma,$$

$$\left[\sum_{i,j=1}^{n} k_{ij} \frac{\partial p}{\partial x_j} \cos(v, x_i) \right] = z_g - y, \quad x \in \gamma.$$

Definition 5.1. A generalized (weak) solution to boundary-value problem (5.9) is called a vector function $(y, p)^T \in H$ that satisfies the following integral equation $\forall z \in H$:

$$\int_{\Omega} \left\{ \sum_{i,j=1}^{n} k_{ij} \frac{\partial y}{\partial x_j} \frac{\partial z_1}{\partial x_i} + \sum_{i,j=1}^{n} k_{ij} \frac{\partial p}{\partial x_j} \frac{\partial z_2}{\partial x_i} \right\} dx +$$

$$+ \int_{\Omega} y \, dx \int_{\Omega} z_1 \, dx + \int_{\Omega} p \, dx \int_{\Omega} z_2 \, dx = \int_{\Omega} (f + Q) z_1 \, dx +$$

$$+ \int_{\Gamma} (g - p/\bar{a}) z_1 \, d\Gamma - \int_{\gamma} \omega z_1 \, d\gamma - \int_{\gamma} (z_g - y) z_2 \, d\gamma. \tag{5.10}$$

Let $u = (u_1, u_2)^T$ and $v = (v_1, v_2)^T$ be arbitrary elements of the complete Hilbert space H. Specify the bilinear form

$$a(u, v) = \int_{\Omega} \left\{ \sum_{l=1}^{2} \sum_{i,j=1}^{n} k_{ij} \frac{\partial u_l}{\partial x_j} \frac{\partial v_l}{\partial x_i} \right\} dx +$$

$$+ \sum_{l=1}^{2} \int_{\Omega} u_l \, dx \int_{\Omega} v_l \, dx + \int_{\Gamma} u_2 v_1 / \bar{a} \, d\Gamma - \int_{\gamma} u_1 v_2 \, d\gamma$$

and linear functional

$$l(v) = \int_{\Omega} (f + Q) v_1 \, dx + \int_{\Gamma} g \, v_1 \, d\Gamma - \int_{\gamma} \omega v_1 \, d\gamma - \int_{\gamma} z_g v_2 \, d\gamma$$

on H.

If constraint like (4.16′) is met, then the unique solution $(y, p)^{\mathrm{T}}$ to problem (5.10) exists in H. Estimate like (1.35) is true for its approximate solution $U_k^N \in H_k^N$ and the estimate

$$\left\| u - u_k^N \right\|_{L_2(\Gamma)} \le c h^k$$

takes place for the approximation $u_k^N(x)$ of the control u.

CONTROL OF A SYSTEM DESCRIBED BY A ONE-DIMENSIONAL QUARTIC EQUATION UNDER CONJUGATION CONDITIONS

3.1 DISTRIBUTED CONTROL WITH OBSERVATION THROUGHOUT A WHOLE DOMAIN

Assume that the equation

$$\frac{d^2}{dx^2}\left(k\frac{d^2y}{dx^2}\right) = f(x) \tag{1.1}$$

is specified in a domain $\Omega = \Omega_1 \cup \Omega_2$ ($\Omega_1 = (0,\xi)$, $\Omega_2 = (\xi, l)$, $0 < \xi < l$), where $k\big|_{\bar{\Omega}_l} = k(x)\big|_{\bar{\Omega}_l} \in C^1(\bar{\Omega}_l) \cap C^2(\Omega_l)$, $0 < k_0 \le k(x) \le k_1 < \infty$, $k_0, k_1 = \text{const}$, $f\big|_{\Omega_l} \in C(\Omega_l)$, $l = 1,2$, $|f| < \infty$.

The conditions

$$y = y' = 0 \tag{1.2}$$

are specified, in their turn, at the ends of a line segment $[0, l]$.

At a point $x = \xi$, the conjugation conditions are

$$y^- = y^+ = 0 \tag{1.3}$$

and

$$[ky''] = 0, \quad \{ky''\}^\pm = \alpha[y']. \tag{1.4}$$

Problem (1.1)–(1.4) describes deflections of a complicated rod that is rigidly fixed at its ends and has a hinge of a final rigidity $\alpha > 0$, and such a hinge is absolutely rigidly supported at the point $x = \xi$; in this case, $y = y(x)$ is a deflection of a rod at a point with a coordinate x, $[\varphi] = = \varphi^+ - \varphi^-$, $\varphi^\pm = \{\varphi\}^\pm = \varphi(\xi \pm 0)$.

Let there be a control Hilbert space $\mathcal{U}$ and mapping $B \in \mathcal{L}(\mathcal{U}; V')$, where V' is a space dual with respect to a state Hilbert space V. Assume the following: $\mathcal{U} = L_2(\Omega)$.

For every control $u \in \mathcal{U}$, determine a system state y as a generalized solution to the boundary-value problem specified by the equation

$$\frac{d^2}{dx^2}\left(k \frac{d^2 y}{dx^2} \right) = f(x) + Bu, \quad y \in V, \tag{1.5}$$

and by conditions (1.2)–(1.4).

Specify the observation

$$Z(u) = C\,y(u), \tag{1.6}$$

where $C \in \mathcal{L}(V; \mathcal{H})$ and $\mathcal{H}$ is some Hilbert space. Assume the following:

$$Cy(u) \equiv y(u), \quad \mathcal{H} = V \subset L_2(\Omega). \tag{1.7}$$

Bring a value of the cost functional

$$J(u) = \left\| Cy(u) - z_g \right\|_{\mathcal{H}}^2 + (\mathcal{N}u, u)_{\mathcal{U}} \tag{1.8}$$

in correspondence with every control $u \in \mathcal{U}$; in this case, z_g is a known element of $\mathcal{H}$, and

$$\mathcal{N} \in \mathcal{L}(\mathcal{U}; \mathcal{U}), \quad (\mathcal{N}u, u)_{\mathcal{U}} \geq \nu_0 \|u\|_{\mathcal{U}}^2, \quad \nu_0 = \mathrm{const} > 0, \quad \forall u \in \mathcal{U}. \tag{1.9}$$

Assume the following: $f = L_2(\Omega)$, $Bu \equiv u \in L_2(\Omega)$, $\mathcal{N}u = \bar{a}(x)u$, and, in this case, $0 < a_0 \leq \bar{a}(x) \leq a_1 < \infty$, $\bar{a}(x)|_{\Omega_l} \in C(\Omega_l)$, $l = 1, 2$, $a_0, a_1 = \mathrm{const}$, $(\varphi, \psi)_{\mathcal{U}} = (\varphi, \psi) = \int\limits_{\Omega} \varphi\psi\,dx$. A unique state, namely, a function $y(u) \in V = \{v: v|_{\Omega_l} \in W_2^2(\Omega_l), \quad l = 1, 2; \quad v(0) = v'(0) = v(l) = $

$= v'(l) = 0; \quad y^- = y^+ = 0 \Big\}$ corresponds to every control $u \in \mathcal{U}$, delivers the minimum to the functional

$$\mathscr{P}(v) = a(v, v) - 2l(v) \tag{1.10}$$

on V, and it is the unique solution in V to the weakly stated problem: Find an element $y \in V$ that meets the equation

$$a(y, v) = l(v), \quad \forall v \in V, \tag{1.11}$$

where $a(u, v) = \int_0^l k u'' v'' dx + \alpha [u'][v'], \quad l(v) = \int_0^l (f + u) v dx$.

Introduce the following denotation: $\bar{H}_2^k = \{v(x): v|_{\Omega_l} \in W_2^k(\Omega_l), l = 1, 2\}$. The estimates

$$|a(u, v)| \le c_1 \|u\|_H \|v\|_H \text{ and } |l(v)| \le c_2 \|v\|_H \tag{1.12}$$

are true for the bilinear form $a(\cdot, \cdot): \bar{H}_2^2 \times \bar{H}_2^2 \to R^1$ and linear functional $l(\cdot): \bar{H}_2^2 \to R^1$. In this case, $\|v\|_H^2 = \sum_{i=1}^2 \|v\|_{W_2^2(\Omega_i)}^2$, where $\|\cdot\|_{W_2^2(\Omega_i)}$ is the norm of the Sobolev space $W_2^2(\Omega_i)$, i.e. the bilinear form $a(\cdot, \cdot)$ and linear functional $l(\cdot)$ are continuous on a complete Hilbert space $\bar{H}_2^2$ with the norm $\|\cdot\|_H$. Illustrate the H-ellipticity of the bilinear form $a(\cdot, \cdot)$ on the subspace $V \subset \bar{H}_2^2$. Take the Friedrichs inequality into account, and the following inequality is derived:

$$a(v, v) \ge \mu \int_0^l (v')^2 dx, \quad \mu = \text{const} > 0. \tag{1.13}$$

Consider the line segment $[\alpha, \beta]$. For an arbitrary element $v \in W_2^2(\alpha, \beta)$, that meets the conditions $v(\alpha) = v(\beta) = 0$, the equality

$$\int_{\alpha}^{\beta}(v')^2 dx = vv'\Big|_{\alpha}^{\beta} - \int_{\alpha}^{\beta} vv''dx = -\int_{\alpha}^{\beta} vv''dx \qquad (1.14)$$

is obtained. On the basis of the ε-, Cauchy-Bunyakovsky and Friedrichs inequalities, the inequality

$$\mu_1 \int_{\alpha}^{\beta} v^2 dx \le \varepsilon \int_{\alpha}^{\beta} v^2 dx + \frac{1}{4\varepsilon} \int_{\alpha}^{\beta}(v'')^2 dx, \quad \forall v \in V,$$

follows from equality (1.14), viz.:

$$\mu_2 \int_{\alpha}^{\beta} v^2 dx \le \int_{\alpha}^{\beta}(v'')^2 dx, \quad \mu_2 = const > 0. \qquad (1.15)$$

Consider inequalities (1.13) and (1.15), and

$$a(v,v) \ge \mu_3 \|v\|_H^2, \quad \mu_3 = const > 0. \qquad (1.16)$$

Use the Lax-Milgramm lemma [16], and it is concluded that problem (1.11) has the unique solution $\forall f, u \in L_2(\Omega)$ in V. It is easy to state the equivalence of problems (1.10) and (1.11).

Therefore, there exists such an operator A acting from V into $L_2(\Omega)$, that

$$y(u) = A^{-1}(f + Bu), \quad \forall u \in L_2(\Omega). \qquad (1.17)$$

Rewrite the cost functional as

$$J(u) = \pi(u,u) - 2L(u) + \|z_g - y(0)\|^2, \qquad (1.18)$$

where the bilinear form $\pi(\cdot,\cdot)$ and linear functional $L(\cdot)$ are expressed as

$$\pi(u,v) = \big(y(u) - y(0), y(v) - y(0)\big) + (\bar{a}u,v)$$

and

$$L(v) = \big(z_g - y(0), y(v) - y(0)\big); \qquad (1.19)$$

in this case: $(\varphi, \psi) = \int\limits_0^l \varphi \psi \, dx$, $\|\varphi\| = (\varphi, \varphi)^{1/2}$.

The form $\pi(\cdot, \cdot)$ is coercive on $\mathcal{U}$, i.e.:

$$\pi(u, u) = \big(y(u) - y(0),\ y(u) - y(0)\big) + \big(\bar{a}u, u\big) \geq a_0(u, u).$$

Let $\tilde{y}' = \tilde{y}(u')$ and $\tilde{y}'' = \tilde{y}(u'')$ be solutions from V to problem (1.11) under $f = 0$ and under a function $u = u(x)$ that is equal, respectively, to u' and u''. Then, the inequality

$$c_1 \|\tilde{y}' - \tilde{y}''\|^2 \leq c_1 \|\tilde{y}' - \tilde{y}''\|_H^2 \leq a\big(\tilde{y}' - \tilde{y}'', \tilde{y}' - \tilde{y}''\big) \leq \|u' - u''\| \|\tilde{y}' - \tilde{y}''\|,$$

$$c_1 = \text{const} > 0,$$

is derived that provides the continuity of the linear functional $L(\cdot)$ and bilinear form $\pi(\cdot, \cdot)$ on $\mathcal{U}$.

On the basis of [58, Chapter 1, Theorem 1.1], the validity of the following statement is proved.

Theorem 1.1. *Let a system state be determined as a solution to equivalent problems (1.10) and (1.11). Then, there exists a unique element u of a convex set $\mathcal{U}_\partial$ that is closed in $\mathcal{U}$, and*

$$J(u) = \inf_{v \in \mathcal{U}_\partial} J(v) \tag{1.20}$$

takes place for u.

Definition 1.1. If an element $u \in \mathcal{U}_\partial$ meets condition (1.20), it is called an optimal control.

If $u \in \mathcal{U}_\partial$ is the optimal control, then the following inequality is true:

$$\pi(u, v - u) \geq L(v - u), \quad \forall v \in \mathcal{U}_\partial. \tag{1.21}$$

Proceed from expressions (1.19), and the inequality

$$\big(y(u) - z_g,\ y(v) - y(u)\big) + \big(\bar{a}u, v - u\big) \geq 0, \quad \forall v \in \mathcal{U}_\partial, \tag{1.22}$$

follows from inequality (1.21), and it is the necessary and sufficient condition under which $u \in \mathcal{U}_\partial$ is the optimal control for the considered problem.

As for the control $v \in \mathcal{U}$, the conjugate state is specified by the relations

$$A^* p(v) = y(v) - z_g ,$$ (1.23)

$$p = 0, \quad p' = 0, \quad x = 0, l,$$ (1.24)

$$p^- = p^+ = 0$$ (1.25)

and

$$[kp''] = 0, \quad \{kp''\}^{\pm} = \alpha [p'],$$ (1.26)

where V^* is a space conjugate to V, $V^* = V$, and

$$A^* p = (kp'')''.$$ (1.27)

Further on, use the formula of integration by parts, and the equality

$$\left(A^* p(u), y(v) - y(u) \right) = \left(y(u) - z_g, y(v) - y(u) \right) = a \left(p, y(v) - y(u) \right) =$$

$$= \left(p(u), A(y(v) - y(u)) \right) = \left(p(u), v - u \right),$$ (1.28)

where

$$a(u,v) = \int_0^l k u'' v'' dx + \alpha [u'][v'],$$ (1.29)

is obtained. Take equality (1.28) into account, and it is stated that inequality (1.22) is equivalent to the inequality

$$(p(u) + \bar{a} u, v - u) \geq 0, \quad \forall v \in \mathcal{U}_\partial.$$ (1.30)

Therefore, the necessary condition for the existence of the optimal control $u \in \mathcal{U}_\partial$ is the one under which the relations

$$A y(u) = f + u,$$ (1.31)

$$A^* p(u) = y(u) - z_g$$ (1.32)

and

$$(p(u) + \bar{a}u, v - u) \geq 0, \ \forall v \in \mathcal{U}_\partial, \tag{1.33}$$

are met. If the constraints are absent, i.e. when $\mathcal{U}_\partial = \mathcal{U}$, then the equality

$$p(u) + \bar{a}u = 0 \tag{1.34}$$

follows from condition (1.33). Therefore, when the constraints are absent, the control $u(x)$ can be excluded from equality (1.31) by means of equality (1.34). On the basis of equalities (1.31) and (1.32), the problem

$$Ay + p/\bar{a} = f, \ y \in V, \tag{1.35}$$

$$A^*p - y = -z_g, \ p \in V^*, \tag{1.36}$$

is derived, and the vector solution $(y, p)^{\mathrm{T}}$ is found from this problem along with the optimal control

$$u = -p/\bar{a}, \tag{1.37}$$

where $\quad V^* = \left\{ v : v|_{\Omega_l} \in W_2^2(\Omega_l), \ l = 1, 2; \ v(0) = v'(0) = v(l) = v'(l) = 0; \right.$

$\left. v^+ = v^- = 0 \right\}.$

If the vector solution $(y, p)^{\mathrm{T}}$ to problem (1.35), (1.36) is smooth enough on $\bar{\Omega}_l$, viz. $y|_{\bar{\Omega}_l}, \ p|_{\bar{\Omega}_l} \in C^3(\bar{\Omega}_l) \cap C^4(\Omega_l), \ l = 1, 2,$ then the differential problem of finding the vector-function $(y, p)^{\mathrm{T}}$, that satisfies the relations

$$\left(ky'' \right)'' + p/\bar{a} = f, \ x \in \Omega_1 \cup \Omega_2,$$

$$\left(kp'' \right)'' - y = -z_g, \ x \in \Omega_1 \cup \Omega_2,$$

$$y(0) = y'(0) = y(l) = y'(l) = 0,$$

$$p(0) = p'(0) = p(l) = p'(l) = 0, \tag{1.37$'$}$$

$$y^- = y^+ = 0, \ p^- = p^+ = 0,$$

$$\left[ky'' \right] = 0, \ \left\{ ky'' \right\}^\pm = \alpha \left[y' \right],$$

$$[kp''] = 0, \quad \{kp''\}^{\pm} = \alpha[p'],$$

corresponds to problem (1.35), (1.36).

Definition 1.2. A generalized (weak) solution to boundary-value problem (1.37′) is called a vector-function $(y,p)^{\mathrm{T}} \in H = = \{v = (v_1, v_2)^{\mathrm{T}} : v_1, v_2 \in V\}$ that satisfies the following integral equation $\forall z \in H$:

$$\int_0^l \{ k\, y''z_1'' + p\, z_1/\bar{a} + k\, p''z_2'' - yz_2 \} dx +$$

$$+ \alpha\, [y'][z_1'] + \alpha\, [p'][z_2'] = \int_0^l \left(f z_1 - z_g z_2 \right) dx. \tag{1.38}$$

Let $u = (u_1, u_2)^{\mathrm{T}}$ and $v = (v_1, v_2)^{\mathrm{T}}$ be arbitrary elements of the complete Hilbert space H with the norm $\|v\|_H = \left\{ \sum_{i=1}^{2} \|v\|_{W_2^2(\Omega_i)}^2 \right\}^{1/2}$. Specify the bilinear form

$$a(u,v) = \int_0^l \{ k(u_1''v_1'' + u_2''v_2'') + u_2\, v_1/\bar{a} - u_1 v_2 \} dx +$$

$$+ \alpha \sum_{i=1}^{2} [u_i'][v_i'] \tag{1.39}$$

and linear functional

$$l(v) = \int_0^l \left(f v_1 - z_g v_2 \right) dx \tag{1.40}$$

on H.

Let the costraint

$$\mu_3 - \frac{1}{2}\left(1 + \frac{1}{a_0}\right) > 0 \tag{1.41}$$

be met, where $\mu_3 = \text{const} > 0$ is the constant in inequality (1.16).

Proceed from the Cauchy-Bunyakovsky and Friedrichs inequalities and from [21]

$$\left|v_l^-\right| \le \bar{c}_1 \|v_l\|_{W_2^1(\Omega_1)} \quad \text{and} \quad \left|v_l^+\right| \le \bar{c}_2 \|v_l\|_{W_2^1(\Omega_2)}, \quad l = 1,2, \tag{1.42}$$

and the inequalities

$$a(v,v) \ge \bar{\alpha}_1 \|v\|_H^2, \quad \forall v \in H, \quad \bar{\alpha}_1 = \text{const} > 0,$$

$$|a(u,v)| \le c_1 \|u\|_H \|v\|_H, \quad \forall u,v \in H, \quad c_1 = \text{const} > 0, \tag{1.41'}$$

are obtained for the bilinear form $a(\cdot,\cdot)$, i.e. this form is H-elliptic and continuous on H [49].

Consider also the Cauchy-Bunyakovsky inequality, and

$$|l(v)| \le c_2 \|v\|_H, \quad c_2 = \text{const} > 0.$$

Use the Lax-Milgramm lemma [16], and it is concluded that the unique solution $(y,p)^T$ to problem (1.38) exists in H.

Problem (1.38) can be solved approximately by means of the finite-element method. For this purpose, divide the line segments $\bar{\Omega}_i$ into the elementary ones, i.e. $\bar{e}_i^j, j = \overline{1,N_i}, i = 1,2$. Specify the subspace $H_k^N \subset H$ ($N = N_1 + N_2$) of the vector-functions $V_k^N(x)$. The components $v_{1k}^N, v_{2k}^N\big|_{\bar{\Omega}_l} \in C^1(\bar{\Omega}_i)$ ($i = 1,2$) of $V_k^N(x)$ are the complete polynomials of the power $k \ge 3$ that contain the variable x at every elementary line segment $\bar{e}_i^j$. Then, the linear algebraic equation system

$$A\bar{U} = B \tag{1.43}$$

follows from equation (1.38), and the solution $\bar{U}$ to system (1.43) exists and such solution is unique. The vector $\bar{U}$ specifies the unique approximate solution $U_k^N \in H_k^N$ to problem (1.38) as the unique one to the equation

$$a\left(U_k^N, V_k^N\right) = l\left(V_k^N\right), \ \forall V_k^N \in H_k^N. \tag{1.44}$$

Let $U = U(x) \in H$ be the solution to problem (1.38). Then:

$$a\left(U - U_k^N, V_k^N\right) = 0, \ \forall V_k^N \in H_k^N. \tag{1.45}$$

Therefore $\forall \tilde{U} \in H_k^N$, the inequality

$$\bar{\alpha}_1 \left\| U - U_k^N \right\|_H^2 \le a(U - U_k^N, U - U_k^N) \le c_1 \left\| U - U_k^N \right\|_H \left\| U - \tilde{U} \right\|_H$$

is obtained, i.e.:

$$\left\| U - U_k^N \right\|_H \le \frac{c_1}{\bar{\alpha}_1} \left\| U - \tilde{U} \right\|_H. \tag{1.46}$$

Suppose that $\tilde{U} \in H_k^N$ is a complete interpolation polynomial for the solution U to problem (1.38) at every $\bar{e}_i^j$. Take the interpolation estimates into account, assume that every component U_1 and U_2 of the solution U on Ω_l has the continuous limited $(k+1)$-th-order derivative, and the estimate

$$\left\| U - U_k^N \right\|_H \le ch^{k-1}, \tag{1.47}$$

where h is a length of the largest finite element $\bar{e}_i^j$, follows from inequality (1.46).

Take estimate (1.47) into consideration, and the estimate

$$\left\| u - u_k^N \right\|_{W_2^2} \le c_2 \left\| p - p_k^N \right\|_{W_2^2} \le c_3 \, h^{k-1},$$

where $\|\cdot\|_{W_2^2} = \left\{ \sum_{i=1}^{2} \|\cdot\|_{W_2^2(\Omega_i)}^2 \right\}^{1/2}$, takes place for the approximation

$u_k^N(x) = -p_k^N(x)/\bar{a}(x)$ of the control $u = u(x)$.

3.2 CONTROL UNDER CONJUGATION CONDITION

Assume that equation (1.1) is specified on intervals $(0, \xi)$ and (ξ, l). The boundary conditions

$$y(0) = 0, \quad y(l) = 0, \quad y'(l) = 0 \tag{2.1}$$

and

$$y''(0) = 0 \tag{2.2}$$

are specified, in their turn, at the ends of the line segment $[0, l]$. At the point $x = \xi$, the conjugation conditions are

$$[y] = 0,$$

$$[ky''] = 0, \quad \{ky''\}^{\pm} = \alpha[y'], \quad \left[(ky'')'\right] = -\beta y + r + u, \tag{2.3}$$

where $\alpha, \beta = \text{const} > 0$, $u, r \in R^1$.

Specify the observation as

$$Cy(u) \equiv y(u), \quad x \in \Omega. \tag{2.4}$$

Bring a value of the cost functional

$$J(u) = \int_0^l \left(y(u) - z_g\right)^2 dx + (\mathcal{N}u, u)_{\mathcal{U}} \tag{2.5}$$

in correspondence with every control $u \in \mathcal{U}$; in this case, z_g is a known element from $L_2(\Omega)$, $\mathcal{N}u = \bar{a}u$, $\bar{a} \in R^1$, $\bar{a} > 0$, $(\varphi, \psi)_{\mathcal{U}} = \varphi\psi$, $\forall \varphi, \psi \in R^1$, $\mathcal{U} = R^1$.

A unique state, namely, a function $y(u) \in V$ corresponds to every control $u \in \mathcal{U}$, minimizes the energy functional

$$\mathcal{P}(v) = a(v,v) - 2l(v) \tag{2.6}$$

on V, and it is the unique solution in V to the weakly stated problem: Find an element $y(u) \in V$ that meets the equation

$$a(y,v) = l(v), \quad \forall v \in V, \tag{2.7}$$

where $V = \left\{ v \in \bar{H}_2^2 : v(0) = 0, \ v(l) = v'(l) = 0, \ [v] = 0 \right\}$,

$$a(y,v) = \int_0^l k y'' v'' dx + \beta y(\xi) v(\xi) + \alpha [y'][v'],$$

$$l(v) = \int_0^l f v dx + (r+u) v(\xi). \tag{2.8}$$

To establish the fact that the solution $y(u)$ to equivalent problems (2.6) and (2.7) exists and that this solution is unique $\forall u \in R^1$, some additional investigations are needed. Consider the following expression $\forall y \in V$:

$$\int_0^l (y')^2 dx = y \, y'\Big|_0^{\xi-0} + y \, y'\Big|_{\xi+0}^l - \int_0^l y y'' dx = -y(\xi)[y'] - \int_0^l y y'' dx.$$

Take the ε-, Friedrichs and Cauchy-Bunyakovsky inequalities into account, and the inequalities

$$\mu \int_0^l y^2 dx \le \int_0^l (y')^2 dx \le \varepsilon_1 y^2(\xi) + \frac{1}{4\varepsilon_1}[y']^2 +$$

$$+ \varepsilon_2 \int_0^l y^2 dx + \frac{1}{4\varepsilon_2} \int_0^l (y'')^2 dx \tag{2.9}$$

and

$$(\mu - \varepsilon_2) \int_0^l y^2 dx \le \varepsilon_1 y^2(\xi) + \frac{1}{4\varepsilon_1}[y']^2 + \frac{1}{4\varepsilon_2} \int_0^l (y'')^2 dx, \tag{2.10}$$

where $\varepsilon_2 \in (0, \mu)$, are derived. The inequality

$$\|y\|_{1,\Omega}^2 \le c\left(y^2(\xi) + [y']^2 + \int_0^l (y'')^2\, dx \right), \quad \forall y \in V, \qquad (2.11)$$

where $\quad \|y\|_{k,\Omega} = \left\{ \sum_{i=1}^2 \|y\|_{W_2^k(\Omega_i)}^2 \right\}^{1/2}, \quad k = 1,2, \quad \|y\|_{0,\Omega} = \int_0^l y^2 dx, \quad$ follows

from inequalities (2.9) and (2.10).

Therefore, the validity of the following statement is proved.

Lemma 2.1. *Inequality (2.11) takes place for an arbitrary function*

$y \in V_{2,0}^2 = \left\{ v \in \bar{H}_2^2 : v(0) = v(l) = 0, \ [v] = 0 \right\}.$

Consider inequalities (1.42) and (2.11) and the Cauchy-Bunyakovsky and Friedrichs ones, and it is easy to show that inequalities like (1.41′) are true under the fixed $\forall u \in R^1$, where $a(\cdot,\cdot)$ and $l(\cdot)$ have the form of expressions (2.8). Use the Lax-Milgramm lemma, and it is concluded that, at every $u \in R^1$, the solution to equivalent problems (2.6) and (2.7) exists and that such solution is unique.

Expression like (1.18) is obtained from expression (2.5). For expression like (1.18), the bilinear form $\pi(\cdot,\cdot)$ is

$$\pi(u,v) = \big(y(u) - y(0), y(v) - y(0) \big) + \bar{a}\, uv \qquad (2.12)$$

that meets the inequality

$$\pi(u,u) \ge \bar{a}\, u^2, \qquad (2.13)$$

and the linear functional $L(\cdot)$ is specified by expression (1.19).

Let $\tilde{y}' = \tilde{y}(u')$ and $\tilde{y}'' = \tilde{y}(u'')$ be solutions from V to problem (2.7) under $f = 0$ and $r = 0$ and under a value u that is equal, respectively, to u' and u''. Then:

$$\bar{\alpha}_1 \|\tilde{y}' - \tilde{y}''\|_{0,\Omega}^2 \le \bar{\alpha}_1 \|\tilde{y}' - \tilde{y}''\|_H^2 \le a\big(\tilde{y}' - \tilde{y}'', \tilde{y}' - \tilde{y}''\big) =$$

$$= (u' - u'')(\tilde{y}' - \tilde{y}'')\big|_{x=\xi} \le c_1 |u' - u''| \|\tilde{y}' - \tilde{y}''\|_H.$$

Therefore, the inequality $\|\tilde{y}' - \tilde{y}''\|_{0,\Omega} \le c|u' - u''|$ is derived that provides the continuity of the linear functional $L(\cdot)$ and bilinear form $\pi(\cdot,\cdot)$ on $\mathcal{U}$.

On the basis of [58, Chapter 1, Theorem 1.1], the validity of the following statement is proved.

Theorem 2.1. *If a system state is determined as a solution to equivalent problems (2.6) and (2.7), then there exists a unique element u of a convex set $\mathcal{U}_\partial$ that is closed in $\mathcal{U}$, and relation like (1.20) takes place for u, where the cost functional J(u) is specified by expression (2.5).*

If $u \in \mathcal{U}_\partial$ is the optimal control, then the following inequality is true:

$$\left(y(u) - z_g, \, y(v) - y(u)\right) + \bar{a}u(v - u) \ge 0, \quad \forall v \in \mathcal{U}_\partial. \tag{2.14}$$

As for the control $v \in \mathcal{U}$, the conjugate state $p(v) \in V^*$ is specified by the relations

$$A^* p(v) = y(v) - z_g,$$

$$p(0) = p(l) = p'(l) = 0,$$

$$p''(0) = 0, \tag{2.15}$$

$$[p] = 0,$$

$$[kp''] = 0, \quad \{kp''\}^{\pm} = \alpha[p'], \quad \left[(kp'')'\right] = -\beta p,$$

where $V^* = V$ and $A^* p = (kp'')''$.

Further on, use the formula of integration by parts, and the equality

$$\left(A^* p(u), \, y(v) - y(u)\right) = \left(y(u) - z_g, y(v) - y(u)\right) =$$

$$= a\left(p, y(v) - y(u)\right) = p\big|_{x=\xi}(v - u),$$

$$\left(y(u) - z_g, y(v) - y(u)\right) = p\big|_{x=\xi}(v - u) \tag{2.16}$$

is obtained. Take it into account, and it is stated that inequality (2.14) is equivalent to the inequality

$$\left(p\big|_{x=\xi} + \bar{a}u\right)(v-u) \geq 0, \ \forall v \in \mathcal{U}_{\partial}. \tag{2.17}$$

The necessary and sufficient condition for $u \in \mathcal{U}_{\partial}$ to be the optimal control is the one under which inequality (2.17) and the relations

$$a(y,v) = l_1(u,v), \ y \in V, \ \forall v \in V, \tag{2.18}$$

and

$$a(p,v) = l_2(y,v), \ p \in V, \ \forall v \in V, \tag{2.19}$$

are met, where $l_1(u,v) = \int_0^l f v \, dx + (r+u)v(\xi)$ and $l_2(y,v) = \int_0^l \left(y - z_g\right)v \, dx$.

If the constraints are absent, i.e. when $\mathcal{U}_{\partial} = \mathcal{U}$, then the equality

$$p\big|_{x=\xi} + \bar{a}u = 0 \tag{2.20}$$

follows from condition (2.17). Therefore, when the constraints are absent, the control u can be excluded from equality (2.18) by means of equality (2.20), and it is possible to obtain problem (2.18), (2.19), where $l_1(u,v) = l_1(u(p),v)$. The solution to problem (2.18), (2.19) is $(y,p)^T$ and the optimal control is

$$u = -p\big|_{x=\xi}/\bar{a}. \tag{2.21}$$

Let the vector solution $(y,p)^T$ to problem (2.18), (2.19), (2.21) be smooth enough on $\bar{\Omega}_l$, $l=1,2$. Then, the differential problem of finding the vector-function $(y,p)^T$, that satisfies the relations

$$(ky'')'' = f(x), \ x \in \Omega_1 \cup \Omega_2,$$

$$(kp'')'' - y = -z_g, \ x \in \Omega_1 \cup \Omega_2,$$

$$y(0) = 0, \ y(l) = 0, \ y'(l) = 0, \ y''(0) = 0,$$

$$p(0) = 0, \ p(l) = 0, \ p'(l) = 0, \ p''(0) = 0, \tag{2.22}$$

$$[y] = 0, \ [k\,y''] = 0, \ \{k\,y''\}^{\pm} = \alpha[y'],$$

$$\left[(k\,y'')'\right] = -\beta\, y + r - p\big|_{x=\xi}/\bar{a},$$

$$[p] = 0, \ [k\,p''] = 0, \ \{k\,p''\}^{\pm} = \alpha[p'],$$

$$\left[(k\,p'')'\right] = -\beta\, p,$$

corresponds to problem (2.18), (2.19), (2.21).

Definition 2.1. A generalized (weak) solution to boundary-value problem (2.22) is called a vector-function $(y,p)^{\mathrm{T}} \in H = \left\{v = (v_1, v_2)^{\mathrm{T}} : v_1, v_2 \in V\right\}$ that satisfies the following integral equation $\forall z \in H$:

$$\int_0^l \{k\,y''z_1'' + k\,p''z_2'' - y z_2\}\,dx + \beta\, y(\xi)\, z_1(\xi) - r z_1(\xi) +$$

$$+ p\big|_{x=\xi}\, z_1(\xi)/\bar{a} + \alpha[y'][z_1'] + \beta\, p(\xi)\, z_2(\xi) +$$

$$+ \alpha[p'][z_2'] = \int_0^l \left(f z_1 - z_g z_2\right) dx. \tag{2.23}$$

Let $u = (u_1, u_2)^{\mathrm{T}}$ and $v = (v_1, v_2)^{\mathrm{T}}$ be arbitrary elements of the complete Hilbert space H with the norm $\|\cdot\|_H$ introduced in point 3.1. Specify the bilinear form

$$a(u,v) = \int_0^l \left\{ k \sum_{i=1}^2 u_i'' v_i'' - u_1 v_2 \right\} dx + \beta \sum_{i=1}^2 u_i(\xi) v_i(\xi) +$$

$$+ \alpha \sum_{i=1}^2 [u_i'][v_i'] + u_2(\xi)\, v_1(\xi)/\bar{a}$$

and linear functional

$$l(v) = \int_0^l \left(f v_1 - z_g v_2 \right) dx + r v_1(\xi)$$

on H.

If the constraints $0 < 1/\bar{a} < 2\beta$ and $2\bar{\alpha}_1 > 1$ are met, then, use the Lax-Milgramm lemma, and it is concluded that the unique solution $U = (U_1, U_2)^{\mathrm{T}}$ to problem (2,23) exists in H. Problem (2.23) can be solved by means of the finite-element method. Estimate like (1.47) is true for its approximate solution $U_k^N \in H_k^N$. Then, the estimate

$$\left| u - \tilde{u}_{2k}^N \right| \le c_0 \, h^{k-1}, \quad c_0 = \text{const} > 0,$$

takes place for the approximation $\tilde{u}_{2k}^N = -p_k^N(\xi)/\bar{a}$ of the control $u = -p(\xi)/\bar{a}$.

3.3   BOUNDARY CONTROL UNDER A FIXED ROD END

Assume that equation (1.1) is specified in the domain $\Omega = (0, \xi) \cup (\xi, l)$. Boundary conditions (2.1) and the constraint

$$-k\, y''(0) = Q + u, \quad Q, u \in R^1, \tag{3.1}$$

are specified, in their turn, at the ends of the line segment $[0, l]$. At the point $x = \xi$, the conjugation conditions are

$$[y] = 0, \ [k y''] = 0, \ \{k y''\}^{\pm} = \alpha [y'],$$

$$\left[(k y'')' \right] = -\beta \, y, \tag{3.2}$$

where $\alpha, \beta = \text{const} > 0$.

Specify the observation in the form of expression (1.7). Bring a value of the cost functional

$$J(u) = \int_0^l \left(y(u) - z_g \right)^2 dx + \bar{a} u^2 \qquad (3.3)$$

in correspondence with every control $u \in \mathcal{U} = R^1$; in this case, z_g is a known element of $L_2(\Omega)$, $\bar{a} = \text{const} > 0$.

A unique state, namely, a function $y(u) \in V$ corresponds to every control $u \in \mathcal{U}$, minimizes energy functional like (2.6) on V, and it is the solution in V to weakly stated problem like (2.7). The space V is specified in point 3.2. In this case:

$$a(y,v) = \int_0^l k y'' v'' dx + \beta y(\xi) v(\xi) + \alpha [y'][v'],$$

$$l(v) = \int_0^l f v dx + (Q + u) v'(0). \qquad (3.4)$$

Use inequality (2.11), and it is easy to show the validity of the following statement.

Lemma 3.1. *Variational problem like (2.6) and weakly stated problem (2.7), that correspond to boundary-value problem (1.1), (2.1), (3.1), (3.2), are equivalent and have a unique solution $y \in V$. The bilinear form $a(\cdot, \cdot)$ and linear functional $l(\cdot)$ are specified by formulas (3.4).*

Expression like (1.18) is obtained from expression (3.3), where the bilinear form $\pi(\cdot, \cdot)$ and linear functional $L(\cdot)$ have the form of expressions, respectively, (2.12) and (1.19). In this case, inequality (2.13) is met for $\pi(\cdot, \cdot)$.

Let $\tilde{y}' = \tilde{y}(u')$ and $\tilde{y}'' = \tilde{y}(u'')$ be solutions from V to problem like (2.7) that corresponds to boundary-value problem (1.1), (2.1), (3.1), (3.2) under $f = 0$ and $Q = 0$ and under $u = u'$ and $u = u''$. Then:

$$\bar{\alpha}_1 \| \tilde{y}' - \tilde{y}'' \|_{0,\Omega}^2 \le \bar{\alpha}_1 \| \tilde{y}' - \tilde{y}'' \|_H^2 \le a \left(\tilde{y}' - \tilde{y}'', \tilde{y}' - \tilde{y}'' \right) \le$$

$$\leq |u' - u''| \left| \left(\frac{d}{dx} \tilde{y}' - \frac{d}{dx} \tilde{y}'' \right) \right|_{x=0} \right| \leq c_1 |u' - u''| \, \|\tilde{y}' - \tilde{y}''\|_H . \qquad (3.5)$$

Therefore, the inequality $\|\tilde{y}' - \tilde{y}''\|_{0,\Omega} \leq c_2 |u' - u''|$, $c_1, c_2 = \text{const} > 0$, is derived that provides the continuity of the linear functional $L(\cdot)$ and bilinear form $\pi(\cdot,\cdot)$ on $\mathcal{U}$.

On the basis of [58, Chapter 1, Theorem 1.1], the validity of the following statement is proved.

Theorem 3.1. *Let a system state be determined as a solution to equivalent problems (2.6) and (2.7), where the bilinear form $a(\cdot,\cdot)$ and linear functional $l(\cdot)$ have the form of expressions (3.4). Then, there exists a unique element u of a convex set $\mathcal{U}_\partial$ that is closed in $\mathcal{U}$, and relation like (1.20) takes place for u, where the cost functional J(u) is specified by expression (3.3).*

If $u \in \mathcal{U}_\partial$ is the optimal control, then inequality like (2.14) is true $\forall v \in \mathcal{U}_\partial$. As for the control $v \in \mathcal{U}$, the conjugate state $p(v) \in V^*$ is specified by relations (2.15). The equality

$$\left(A^* p(u), \; y(v) - y(u) \right) = \left(y(u) - z_g, y(v) - y(u) \right) =$$

$$= a\left(p, y(v) - y(u) \right) = \frac{dp}{dx}\bigg|_{x=0} (v - u),$$

viz.

$$\left(y(u) - z_g, \; y(v) - y(u) \right) = \frac{dp}{dx}\bigg|_{x=0} (v - u) \qquad (3.6)$$

is obtained. Take it into account, and it is stated that inequality like (2.14) corresponds to the optimal control for the problem of the present point and that such inequality is equivalent to the inequality

$$\left(\frac{dp}{dx}\bigg|_{x=0} + \bar{a}\, u \right)(v - u) \geq 0, \; \forall v \in \mathcal{U}_\partial. \qquad (3.7)$$

Therefore, the necessary condition for the existence of the optimal control $u \in \mathcal{U}_{\partial}$ is the one under which inequality (3.7) and the relations

$$a(y,v) = l_1(u,v), \quad y \in V, \quad \forall v \in V, \tag{3.8}$$

and

$$a(p,v) = l_2(y,v), \quad p \in V, \quad \forall v \in V, \tag{3.9}$$

are met. In this case, the bilinear form $a(\cdot,\cdot)$ is specified by the first formula from expressions (2.8) and functionals $l_1(\cdot,\cdot)$ and $l_2(\cdot,\cdot)$ are expressed, respectively, as

$$l_1(u,v) = \int_0^l f v \, dx + (Q+u)v'(0)$$

and

$$l_2(y,v) = \int_0^l \left(y - z_g\right) v \, dx.$$

If the constraints are absent, i.e. when $\mathcal{U}_{\partial} = \mathcal{U}$, then the equality

$$\left.\frac{dp}{dx}\right|_{x=0} + \bar{a}\, u = 0$$

follows from condition (3.7). Therefore, when the constraints are absent, the control u can be excluded from equality (3.8), and it is possible to obtain problem (3.8), (3.9), where $l_1(u,v) = l_1\left(u(p'),v\right)$ The solution to problem (3.8), (3.9) is $(y,p)^{\mathrm{T}}$ and the optimal control is

$$u = -\left.\frac{dp}{dx}\right|_{x=0} \Big/ \bar{a}. \tag{3.10}$$

If the vector solution $(y,p)^{\mathrm{T}}$ to problem (3.8), (3.9), (3.10) is smooth enough on $\bar{\Omega}_l$ $(l=1,2)$, then the differential problem of finding the vector-function $(y,p)^{\mathrm{T}}$, that satisfies the relations

$$(ky'')'' = f(x), \quad x \in \Omega_1 \cup \Omega_2,$$

$$(kp'')'' - y = -z_g, \quad x \in \Omega_1 \cup \Omega_2,$$

$$y(0) = y(l) = y'(l) = 0, \quad -ky''(0) = Q - \frac{dp}{dx}\bigg|_{x=0} \bigg/ \bar{a},$$

$$p(0) = p(l) = p'(l) = 0, \quad p''(0) = 0, \tag{3.11}$$

$$[y] = 0, \quad [ky''] = 0, \quad \{ky''\}^{\pm} = \alpha[y'],$$

$$\left[(ky'')' \right] = -\beta y, \quad [p] = 0, \quad [kp''] = 0,$$

$$\{kp''\}^{\pm} = \alpha[p'], \quad \left[(kp'')' \right] = -\beta p,$$

corresponds to problem (3.8), (3.9), (3.10).

Definition 3.1. A generalized (weak) solution to boundary-value problem (3.11) is called a vector-function $(y, p)^{\mathrm{T}} \in H = \{ v = (v_1, v_2)^{\mathrm{T}} : v_1, v_2 \in V \}$ that satisfies the following integral equation $\forall z \in H$:

$$\int_0^l \{ ky'' z_1'' + kp'' z_2'' - yz_2 \} dx + \beta y(\xi) z_1(\xi) +$$

$$+ p'(0) z_1'(0) / \bar{a} + \alpha[y'][z_1'] + \beta p(\xi) z_2(\xi) +$$

$$+ \alpha[p'][z_2'] = \int_0^l \left(f z_1 - z_g z_2 \right) dx + Q z_1'(0). \tag{3.12}$$

Let $u = (u_1, u_2)^T$ and $v = (v_1, v_2)^T$ be arbitrary elements of the complete Hilbert space H with the norm $\|\cdot\|_H$ introduced in point 3.1. Specify the bilinear form

$$a(u,v) = \int_0^l \left\{ k \sum_{i=1}^2 u_i'' v_i'' - u_1 v_2 \right\} dx + \beta \sum_{i=1}^2 u_i(\xi) v_i(\xi) +$$

$$+ \alpha \sum_{i=1}^2 [u_i'][v_i'] + u_2'(0) v_1'(0)/\bar{a}$$

and linear functional

$$l(v) = \int_0^l \left(f v_1 - z_g v_2 \right) dx + Q v_1'(0)$$

on H. If the constraint $\bar{\alpha}_1 - \dfrac{1}{2} - \dfrac{c_1}{2\bar{a}} > 0$, where $\bar{\alpha}_1$ and c_1 are, respectively, the constants in inequalities (3.5) and the embedding theorems, is met, then problem (3.12) has the unique solution in H. Problem (3.12) can be solved by means of the finite-element method, and the approximate solution $U_k^N \in H_k^N$ to this problem is obtained on its basis. Estimate like (1.47) is true for $U_k^N \in H_k^N$. Then, the estimate

$$\left| u - \tilde{u}_{2k}^N \right| \le c_0 h^{k-1}, \quad c_0 = \text{const} > 0,$$

can be written for the approximation $\tilde{u}_{2k}^N = -\dfrac{dp_k^N}{dx}\bigg|_{x=0} \bigg/ \bar{a}$ of the control

$$u = -\dfrac{dp}{dx}\bigg|_{x=0} \bigg/ \bar{a}.$$

3.4.   BOUNDARY CONTROL UNDER AN ELASTIC ROD END SUPPORT

Assume that equation (1.1) is specified on the domain $\Omega = (0,\xi) \cup (\xi,l)$. The boundary conditions

$$y''(0) = 0, \quad (ky'')' \big|_{x=0} + \beta_0\, y(0) = 0, \quad \beta_0 = \text{const} > 0,$$

$$y(l) = 0, \quad k\, y''(l) = Q + u, \quad Q, u \in R^1, \tag{4.1}$$

are specified, in their turn, at the ends of the line segment $[0,l]$. At the point $x=\xi$, the conjugation conditions are

$$[y] = 0, \quad [ky''] = 0, \quad \{ky''\}^{\pm} = \alpha[y'],$$

$$\left[(ky'')'\right] = -\beta\, y, \tag{4.2}$$

where α, $\beta = \text{const} > 0$.

Specify the observation in the form of expression (1.7). Bring a value of cost functional like (3.3) in correspondence with every control $u \in \mathcal{U} = R^1$; in this case, z_g is a known element of $L_2(\Omega)$, $\bar{a} = \text{const} > 0$.

A unique state, namely, a function $y(u) \in V$ corresponds to every control $u \in \mathcal{U}$, delivers the minimum to energy functional like (2.6) on V, and it is the solution in V to weakly stated problem like (2.7); in this case, $V = \left\{ v \in \bar{H}_2^2 : v(l) = 0, \ [v] = 0 \right\}$,

$$a(y,v) = \int_0^l k\, y'' v''\, dx + \beta y(\xi) v(\xi) + \beta_0\, y(0) v(0) + \alpha[y'][v'],$$

$$l(v) = \int_0^l f\, v\, dx + (Q + u) v'(l). \tag{4.3}$$

Lemma 4.1. *The inequality*

$$\|y\|_{1,\Omega}^2 \le c\left(y^2(0) + y^2(\xi) + [y']^2 + \int_0^l (y'')^2\, dx \right) \qquad (4.4)$$

is true for an arbitrary function $y \in V$.

Proof. Suppose that y is an arbitrary element from V. Consider the expression

$$\int_0^l (y')^2\, dx = yy'\Big|_0^{\xi-0} + yy'\Big|_{\xi+0}^l - \int_0^l yy''dx =$$

$$= -y(\xi)[y'] - y(0)y'(0) - \int_0^l yy''dx \le \varepsilon_1 y^2(\xi) + \frac{1}{4\varepsilon_1}[y']^2 +$$

$$+\varepsilon_2 (y'(0))^2 + \frac{1}{4\varepsilon_2} y^2(0) + \varepsilon_3 \|y\|_{0,\Omega}^2 + \frac{1}{4\varepsilon_3}\|y''\|_{0,\Omega}^2, \qquad (4.5)$$

from which the inequality

$$y^2(0) + (1 - \varepsilon_2 c_1) \int_0^l (y')^2\, dx \le \varepsilon_1\, y^2(\xi) + \frac{1}{4\varepsilon_1}[y']^2 +$$

$$+\left(1 + \frac{1}{4\varepsilon_2}\right) y^2(0) + \varepsilon_3 \|y\|_{0,\Omega}^2 + \left(\varepsilon_2 c_1 + \frac{1}{4\varepsilon_3}\right)\|y''\|_{0,\Omega}^2$$

follows since the inequality $(y'(0))^2 \le c_1\|y'\|_{1,\Omega}^2$ is true [21]. Hence, it is easy to obtain inequality (4.4) when proceeding from the generalized Friedrichs inequality [21].

Lemma is proved.

Lemma 4.2. *Variational problem like (2.6) and weakly stated problem (2.7), where the bilinear form $a(\cdot,\cdot)$ and linear functional $l(\cdot)$ have the form of expressions (4.3), are equivalent and have a unique solution* $y(u) \in V$.

The validity of Lemma 4.2 is stated according to inequality (4.4) and the Lax-Milgramm lemma.

Let $\tilde{y}' = \tilde{y}(u')$ and $\tilde{y}'' = \tilde{y}(u'')$ be solutions from V to problem like (2.7) that corresponds to boundary-value problem (1.1), (4.1), (4.2) under $f = 0$ and $Q = 0$. Then:

$$\bar{\alpha}_1 \|\tilde{y}' - \tilde{y}''\|_{0,\Omega}^2 \le \bar{\alpha}_1 \|\tilde{y}' - \tilde{y}''\|_H \le a\left(\tilde{y}' - \tilde{y}'', \tilde{y}' - \tilde{y}''\right) \le$$

$$\le \left|u' - u''\right| \left|\frac{d}{dx}\left(\tilde{y}' - \tilde{y}''\right)\right|_{x=l} \le c_2 \left|u' - u''\right| \|\tilde{y}' - \tilde{y}''\|_H.$$

Therefore, the inequality $\|\tilde{y}' - \tilde{y}''\|_{0,\Omega} \le c_3 \left|u' - u''\right|$, $\bar{\alpha}_1, c_2, c_3 =$ $= \text{const} > 0$, is derived that provides the continuity of the linear functional $L(\cdot)$ and bilinear form $\pi(\cdot, \cdot)$ on $\mathcal{U}$.

On the basis of [58, Chapter 1, Theorem 1.1], the validity of the following statement is proved.

Theorem 4.1. *If a system state is determined as a solution to equivalent problems (2.6) and (2.7), where the bilinear form $a(\cdot, \cdot)$ and linear functional $l(\cdot)$ have the form of expressions (4.3), then there exists a unique element u of a convex set $\mathcal{U}_{\partial}$ that is closed in $\mathcal{U}$, and relation like (1.20) takes place for u, where the cost functional J(u) is specified by expression (3.3).*

If $u \in \mathcal{U}_{\partial}$ is the optimal control, then inequality like (2.14) is true $\forall v \in \mathcal{U}_{\partial}$. As for the control $v \in \mathcal{U}$, the conjugate state $p(v) \in V^*$ is specified by the relations

$$A^* p(v) = y(v) - z_g,$$

$$p''(0) = 0, \quad \left(k p''\right)'\Big|_{x=0} + \beta_0\, p(0) = 0,$$

$$p(l) = 0, \quad k\, p''(l) = 0, \tag{4.6}$$

$$[p] = 0, \quad [kp''] = 0, \quad \{k\, p''\}^{\pm} = \alpha [p'],$$

$$\left[(kp'')'\right] = -\beta p.$$

The equality

$$\left(A^*p(u),\, y(v)-y(u)\right) = \left(y(u)-z_g,\, y(v)-y(u)\right) =$$

$$= a\left(p,(y(v)-y(u))\right) = \frac{dp}{dx}\bigg|_{x=l} (v-u),$$

viz.

$$\left(y(u)-z_g,\, y(v)-y(u)\right) = \frac{dp}{dx}\bigg|_{x=l} (v-u) \qquad (4.7)$$

is obtained. Take it into account, and it is stated that inequality like (2.14), that corresponds to the optimal control for the problem of the present point, is equivalent to the inequality

$$\left(\frac{dp}{dx}\bigg|_{x=l} + \bar{a}\,u\right)(v-u) \geq 0,\ \forall v \in \mathcal{U}_\partial. \qquad (4.8)$$

Therefore, the necessary and sufficient condition for the existence of the optimal control $u \in \mathcal{U}_\partial$ is the one under which inequality like (4.8) and relations like (3.8) and (3.9) are met, where the bilinear form $a(\cdot,\cdot)$ is specified by expression (4.3) and the functionals $l_1(\cdot,\cdot)$ and $l_2(\cdot,\cdot)$ are expressed, respectively, as

$$l_1(u,v) = \int_0^l f v\, dx + (Q+u)v'(l)$$

and

$$l_2(u,v) = \int_0^l \left(y-z_g\right)v\, dx.$$

If the constraints are absent, i.e. when $\mathcal{U}_\partial = \mathcal{U}$, then the equality

$$\left.\frac{dp}{dx}\right|_{x=l} + \bar{a}\,u = 0$$

follows from condition (4.8).

Therefore, when the constraints are absent, the control u can be excluded from equality (3.8), and it is possible to obtain the problem like (3.8), (3.9), where $l_1(u,v) = l_1\big(u(p'),v\big)$. The solution to problem like (3.8), (3.9) is $(y, p)^T$ and the optimal control is

$$u = -\left.\frac{dp}{dx}\right|_{x=l} \Big/ \bar{a}\,. \tag{4.9}$$

Let the vector solution $(y, p)^T$ to problem like (3.8), (3.9), (4.9) be smooth enough on $\bar{\Omega}_l$ $(l = 1,2)$. Then, the differential problem of finding the vector-function $(y, p)^T$, that satisfies the relations

$$\big(ky''\big)'' = f(x), \quad x \in \Omega_1 \cup \Omega_2,$$

$$\big(kp''\big)'' - y = -z_g, \quad x \in \Omega_1 \cup \Omega_2,$$

$$y''(0) = 0, \quad \big(ky''\big)'\Big|_{x=0} + \beta_0\, y(0) = 0,$$

$$y(l) = 0, \quad ky''(l) = Q - \left.\frac{dp}{dx}\right|_{x=l} \Big/ \bar{a}\,, \tag{4.10}$$

$$[y] = 0, \quad [ky''] = 0, \quad \{ky''\}^{\pm} = \alpha[y'], \quad \Big[\big(ky''\big)'\Big] = -\beta\, y,$$

$$p''(0) = 0, \quad \big(kp''\big)'\Big|_{x=0} + \beta_0\, p(0) = 0, \quad p(l) = 0, \quad kp''(l) = 0,$$

$$[p] = 0, \quad [kp''] = 0, \quad \{kp''\}^{\pm} = \alpha[p'], \quad \Big[\big(kp''\big)'\Big] = -\beta\, p,$$

corresponds to problem like (3.8), (3.9), (4.9).

Definition 4.1. A generalized (weak) solution to boundary-value problem (4.10) is called a vector function $(y, p)^{\mathrm{T}} \in H = \{ v = (v_1, v_2)^{\mathrm{T}} : v_1, v_2 \in V \}$ that satisfies the following integral equation $\forall z \in H$:

$$\int_0^l \{ k y'' z_1'' + k p'' z_2'' - y z_2 \} dx + \beta_0 y(0) z_1(0) + \beta_0 p(0) z_2(0) +$$

$$+ \beta y(\xi) z_1(\xi) + \beta p(\xi) z_2(\xi) + \alpha [y'][z_1'] + \alpha [p'][z_2'] +$$

$$+ p'(l) z_1'(l) / \bar{a} = \int_0^l \left(f z_1 - z_g z_2 \right) dx + Q z_1'(l). \tag{4.11}$$

Let $u = (u_1, u_2)^{\mathrm{T}}$ and $v = (v_1, v_2)^{\mathrm{T}}$ be arbitrary elements of the complete Hilbert space H with the norm $\| \cdot \|_H$. Specify the bilinear form

$$a(u, v) = \int_0^l \left\{ k \sum_{i=1}^2 u_i'' v_i'' - u_1 v_2 \right\} dx + \beta_0 \sum_{i=1}^2 u_i(0) v_i(0) +$$

$$+ \beta \sum_{i=1}^2 u_i(\xi) v_i(\xi) + \alpha \sum_{i=1}^2 [u_i'][v_i'] + u_2'(l) v_1'(l) / \bar{a}$$

and linear functional

$$l_1(v) = \int_0^l \left(f v_1 - z_g v_2 \right) dx + Q v_1'(l)$$

on H.

If the constraint $\bar{\alpha}_1 - \dfrac{1}{2} - \dfrac{c_1^2}{2\bar{a}} > 0$, where c_1 is the constant from the embedding theorem, is met, then problem (4.11) has the unique solution in H. Problem (4.11) can be solved by means of the finite-element method, and the approximate solution $U_k^N \in H_k^N$ to it is obtained. Estimate like (1.47) is true for $U_k^N \in H_k^N$. Then, the estimate

$$\left| u - \tilde{u}_k^N \right| \le c_0 \, h^{k-1} \qquad (4.12)$$

takes place for the approximation $\tilde{u}_k^N = -\dfrac{dp_k^N}{dx}\bigg|_{x=l} \bigg/ \bar{a}$ of the control

$$u = -\frac{dp}{dx}\bigg|_{x=l} \bigg/ \bar{a} \,.$$

3.5   CONTROL UNDER CONJUGATION CONDITION WITH OBSERVATION AT THEIR SPECIFICATION POINT

Assume that equation (1.1) is specified in the domain $\Omega = (0, \xi) \cup (\xi, l)$. The boundary conditions

$$y''(0) = 0, \ \left(k y'' \right)' \bigg|_{x=0} + \beta_0 \, y(0) = 0 \,,$$

$$y(l) = 0, \ k y''(l) = 0 \qquad (5.1)$$

are specified, in their turn, at the ends of the line segment $[0, l]$. At the point $x = \xi$, the conjugation conditions are conditions (2.3).

Specify the observation

$$Z(u) = Cy(u) \equiv y(u)\big|_{x=\xi} \,.$$

Bring a value of the cost functional

$$J(u) = \left(y(u)\big|_{x=\xi} - z_g \right)^2 + \bar{a} \, u^2 \qquad (5.2)$$

in correspondence with every control $u \in \mathcal{U} = R^1$; in this case, z_g is some real number, $\bar{a} = \text{const} > 0$.

A unique state, namely, a function $y(u) \in V$ corresponds to every control $u \in \mathcal{U}$, minimizes functional like (2.6) on V, and it is the solution in V to weakly stated problem like (2.7). The space V is specified in point 3.4. In this case:

$$a(y,v) = \int_0^l k y'' v'' dx + \beta y(\xi) v(\xi) + \beta_0 y(0) v(0) + \alpha [y'][v'],$$

$$l(v) = \int_0^l f v dx + (r+u) v(\xi). \tag{5.3}$$

Lemma 5.1. *Variational problem like (2.6) and weakly stated problem like (2.7), where the bilinear form $a(\cdot,\cdot)$ and linear functional $l(\cdot)$ have the form of expressions (5.3), are equivalent and have a unique solution $y(u) \in V$.*

The validity of Lemma 5.1 is stated according to the Lax-Milgramm lemma.

Rewrite cost functional (5.2) as

$$J(u) = \pi(u,u) - 2L(u) + \left(z_g - y(0) \big|_{x=\xi} \right)^2,$$

where

$$\pi(u,v) = \left(y(u) - y(0) \right) \big|_{x=\xi} \left(y(v) - y(0) \right) \big|_{x=\xi} + \bar{a} u v$$

and

$$L(v) = \left(z_g - y(0) \right) \big|_{x=\xi} \left(y(v) - y(0) \right) \big|_{x=\xi}.$$

Let $\tilde{y}' = \tilde{y}(u')$ and $\tilde{y}'' = \tilde{y}(u'')$ be solutions from V to problem like (2.7), where the bilinear form $a(\cdot,\cdot)$ and the linear functional $l(\cdot)$ are specified by formulas (5.3) under $f = 0$ and $r = 0$. Then:

$$\alpha_0 \left| (\tilde{y}' - \tilde{y}'') \right|_{x=\xi} \Big|^2 \le \bar{\alpha}_1 \| \tilde{y}' - \tilde{y}'' \|_{1,\Omega}^2 \le \bar{\alpha}_1 \| \tilde{y}' - \tilde{y}'' \|_H^2 \le$$

$$\leq a(\tilde{y}' - \tilde{y}'', \tilde{y}' - \tilde{y}'') \leq |u' - u''| \left| (\tilde{y}' - \tilde{y}'') \right|_{x=\xi} \leq c_1 |u' - u''| \|\tilde{y}' - \tilde{y}''\|_H.$$

Therefore, the inequality $\left| (\tilde{y}' - \tilde{y}'') \right|_{x=\xi} \leq c_2 |u' - u''|$, $\alpha_0, \bar{\alpha}_1, c_1, c_2 =$ $= \text{const} > 0$, is derived that provides the continuity of the functional $L(\cdot)$ and bilinear form $\pi(\cdot, \cdot)$ on $\mathcal{U}$.

On the basis of [58, Chapter 1, Theorem 1.1], the validity of the following statement is proved.

Theorem 5.1. *Let a system state be determined as a solution to equivalent problems (2.6) and (2.7), where the bilinear form $a(\cdot, \cdot)$ and linear functional have the form of expressions (5.3). Then, there exists a unique element u of a convex set $\mathcal{U}_\partial$ that is closed in $\mathcal{U}$, and relation like (1.20) takes place for u, where the functional $J(u)$ is specified by expression (5.2).*

If $u \in \mathcal{U}_\partial$ is the optimal control, then the following inequality is true:

$$\left(y(u) - z_g \right)\big|_{x=\xi} \left(y(v) - y(u) \right)\big|_{x=\xi} + \bar{a} u (v - u) \geq 0, \ \forall v \in \mathcal{U}_\partial.$$

As for the control $v \in \mathcal{U}$, the conjugate state $p(v) \in V^*$ is specified by the relations

$$A^* p(v) = 0,$$

$$p''(0) = 0, \ \left(k p'' \right)'\big|_{x=0} + \beta_0 \, p(0) = 0,$$

$$p(l) = 0, \ k p''(l) = 0, \tag{5.4}$$

$$[p] = 0, \ [k p''] = 0, \ \{k p''\}^{\pm} = \alpha [p'],$$

$$\left[(k p'')' \right] = -\beta \, p + y(v) - z_g.$$

The equality

$$0 = \left(A^* p(u), y(v) - y(u) \right) = a \left(p, y(v) - y(u) \right) -$$

$$-\left(y(u)\big|_{x=\xi} - z_g\right)\left(y(v) - y(u)\right)\big|_{x=\xi} =$$

$$= -\left(y(u)\big|_{x=\xi} - z_g\right)\left(y(v) - y(u)\right)\big|_{x=\xi} + p(\xi)(v-u),$$

viz.

$$\left(y(u)\big|_{x=\xi} - z_g\right)\left(y(v) - y(u)\right)\big|_{x=\xi} = p(\xi)(v-u)$$

is obtained.

Therefore:

$$(p(\xi) + \bar{a}\, u)(v - u) \geq 0, \quad \forall v \in \mathscr{U}_\partial. \tag{5.5}$$

If the constraints are absent, i.e. when $\mathscr{U}_\partial = \mathscr{U}$, then the equality

$$u = -p(\xi)/\bar{a} \tag{5.6}$$

follows from condition (5.5).

Therefore, the necessary and sufficient condition for the existence of the optimal control $u \in \mathscr{U}_\partial$ is the one under which inequality (5.5) and the relations

$$a(y,v) = l_1(u,v), \quad y \in V, \ \forall v \in V, \tag{5.7}$$

and

$$a(p,v) = l_2(y,v), \quad p \in V, \ \forall v \in V, \tag{5.8}$$

are met. In this case, the bilinear form $a(\cdot,\cdot)$ is specified by expression (5.3) and the functionals $l_1(\cdot,\cdot)$ and $l_2(\cdot,\cdot)$ are expressed, respectively, as

$$l_1(u,v) = \int_0^l fv dx + (r+u)v(\xi)$$

and

$$l_2(y,v) = \left(y\big|_{x=\xi} - z_g\right)v(\xi).$$

If the vector solution $(y, p)^T$ to problem like (5.6)–(5.8) is smooth enough on $\bar{\Omega}_l$ $(l = 1, 2)$, then the differential problem of finding the vector function $(y, p)^T$, for which relations (1.1), (2.3), (5.1), (5.4) and (5.6), where $A^* p = (k p'')''$, $(y, p)^T \in H = \{ v = (v_1, v_2)^T : v_1, v_2 \in V \}$, are met, corresponds to problem (5.6)–(5.8).

Definition 5.1. A generalized (weak) solution to boundary-value problem (1.1), (2.3), (5.1), (5.4), (5.6) is called a vector-function $(y, p) \in H$ that satisfies the following equation $\forall z \in H$:

$$\int_0^l \{ k\, y'' z_1'' + k\, p'' z_2'' \}\, dx + \beta_0\, y(0) z_1(0) + \beta_0\, p(0) z_2(0) +$$

$$+ \beta\, y(\xi) z_1(\xi) + \beta\, p(\xi) z_2(\xi) + \alpha [y'][z_1'] + \alpha [p'][z_2'] =$$

$$= \int_0^l f z_1\, dx + (r - p(\xi)/\bar{a}) z_1(\xi) + \left(y\big|_{x=\xi} - z_g \right) z_2(\xi). \tag{5.9}$$

Let $u = (u_1, u_2)^T$ and $v = (v_1, v_2)^T$ be arbitrary elements of the complete Hilbert space H. Specify the bilinear form

$$a(u, v) = \int_0^l k \sum_{i=1}^2 u_i'' v_i''\, dx + \beta_0 \sum_{i=1}^2 u_i(0) v_i(0) + \beta \sum_{i=1}^2 u_i(\xi) v_i(\xi) +$$

$$+ \alpha \sum_{i=1}^2 [u_i'][v_i'] + u_2(\xi) v_1(\xi)/\bar{a} - u_1(\xi) v_2(\xi)$$

and linear functional

$$l(v) = \int_0^l f v_1\, dx + r v_1(\xi) - z_g v_2(\xi)$$

on H.

If the constraint

$$\beta - \frac{1}{2}\left(\frac{1}{\overline{a}} + 1\right) > 0$$

is met, then problem (5.9) has the unique solution in H. If problem (5.9) is solved by means of the finite-element method, then estimate like (4.12) takes place for the approximation $\tilde{u}_k^N = -p_k^N(\xi)/\overline{a}$ of the optimal control $u = -p(\xi)/\overline{a}$.

CONTROL OF A SYSTEM DESCRIBED BY A TWO-DIMENSIONAL QUARTIC EQUATION UNDER CONJUGATION CONDITIONS

4.1   DISTRIBUTED CONTROL WITH OBSERVATION THROUGHOUT A WHOLE DOMAIN

Assume that the quartic equation

$$Ay \equiv \frac{\partial^2}{\partial x_1^2} D\left(\frac{\partial^2 y}{\partial x_1^2} + v\frac{\partial^2 y}{\partial x_2^2}\right) + \frac{\partial^2}{\partial x_2^2} D\left(\frac{\partial^2 y}{\partial x_2^2} + v\frac{\partial^2 y}{\partial x_1^2}\right) +$$

$$+2\frac{\partial^2}{\partial x_1 \partial x_2} D(1-v)\frac{\partial^2 y}{\partial x_1 \partial x_2} = q \qquad (1.1)$$

is specified in a domain Ω that consists of two rectangular domains Ω_1 and Ω_2, where $\Omega_1 = \{x : -\infty < a_1 < x_1 < 0, \ 0 < x_2 < b < \infty\}$, $\Omega_2 = \{x : 0 < x_1 < a_2 < \infty, \ 0 < x_2 < b\}$, $x = (x_1, x_2)$; $D = E h^3/12(1-v^2)$ $(0 < D_0 \le D \le D_1; \ D_0, D_1 = \text{const})$ is the cylindrical rigidity coefficient for a thin plate of a thickness $h = h(x)$; E and v $(0 < v < 1/2)$ are, respectively, the Young modulus and Poisson ratio that are different for the domains Ω_1 and Ω_2; $q = q(x)$ is a transversal load value; $y = y(x)$ is a deflection of a middle plate surface at a point x [92].

The boundary conditions [78]

$$y = 0 \qquad (1.2)$$

and

$$\Delta y - (1 - v)\left(n_2^2 \frac{\partial^2 y}{\partial x_1^2} + n_1^2 \frac{\partial^2 y}{\partial x_2^2}\right) = 0 \qquad (1.3)$$

are specified on a boundary $\Gamma = (\partial\Omega_1 \cup \partial\Omega_2)\backslash\gamma$ $(\gamma = \partial\Omega_1 \cap \partial\Omega_2 \neq \varnothing)$ $\gamma = \{x: x_1 = 0, 0 \le x_2 \le b\}$; in this case, $n_1 = \cos(n, x_1)$, $n_2 = \cos(n, x_2)$ and n is an ort of an outer normal to Γ (called simply an outer normal to Γ). On a section γ of the domain $\bar{\Omega} = \bar{\Omega}_1 \cup \bar{\Omega}_2$, the conjugation conditions are [92]

$$[y] = 0, \qquad (1.4)$$

$$\left[Q_y\right] = \beta y \qquad (1.5)$$

and

$$\left[M_y\right] = 0, \quad \left\{M_y\right\}^{\pm} = -\alpha\left[\frac{\partial y}{\partial x_1}\right], \qquad (1.6)$$

where

$$[\varphi] = \varphi^+ - \varphi^-, \quad \varphi^{\pm} = \{\varphi\}^{\pm} = \varphi(0 \pm 0, x_2),$$

$$Q_y = -\frac{\partial}{\partial x_1}D\left(\frac{\partial^2 y}{\partial x_1^2} + v\frac{\partial^2 y}{\partial x_2^2}\right) - 2\frac{\partial}{\partial x_2}(1 - v)D\frac{\partial^2 y}{\partial x_1 \partial x_2}$$

and

$$M_y = -D\left(\frac{\partial^2 y}{\partial x_1^2} + v\frac{\partial^2 y}{\partial x_2^2}\right).$$

Conjugation conditions (1.4)–(1.6) describe a hinge joint of two thin plates on γ; $0 \le \alpha_0 \le \alpha \le \alpha_1 < \infty$ is the hinge rigidity coefficient; $0 < \beta_0 \le \beta \le \beta_1 < \infty$ is the rigidity coefficient for a support on γ; $\alpha_i, \beta_i = \text{const}, i = 0, 1$.

Let there be a control Hilbert space $\mathcal{U}$ and mapping $B \in \mathcal{L}(\mathcal{U}; V')$, where V' is a space dual with respect to a state Hilbert space V. Assume the following: $\mathcal{U} = L_2(\Omega)$. For every control $u \in \mathcal{U}$, determine a system state y as a generalized solution to the boundary-value problem specified by the equation

$$\frac{\partial^2}{\partial x_1^2} D\left(\frac{\partial^2 y}{\partial x_1^2} + v\frac{\partial^2 y}{\partial x_2^2}\right) + \frac{\partial^2}{\partial x_2^2} D\left(\frac{\partial^2 y}{\partial x_2^2} + v\frac{\partial^2 y}{\partial x_1^2}\right) +$$

$$+2\frac{\partial^2}{\partial x_1 \partial x_2} D(1-v)\frac{\partial^2 y}{\partial x_1 \partial x_2} = q + Bu \qquad (1.7)$$

and by conditions (1.2)–(1.6).

Specify the observation

$$Z(u) = C\,y(u), \qquad (1.8)$$

where $C \in \mathcal{L}(V; \mathcal{H})$ and $\mathcal{H}$ is some Hilbert space. Assume the following:

$$C\,y(u) \equiv y(u), \quad \mathcal{H} = V \subset L_2(\Omega). \qquad (1.9)$$

Bring a value of the cost functional

$$J(u) = \left\|C\,y(u) - z_g\right\|_{\mathcal{H}}^2 + (\mathcal{N}u, u)_{\mathcal{U}} \qquad (1.10)$$

in correspondence with every control $u \in \mathcal{U}$; in this case, z_g is a known element of a space $L_2(\Omega)$, $\|\cdot\|_{\mathcal{H}} = \|\cdot\|_{L_2(\Omega)} = \|\cdot\|$ and

$$\mathcal{N} \in \mathcal{L}(\mathcal{U}; \mathcal{U}), \quad (\mathcal{N}u, u)_{\mathcal{U}} \geq v_0 \|u\|_{\mathcal{U}}^2, \quad v_0 = \text{const} > 0, \quad \forall u \in \mathcal{U}. \qquad (1.11)$$

Assume the following: $q \in L_2(\Omega)$, $Bu \equiv u \in L_2(\Omega)$, $\mathcal{N}u = \bar{a}(x)u$, $0 < \bar{a}_0 \leq$ $\leq \bar{a}(x) \leq \bar{a}_1 < \infty$, $\bar{a}(x)|_{\Omega_l} \in C(\Omega_l)$, $l = 1, 2$; $\bar{a}_0, \bar{a}_1 = \text{const}$, $(\varphi, \psi)_{\mathcal{U}} =$ $= (\varphi, \psi) = \int_{\Omega} \varphi\psi\,dx$.

A unique state, namely, a function $y(u) \in V = \left\{v: v|_{\Omega_l} \in W_2^2(\Omega_l),\right.$ $l = 1, 2; \ v|_{\Gamma} = 0, \ [v]_{\gamma} = 0\right\}$ corresponds to every control $u \in \mathcal{U}$, delivers the minimum to the functional

$$\Phi(v) = a(v,v) - 2l(v) \tag{1.12}$$

on V, and it is the unique solution in V to the weakly stated problem: Find an element $y \in V$ that meets the equation

$$a(y,v) = l(v), \quad \forall v \in V, \tag{1.13}$$

where

$$a(y,v) = \iint_\Omega D \left\{ \Delta y \, \Delta v - (1-v) \left(\frac{\partial^2 y}{\partial x_1^2} \frac{\partial^2 v}{\partial x_2^2} + \frac{\partial^2 y}{\partial x_2^2} \frac{\partial^2 v}{\partial x_1^2} \right) + \right.$$

$$\left. +2(1-v) \frac{\partial^2 y}{\partial x_1 \partial x_2} \frac{\partial^2 v}{\partial x_1 \partial x_2} \right\} dx + \int_\gamma \beta \, y v \, d\gamma + \int_\gamma \alpha \left[\frac{\partial y}{\partial x_1} \right] \left[\frac{\partial v}{\partial x_1} \right] d\gamma, \tag{1.13'}$$

$$l(v) = \iint_\Omega (q+u)v \, dx.$$

Introduce the following denotation: $\bar{H}_2^k = \left\{ v(x) : v \big|_{\Omega_l} \in W_2^k(\Omega_l), \ l = 1,2 \right\}$. The estimates

$$|a(u,v)| \le c_1 \|u\|_V \|v\|_V \quad \text{and} \quad |l(v)| \le c_2 \|v\|_V \tag{1.14}$$

are true for the bilinear form $a(\cdot,\cdot) : \bar{H}_2^2 \times \bar{H}_2^2 \to R^1$ and linear functional $l(\cdot) : \bar{H}_2^2 \to R^1$. In this case, $\|v\|_V = \left\{ \sum_{i=1}^2 \|v\|_{W_2^2(\Omega_i)}^2 \right\}^{1/2}$, where $\|\cdot\|_{W_2^2(\Omega_i)}$ is

the norm of the Sobolev space $W_2^2(\Omega_i)$, i.e. the bilinear form $a(\cdot,\cdot)$ and linear functional $l(\cdot)$ are continuous [49] on the complete Hilbert space $\bar{H}_2^2$ with the norm $\|\cdot\|_V$. Illustrate the V-ellipticity of the bilinear form $a(\cdot,\cdot)$ on the subspace $V \subset \bar{H}_2^2$. The following statement is proved for this purpose.

Lemma 1.1. *The inequality*

$$\|v\|_{1,\Omega}^2 \le c_0 \left\{ \iint_\Omega |\Delta v|^2 \, dx + \int_\gamma \left[\frac{\partial v}{\partial x_1} \right]^2 \, d\gamma \right\}, \tag{1.15}$$

where $\|v\|_{k,\Omega} = \left\{ \sum_{l=1}^{2} \iint_{\Omega_l} \sum_{|i| \le k} \left(D^{|i|} v \right)^2 \, dx \right\}^{1/2}$, $c_0 = \text{const} > 0$, *is true for all the*

functions $v \in V$.

Proof. Since the equality

$$\iint_\Omega |\nabla v|^2 \, dx = -\int_\gamma \left[\frac{\partial v}{\partial x_1} \right] v \, d\gamma - \iint_\Omega v \Delta v \, dx$$

takes place $\forall v \in V$, then, consider the ε- and Cauchy-Bunyakovsky inequalities and embedding theorems [55], and the inequality

$$\iint_\Omega |\nabla v|^2 \, dx \le \varepsilon \|v\|_{0,\Omega}^2 + \frac{1}{4\varepsilon} \iint_\Omega |\Delta v|^2 \, dx +$$

$$+ \frac{1}{4\varepsilon_1} \int_\gamma \left[\frac{\partial v}{\partial x_1} \right]^2 \, d\gamma + \varepsilon_1 \, c_0' \, \|v\|_{1,\Omega}^2$$

is obtained, where c_0' is the positive constant from the inequalities in the embedding theorems and ε and ε_1 are the arbitrary constants. Use the generalized Friedrichs inequality [21], and the validity of inequality (1.15) follows from the obtained one. Lemma is proved.

Proceed from inequality (1.15), and the following can be written:

$$a(v,v) \ge$$

$$\ge \iint_\Omega D \left(\left(\frac{\partial^2 v}{\partial x_1^2} \right)^2 + \left(\frac{\partial^2 v}{\partial x_2^2} \right)^2 + 2v \frac{\partial^2 v}{\partial x_1^2} \frac{\partial^2 v}{\partial x_2^2} + 2(1-v) \left(\frac{\partial^2 v}{\partial x_1 \partial x_2} \right)^2 \right) dx \ge$$

$$\geq c_0 \iint\limits_{\Omega} \left(\left(\frac{\partial^2 v}{\partial x_1^2} \right)^2 + \left(\frac{\partial^2 v}{\partial x_2^2} \right)^2 + \left(\frac{\partial^2 v}{\partial x_1 \partial x_2} \right)^2 \right) dx$$

and

$$a(v,v) \geq D_0 \, v_0 \iint\limits_{\Omega} |\Delta v|^2 \, dx + \alpha_0 \int\limits_{\gamma} \left[\frac{\partial v}{\partial x_1} \right]^2 d\gamma \geq c_0' \|v\|_{1,\Omega}^2 \, ;$$

in this case: $c_0' = \text{const} > 0$, $v_0 = \min\limits_{x \in \bar{\Omega}_1 \cup \bar{\Omega}_2} v = \text{const} > 0$, $c_0 = D_0(1 - v_1)$,

$v_1 = \max\limits_{x \in \bar{\Omega}_1 \cup \bar{\Omega}_2} v$, $v_1 < 1/2$.

Take the derived inequalities into account, and the inequality

$$a(v,v) \geq c_1' \|v\|_V^2 \tag{1.16}$$

is true. Hence, the V-ellipticity of the bilinear form $a(\cdot,\cdot)$ on V is proved.

Use the Lax-Milgramm lemma [16], and it is concluded that there exists the unique element y that meets equation (1.13) and implements the minimum of functional (1.12) $\forall q, u \in L_2(\Omega)$ on V. Therefore, there is such an operator A acting from V into $L_2(\Omega)$, that

$$y(u) = A^{-1}(q + Bu), \quad \forall u \in L_2(\Omega). \tag{1.17}$$

Rewrite the cost functional as

$$J(u) = \pi(u,u) - 2L(u) + \|z_g - y(0)\|^2, \tag{1.18}$$

where the bilinear form $\pi(\cdot,\cdot)$ and linear functional $L(\cdot)$ are expressed as

$$\pi(u,v) = \big(y(u) - y(0), \, y(v) - y(0) \big) + (\bar{a} \, u, v),$$

$$L(v) = \big(z_g - y(0), \, y(v) - y(0) \big); \tag{1.19}$$

in this case: $(\varphi, \psi) = \iint\limits_{\Omega} \varphi \psi \, dx$, $\|\varphi\| = (\varphi, \varphi)^{1/2}$. The form $\pi(\cdot,\cdot)$ is coercive on $\mathcal{U}$, i.e.:

$$\pi(u,u) = \big(y(u) - y(0),\ y(u) - y(0)\big) + \big(\bar{a}\,u, u\big) \ge \bar{a}_0(u,u).$$

Let $\tilde{y}' = \tilde{y}(u')$ and $\tilde{y}'' = \tilde{y}(u'')$ be solutions from V to problem (1.13) under $q = 0$ and under a function $u = u(x)$ that is equal respectively, to u' and u''. Then, the inequality

$$c_1 \big\| \tilde{y}' - \tilde{y}'' \big\|^2 \le c_1 \big\| \tilde{y}' - \tilde{y}'' \big\|_V^2 \le a\big(\tilde{y}' - \tilde{y}'',\ \tilde{y}' - \tilde{y}''\big) \le$$

$$\le \big\| u' - u'' \big\|\ \big\| \tilde{y}' - \tilde{y}'' \big\|,\quad c_1 = \text{const} > 0,$$

is derived that provides the continuity of the linear functional $L(\cdot)$ and bilinear form $\pi(\cdot,\cdot)$ on $\mathcal{U}$.

On the basis of [58, Chapter 1, Theorem 1.1], the validity of the following statement is proved.

Theorem 1.1. *Let a system state be determined as a solution to equivalent problems (1.12) and (1.13). Then, there exists a unique element u of a convex set $\mathcal{U}_\partial$ that is closed in $\mathcal{U}$, and*

$$J(u) = \inf_{v \in \mathcal{U}_\partial} J(v) \tag{1.20}$$

takes place for u.

If $u \in \mathcal{U}_\partial$ is the optimal control, then the following inequality is true:

$$\pi(u, v - u) \ge L(v - u),\quad \forall v \in \mathcal{U}_\partial. \tag{1.21}$$

Proceed from expressions (1.19), and the inequality

$$\big(y(u) - z_g,\ y(v) - y(u)\big) + \big(\bar{a}\,u, v - u\big) \ge 0,\quad \forall v \in \mathcal{U}_\partial, \tag{1.22}$$

follows from inequality (1.21), and it is the necessary and sufficient condition under which $u \in \mathcal{U}_\partial$ is the optimal control for the considered problem.

As for the control $v \in \mathcal{U}$, the conjugate state $p(v) \in V^*$ is specified by the relations

$$\frac{\partial^2}{\partial x_1^2} D\left(\frac{\partial^2 p}{\partial x_1^2} + v\frac{\partial^2 p}{\partial x_2^2}\right) + \frac{\partial^2}{\partial x_2^2} D\left(\frac{\partial^2 p}{\partial x_2^2} + v\frac{\partial^2 p}{\partial x_1^2}\right) +$$

$$+2\frac{\partial^2}{\partial x_1 \partial x_2} D(1-v)\frac{\partial^2 p}{\partial x_1 \partial x_2} = y(v) - z_g, \quad x \in \Omega,$$

$$p = 0, \quad x \in \Gamma,$$

$$\Delta p - (1-v)\left(n_2^2 \frac{\partial^2 p}{\partial x_1^2} + n_1^2 \frac{\partial^2 p}{\partial x_2^2}\right) = 0, \quad x \in \Gamma, \tag{1.23}$$

$$[p] = 0, \quad x \in \gamma,$$

$$[Q_p] = \beta p, \quad x \in \gamma,$$

$$[M_p] = 0, \quad \{M_p\}^{\pm} = -\alpha\left[\frac{\partial p}{\partial x_1}\right], \quad x \in \gamma,$$

where V^* is a space conjugate to V, $V^* = V$.

To find a generalized solution p to problem (1.23), the generalized problem means to find such $p \in V$ that meets the equation

$$a(p(u), z) = l_1(y(u), z), \quad \forall z \in V, \tag{1.24}$$

where

$$l_1(y(u), z) = \iint_{\Omega}(y(u) - z_g)z\, dx.$$

It is easy to state that the solution $p \in V$ to problem (1.24) exists and that such solution is unique. Therefore, the necessary and sufficient condition for the existence of the optimal control $u \in \mathcal{U}_\partial$ is the one under which relations (1.13), (1.22) and (1.24) are met. Use the difference $y(v) - y(u)$ in equation (1.24) instead of z, consider equation (1.13), and the equality

$$(y(u) - z_g, y(v) - y(u)) =$$

$$= a\left(p(u),\, y(v)-y(u)\right) = (v-u,\, p(u))$$

is obtained, i.e.:

$$\left(y(u)-z_g,\, y(v)-y(u)\right) = (v-u,\, p(u)), \quad \forall v \in \mathscr{U}_\partial. \tag{1.25}$$

Take it into account, and the inequality

$$\left(p(u)+\bar{a}\,u,\, v-u\right) \ge 0, \quad \forall v \in \mathscr{U}_\partial, \tag{1.26}$$

is derived from inequality (1.22).

If the constraints are absent, i.e. when $\mathscr{U}_\partial = \mathscr{U}$, then the equality

$$p(u)+\bar{a}\,u = 0, \quad x \in \bar{\Omega}, \tag{1.27}$$

follows from condition (1.26). Therefore, when the constraints are absent, the control $u(x)$ can be excluded from equality (1.7) by means of equality (1.27). On the basis of equation (1.13) and relations (1.23), the problem of finding the vector-function $(y,p)^{\mathrm{T}}$ $(y,p \in V)$, that satisfies the equality system

$$a(y,v)+\left(p/\bar{a},v\right) = (q,v), \quad \forall v \in V, \tag{1.28}$$

$$a(p,v)-(y,v) = -(z_g,v), \quad \forall v \in V, \tag{1.29}$$

is derived, and the vector solution $(y,p)^{\mathrm{T}}$ is found from this problem along with the optimal control

$$u = -\,p/\bar{a}, \quad x \in \Omega. \tag{1.30}$$

If the vector solution $(y,p)^{\mathrm{T}}$ to problem (1.28), (1.29) is smooth enough on $\bar{\Omega}_l$, viz. $y|_{\bar{\Omega}_l},\ p|_{\bar{\Omega}_l} \in C^3(\bar{\Omega}_l) \cap C^4(\Omega_l),\ \left|D^4 y\right|,$ $\left|D^4 p\right| < \infty,\ l=1,2$, then the differential problem of finding the vector-function $(y,p)^{\mathrm{T}}$, that satisfies the equations

$$Ay + p/\bar{a} = q, \quad x \in \Omega,$$
$$A^* p - y = -z_g, \quad x \in \Omega, \tag{1.31}$$

constraints (1.2)–(1.6) and all the equalities of system (1.23), except the first one, corresponds to problem (1.28), (1.29); in this case, $A^* = A$.

Definition 1.1. A generalized (weak) solution to the obtained boundary-value problem is called a vector-function $(y, p)^T \in H = \{v = (v_1, v_2)^T : v_1, v_2 \in V\}$, that satisfies the following integral equation $\forall z \in H$:

$$
\iint_\Omega \left\{ D \left[\Delta y \, \Delta z_1 + \Delta p \Delta z_2 - (1 - v) \left(\frac{\partial^2 y}{\partial x_1^2} \frac{\partial^2 z_1}{\partial x_2^2} + \frac{\partial^2 y}{\partial x_2^2} \frac{\partial^2 z_1}{\partial x_1^2} + \right. \right. \right.
$$

$$
+ \frac{\partial^2 p}{\partial x_1^2} \frac{\partial^2 z_2}{\partial x_2^2} + \frac{\partial^2 p}{\partial x_2^2} \frac{\partial^2 z_2}{\partial x_1^2} \bigg) +
$$

$$
+ 2(1 - v) \left(\frac{\partial^2 y}{\partial x_1 \partial x_2} \frac{\partial^2 z_1}{\partial x_1 \partial x_2} + \frac{\partial^2 p}{\partial x_1 \partial x_2} \frac{\partial^2 z_2}{\partial x_1 \partial x_2} \right) \bigg) \bigg] + p z_1 / \bar{a} - y z_2 \bigg\} dx +
$$

$$
+ \int_\gamma \beta (y z_1 + p z_2) d\gamma + \int_\gamma \alpha \left(\left[\frac{\partial y}{\partial x_1} \right] \left[\frac{\partial z_1}{\partial x_1} \right] + \left[\frac{\partial p}{\partial x_1} \right] \left[\frac{\partial z_2}{\partial x_1} \right] \right) d\gamma =
$$

$$
= \iint_\Omega (q z_1 - z_g z_2) \, dx . \tag{1.32}
$$

Let $u = (u_1, u_2)^T$ and $v = (v_1, v_2)^T$ be arbitrary elements of the complete Hilbert space H with the norm $\|v\|_H = \left\{ \sum_{i=1}^2 \|v\|_{W_2^2(\Omega_i)}^2 \right\}^{1/2}$. Specify the bilinear forms

$$
a_0(u, v) = \iint_\Omega D \left\{ \sum_{l=1}^2 \left(\Delta u_l \Delta v_l - (1 - v) \left(\frac{\partial^2 u_l}{\partial x_1^2} \frac{\partial^2 v_l}{\partial x_2^2} + \frac{\partial^2 u_l}{\partial x_2^2} \frac{\partial^2 v_l}{\partial x_1^2} \right) + \right. \right.
$$

$$+2(1-v)\left(\frac{\partial^2 u_l}{\partial x_1 \partial x_2}\frac{\partial^2 v_l}{\partial x_1 \partial x_2}\right)\right\}dx + \int_\gamma \beta \sum_{l=1}^{2} u_l v_l d\gamma + \int_\gamma \alpha \sum_{l=1}^{2} \left[\frac{\partial u_l}{\partial x_1}\right]\left[\frac{\partial v_l}{\partial x_1}\right]d\gamma$$

and

$$a(u,v) = a_0(u,v) + \iint_\Omega (u_2 v_1/\bar{a} - u_1 v_2)dx$$

and linear functional

$$l(v) = \iint_\Omega (q v_1 - z_g v_2)dx$$

on H.

The following can be easily shown $\forall u,v \in H$:

$$|a(u,v)| \le c_1 \|u\|_H \|v\|_H,$$

$$|l(v)| \le c_2 \|v\|_H. \tag{1.33}$$

Let the constraint

$$c_1' - \frac{1}{2}\left(1 + \frac{1}{\bar{a}_0}\right) > 0 \tag{1.34}$$

be met, where c_1' is the positive constant from inequality (1.16). Then:

$$a(u,u) \ge c_0 \|u\|_H^2, \quad c_0 = \text{const} > 0, \tag{1.35}$$

i.e. the bilinear form $a(\cdot,\cdot)$ and linear functional $l(\cdot)$ are continuous on H and the form $a(\cdot,\cdot)$ is H-elliptic on H. Use the Lax-Milgramm lemma [16], and it is concluded that $(y,p)^T \in H$ is the unique solution to problem (1.32).

Problem (1.32) can be solved approximately by means of the finite-element method. For this purpose, divide every domain $\bar{\Omega}_1$ and $\bar{\Omega}_2$ into rectangular finite elements that belong to the class $\mathscr{C}^1$ on $\bar{\Omega}_1$ and $\bar{\Omega}_2$.

Such elements are considered [16] for the scalar functions and polygon domains $\overline{\Omega}$. Then, use the estimate [16]

$$\forall v \in W_2^{k+1}(e) \; |v - \Pi_k v|_{m,e} \le c h_e^{k+1-m} \, |v|_{k+1,e} \tag{1.36}$$

of an approximation error for a scalar function v at a finite element e, and it is easy to show the validity of the following statement.

Theorem 1.2. *Let the components y and p of the solution to problem (1.32) on Ω_i belong to the space $W_2^{k+1}(\Omega_i)$ $(i = 1,2; \; k \ge 2)$. Then, the estimate*

$$\left\| u - u^N \right\|_{2,\Omega} \le c h^{k-1}, \tag{1.37}$$

where $h = \max\limits_e h_e$, h_e is a diameter of a finite element e, $\left(y^N, p^N \right)^{\mathrm{T}} \in$

$\in H^N$ and $H^N = \left\{ v^N : v^N \big|_{\Omega_i} \in P_e, \; i = 1,2; \; v^N \big|_{\Gamma} = 0, \; \left[v^N \right]_\gamma = 0 \right\}$, *takes*

place for the approximation $u^N(x) = -p^N(x)/\bar{a}(x)$ of the control $u = u(x)$.

Proof. The unique finite-element approximate solution $U^N =$

$= \left(y^N, p^N \right)^{\mathrm{T}}$ to problem (1.32) is found from the equation

$$a\left(U^N, V^N \right) = l\left(V^N \right), \; \forall V^N \in H^N.$$

If $U = U(x) \in H$ is the solution to problem (1.32), then:

$$a\left(U - U^N, V^N \right) = 0, \; \forall V^N \in H^N.$$

Therefore, the inequality

$$c_1 \left\| U - U^N \right\|_H^2 \le a\left(U - U^N, U - U^N \right) \le c_2 \left\| U - U^N \right\|_H \left\| U - \tilde{U} \right\|_H,$$

viz.

$$\left\| U - U^N \right\|_H \le c_3 \left\| U - \tilde{U} \right\|_H, \quad c_3 = \frac{c_2}{c_1} = \text{const} > 0, \tag{1.38}$$

is derived $\forall \tilde{U} \in H^N$. Take estimate (1.36) into account, and the estimate

$$\left\| U - U^N \right\|_H \le c_3 h^{k-1} \qquad (1.39)$$

follows from inequality (1.38). Use estimate (1.39), and estimate (1.37) is obtained. Theorem is proved.

4.2 CONTROL UNDER CONJUGATION CONDITION WITH OBSERVATION THROUGHOUT A WHOLE DOMAIN

Assume that equation (1.1) is specified on the domain Ω. On the boundary Γ, the boundary conditions have the form of expressions (1.2) and (1.3) and, on γ, the conjugation conditions are specified, in their turn, by equalities (1.4) and (1.6) and by the constraint

$$\left[Q_y \right] = \beta y + u, \quad x \in \gamma, \qquad (2.1)$$

where $u \in L_2(\gamma)$.

Specify the observation as

$$C y(u) \equiv y(u), \quad x \in \Omega.$$

Bring a value of the cost functional

$$J(u) = \iint_\Omega \left(y(u) - z_g \right)^2 dx + (\mathcal{N} u, u)_{\mathcal{U}} \qquad (2.2)$$

in correspondence with every control $u \in \mathcal{U} = L_2(\gamma)$; in this case, z_g is a known element from $L_2(\Omega)$, $\mathcal{N} u = \bar{a} u$, $\bar{a} \in C(\gamma)$, $0 < a_0 \le \bar{a} \le a_1 < \infty$; $a_0, a_1 = \text{const}$, $(\varphi, \psi)_{\mathcal{U}} = \int_\gamma \varphi \psi \, d\gamma$.

According to the Lax-Milgramm lemma, a unique state, namely, a function $y(u) \in V$ corresponds to every control $u \in \mathcal{U}$, minimizes the functional

$$\Phi(v) = a(v,v) - 2l(v) \qquad (2.3)$$

on V, and it is the unique solution in V to the weakly stated problem: Find an element $y(u) \in V$ that meets the equation

$$a(y,v) = l(v), \quad \forall v \in V,$$ (2.4)

where the bilinear form $a(\cdot,\cdot)$ has the form of expression (1.13') and the linear functional is

$$l(v) = \iint_{\Omega} qv\, dx - \int_{\gamma} uv\, d\gamma.$$ (2.5)

Expression (2.2) yields expression like (1.18), where the bilinear form $\pi(\cdot,\cdot)$ is expressed as

$$\pi(u,v) = \big(y(u) - y(0),\ y(v) - y(0)\big) + \int_{\gamma} \bar{a}\, u\, v\, d\gamma$$ (2.6)

and

$$\pi(u,u) \geq a_0 \int_{\gamma} u^2 d\gamma,$$ (2.7)

and the linear functional $L(\cdot)$ is specified by the second formula of expressions (1.19).

Let $\tilde{y}' = \tilde{y}(u')$ and $\tilde{y}'' = \tilde{y}(u'')$ be solutions from V to problem (2.4) under $q = 0$ and under a function $u = u(x)$ that is equal, respectively, to u' and u''. Then:

$$c_1 \|\tilde{y}' - \tilde{y}''\|^2_{0,\Omega} \leq c_1 \|\tilde{y}' - \tilde{y}''\|^2_V \leq a\big(\tilde{y}' - \tilde{y}'',\ \tilde{y}' - \tilde{y}''\big) =$$

$$= -\int_{\gamma} (u' - u'')(\tilde{y}' - \tilde{y}'')\, d\gamma \leq c_2 \|u' - u''\|_{L_2(\gamma)}\ \|\tilde{y}' - \tilde{y}''\|_V.$$

Therefore, the inequality $\|\tilde{y}' - \tilde{y}''\|_{0,\Omega} \leq c \|u' - u''\|_{L_2(\gamma)}$, $c = \mathrm{const} > 0$, is derived that provides the continuity of the functional $L(\cdot)$ and bilinear form $\pi(\cdot,\cdot)$ on $\mathcal{U}$.

On the basis of [58, Chapter 1, Theorem 1.1], the validity of the following statement is proved.

Theorem 2.1. *If a system state is determined as a solution to equivalent problems (2.3) and (2.4), then there exists a unique element u of a convex* set $\mathscr{U}_\partial$ *that is closed in* $\mathscr{U}$, *and relation like (1.20) takes place for u, where the cost functional J(u) is specified by expression (2.2).*

Let $u \in \mathscr{U}_\partial$ be the optimal control. Then, the following inequality is true $\forall v \in \mathscr{U}_\partial$:

$$\left(y(u) - z_g, \ y(v) - y(u) \right) + \int_\gamma \bar{a} u(v - u) d\gamma \geq 0, \ \forall v \in \mathscr{U}_\partial. \tag{2.8}$$

As for the control $v \in \mathscr{U}$, the conjugate state $p(v) \in V^* = V$ is specified as a generalized solution to the boundary-value problem specified, in its turn, by system (1.23), and it is the solution to weakly stated problem (1.24).

Therefore, the necessary and sufficient condition for the existence of the optimal control $u \in \mathscr{U}_\partial$ is the one under which relations (2.4), (1.24) and (2.8) are met. Use the difference $y(v) - y(u)$ in equation (1.24) instead of z. Then, the equality

$$\left(y(u) - z_g, \ y(v) - y(u) \right) = - \int_\gamma (v - u) \, p(u) \, d\gamma$$

is obtained on the basis of equation (2.4). Take the obtained equality into account, and the inequality

$$\int_\gamma (-p + \bar{a} u)(v - u) d\gamma \geq 0, \ \forall v \in \mathscr{U}_\partial, \tag{2.9}$$

is derived from inequality (2.8).

If the constraints are absent, i.e. when $\mathscr{U}_\partial = \mathscr{U}$, then the equality

$$-p(u) + \bar{a} u = 0, \quad x \in \gamma, \tag{2.10}$$

follows from condition (2.9). Therefore, when the constraints are absent and equality (2.10) is used, conjugation condition (2.1) can be written as

$$\left[Q_y \right] = \beta y + p/\bar{a}, \quad x \in \gamma, \tag{2.11}$$

and problem (2.4) is transformed into

$$a(y,v) + \int_\gamma \frac{pv}{\bar{a}} d\gamma = (q,v), \quad \forall v \in V. \tag{2.12}$$

Hence, the problem of finding the vector-function $(y,p)^T$ $(y,p \in V)$, that satisfies equalities (1.29) and (2.12), is derived, and the vector solution $(y,p)^T$ is found from problem (1.29), (2.12) along with the optimal control

$$u = p/\bar{a}, \quad x \in \gamma. \tag{2.13}$$

Let the vector solution $(y,p)^T$ to problem (1.29), (2.12) be smooth enough on $\bar{\Omega}_l$, $l = 1, 2$. Then, the equivalent differential problem of finding the vector-function $(y,p)^T$, that satisfies equalities (1.1)–(1.4), (1.6), (1.23) and (2.11), corresponds to problem (1.29), (2.12).

Definition 2.1. A generalized (weak) solution to boundary-value problem (1.1)–(1.4), (1.6), (1.23), (2.11) is called the vector-function $(y,p)^T \in H$ that satisfies the following equation $\forall z \in H$:

$$\iint_\Omega \left\{ D \left[\Delta y \Delta z_1 + \Delta p \Delta z_2 - (1-v) \left(\frac{\partial^2 y}{\partial x_1^2} \frac{\partial^2 z_1}{\partial x_2^2} + \frac{\partial^2 y}{\partial x_2^2} \frac{\partial^2 z_1}{\partial x_1^2} + \right. \right. \right.$$

$$\left. + \frac{\partial^2 p}{\partial x_1^2} \frac{\partial^2 z_2}{\partial x_2^2} + \frac{\partial^2 p}{\partial x_2^2} \frac{\partial^2 z_2}{\partial x_1^2} \right) +$$

$$\left. \left. + 2(1-v) \left(\frac{\partial^2 y}{\partial x_1 \partial x_2} \frac{\partial^2 z_1}{\partial x_1 \partial x_2} + \frac{\partial^2 p}{\partial x_1 \partial x_2} \frac{\partial^2 z_2}{\partial x_1 \partial x_2} \right) \right) - y z_2 \right\} dx +$$

$$+ \int_\gamma \beta(y z_1 + p z_2) d\gamma + \int_\gamma p z_1/\bar{a}\, d\gamma + \int_\gamma \alpha \left(\left[\frac{\partial y}{\partial x_1} \right] \left[\frac{\partial z_1}{\partial x_1} \right] + \left[\frac{\partial p}{\partial x_1} \right] \left[\frac{\partial z_2}{\partial x_1} \right] \right) d\gamma =$$

$$= \iint_\Omega \left(q z_1 - z_g z_2 \right) dx; \tag{2.14}$$

in this case, the space H is specified in point 4.1.

Let $u = (u_1, u_2)^T$ and $v = (v_1, v_2)^T$ be arbitrary elements of the complete Hilbert space H. Specify the bilinear form

$$a(u, v) = a_0(u, v) - \iint_\Omega u_1 v_2 dx + \int_\gamma u_2 v_1 / \bar{a}\, d\gamma \qquad (2.15)$$

and linear functional

$$l(v) = \int_\Omega \left(q v_1 - z_g v_2 \right) dx$$

on H.

Let the constraints $2\beta > \dfrac{1}{a_0}$ and $2c_1' > 1$ be met, where c_1' is the constant from inequality (1.16). Then, use the Lax-Milgramm lemma, and it is concluded that the unique solution $U = (U_1, U_2)^T$ to problem (2.14) exists in H. Problem (2.14) can be solved by means of the finite-element method. Estimate like (1.39) is true for its approximate solution $U^N \in H^N$. Therefore, the estimate

$$\left\| u - \tilde{u}^N \right\|_{L_2(\gamma)} \le c_0 h^{k-1}, \quad c_0 = \text{const} > 0, \qquad (2.16)$$

takes place for the approximation $\tilde{u}^N = p^N / \bar{a}$ of the control $u = p / \bar{a}$.

4.3   CONTROL UNDER CONJUGATION CONDITION WITH OBSERVATION ON A SECTION γ

Assume that equation (1.1) is specified in the domain Ω. On the boundary Γ, the boundary conditions have the form of expressions (1.2) and (1.3) and, on γ, the conjugation conditions are specified, in their turn, by equalities (1.4), (1.6) and (2.1), where $u \in L_2(\gamma)$.

Specify the observation as

$$C\, y(u) \equiv y(u), \quad x \in \gamma.$$

Bring a value of the cost functional

$$J(u) = \int_\gamma \left(y(u) - z_g \right)^2 d\gamma + (\mathcal{N}u, u)_{L_2(\gamma)} \tag{3.1}$$

in correspondence with every control $u \in \mathcal{U} = L_2(\gamma)$; in this case, z_g is a known element from $L_2(\gamma)$ and the operator $\mathcal{N}$ is specified in point 4.2. According to the lax-Milgramm lemma, a unique state, namely, a function $y(u) \in V$, corresponds to every control $u \in \mathcal{U}$, minimizes functional (2.3) on V, and it is the unique solution to problem (2.4).

Rewrite cost functional (3.1) as

$$J(u) = \pi(u,u) - 2L(u) + \left\| z_g - y(0) \right\|_{L_2(\gamma)}^2, \tag{3.2}$$

where

$$\pi(u,v) = \left(y(u) - y(0), \, y(v) - y(0) \right)_{L_2(\gamma)} + (\bar{a}\,u, v)_{L_2(\gamma)},$$

$$L(v) = \left(z_g - y(0), \, y(v) - y(0) \right)_{L_2(\gamma)}. \tag{3.3}$$

Let $\tilde{y}' = \tilde{y}(u')$ and $\tilde{y}'' = \tilde{y}(u'')$ be solutions from V to problem (2.4) under $q = 0$ and under a function that is equal, respectively, to u' and u''. Then, proceed from the embedding theorems, and the following can be written:

$$c_0 \left\| \tilde{y}' - \tilde{y}'' \right\|_{L_2(\gamma)}^2 \le c_1 \left\| \tilde{y}' - \tilde{y}'' \right\|_V^2 \le a\left(\tilde{y}' - \tilde{y}'', \tilde{y}' - \tilde{y}'' \right) =$$

$$= -\int_\gamma (u' - u'')(\tilde{y}' - \tilde{y}'') d\gamma \le c_2 \left\| u' - u'' \right\|_{L_2(\gamma)} \left\| \tilde{y}' - \tilde{y}'' \right\|_V.$$

Therefore, the inequality $\left\| \tilde{y}' - \tilde{y}'' \right\|_{L_2(\gamma)} \le c \left\| u' - u'' \right\|_{L_2(\gamma)}$ is derived that provides the continuity of the functional $L(\cdot)$ and bilinear form $\pi(\cdot, \cdot)$ on $\mathcal{U}$.

On the basis of [58, Chapter 1, Theorem 1.1], the validity of the following statement is proved.

Theorem 3.1. *If a system state is determined as a solution to equivalent problems (2.3) and (2.4), then there exists a unique element u of a convex set $\mathcal{U}_\partial$ that is closed in $\mathcal{U}$, and relation like (1.20) takes place for u, where the cost functional J(u) is specified by expression (3.1).*

As for the control $v \in \mathcal{U}$, the conjugate state $p(v) \in V$ is specified as a generalized solution to the boundary-value problem specified, in its turn, by the following equality system:

$$\frac{\partial^2}{\partial x_1^2} D \left(\frac{\partial^2 p}{\partial x_1^2} + v \frac{\partial^2 p}{\partial x_2^2} \right) + \frac{\partial^2}{\partial x_2^2} D \left(\frac{\partial^2 p}{\partial x_2^2} + v \frac{\partial^2 p}{\partial x_1^2} \right) +$$

$$+ 2 \frac{\partial^2}{\partial x_1 \partial x_2} D(1-v) \frac{\partial^2 p}{\partial x_1 \partial x_2} = 0, \quad x \in \Omega,$$

$$p = 0, \quad x \in \Gamma,$$

$$\Delta p - (1-v) \left(n_2^2 \frac{\partial^2 p}{\partial x_1^2} + n_1^2 \frac{\partial^2 p}{\partial x_2^2} \right) = 0, \quad x \in \Gamma,$$

$$[p] = 0, \quad x \in \gamma,$$

$$\left[Q_p \right] = \beta p + y - z_g, \quad x \in \gamma, \tag{3.4}$$

$$\left[M_p \right] = 0, \quad \left\{ M_p \right\}^{\pm} = -\alpha \left[\frac{\partial p}{\partial x_1} \right], \quad x \in \gamma.$$

As for the considered problem, the optimality condition (1.21) for the control $u \in \mathcal{U}_\partial$ is

$$\left(y(u) - z_g, y(v) - y(u) \right)_{L_2(\gamma)} + \left(\bar{a} u, v - u \right)_{L_2(\gamma)} \geq 0, \quad \forall v \in \mathcal{U}_\partial. \tag{3.5}$$

To find the generalized solution p to problem (3.4), the weakly stated problem means to derive $p \in V$ that meets equation like (1.24), where the bilinear form $a(\cdot, \cdot)$ is specified by expression (1.13'), and

$$l_1(y, v) = - \int\limits_{\gamma} \left(y - z_g \right) v \, d\gamma. \tag{3.6}$$

It is easy to state that the solution p exists and it is unique in V. Therefore, the necessary and sufficient condition for the existence of the optimal control $u \in \mathscr{U}_\partial$ is the one under which the relations (2.4), (3.5) and (1.24) are met; in this case, the bilinear form $a(\cdot,\cdot)$ and functional $l_1(y,v)$ are specified, respectively, by expressions (1.13′) and (3.6).

Use the difference $y(v) - y(u)$ instead of z in equality like (1.24) that corresponds to problem (3.4), consider equation (2.4), and the inequality

$$\int_\gamma \big(y(u) - z_g\big)\big(y(v) - y(u)\big)\,d\gamma = \int_\gamma p(u)(v-u)\,d\gamma, \quad \forall v \in \mathscr{U}_\partial, \quad (3.7)$$

is obtained. Then, the inequality

$$\int_\gamma (p + \bar{a}\,u)(v-u)\,d\gamma \geq 0, \quad \forall v \in \mathscr{U}_\partial, \quad (3.8)$$

is derived from condition (3.5).

If the constraints are absent, i.e. when $\mathscr{U}_\partial = \mathscr{U}$, then the equality

$$p + \bar{a}\,u = 0, \quad x \in \gamma, \quad (3.9)$$

follows from condition (3.8). Therefore, when the constraints are absent and equality (3.9) is used, conjugation condition (2.1) can be written as

$$\big[Q_y\big] = \beta\, y - p/\bar{a}, \quad x \in \gamma, \quad (3.10)$$

and problem (2.4) is transformed into

$$a(y,v) - \int_\gamma \frac{pv}{\bar{a}}\,d\gamma = (q,v), \quad \forall v \in V. \quad (3.11)$$

Hence, the problem of finding the vector-function $(y,p)^{\mathrm{T}}$ $(y,p \in V)$, that satisfies equalities (3.11) and (1.24), where the functional l_1 is specified by formula (3.6), is derived and the optimal control is

$$u = -p/\bar{a}, \quad x \in \gamma. \quad (3.12)$$

Let the vector solution $(y,p)^{\mathrm{T}}$ be smooth enough on $\bar{\Omega}_l$, $l = 1,2$. Then, for problem (3.11), (1.24), where the functional $l_1(\cdot,\cdot)$ is specified by expression (3.6), the corresponding equivalent differential problem consists in finding the vector-function $(y, p)^T$ that satisfies equalities (1.1)–(1.4), (1.6) and (3.4) and condition (3.10).

For such differential problem, the generalized problem means to find the vector-function $U \in H$ that meets the equality

$$a(U, z) = l(z), \quad \forall z \in H, \tag{3.13}$$

where

$$a(U, z) = a_0(U, z) - \int_\gamma U_2 z_1 / \bar{a} \, d\gamma + \int_\gamma U_1 z_2 \, d\gamma$$

and

$$l(z) = \iint_\Omega q z_1 \, dx + \int_\gamma z_g z_2 \, d\gamma.$$

If the inequality $\beta > \dfrac{1}{2}\left(\dfrac{1}{a_0} + 1\right)$ is met, then problem (3.13) has the unique solution in H. Therefore, the estimate like (2.16) takes place for the finite-element approximation $\tilde{u}^N = -p^N / \bar{a}$ of the optimal control $u = -p / \bar{a}$.

4.4 CONTROL ON A SECTION γ WITH OBSERVATION ON A PART OF A BOUNDARY Γ

Assume that equation (1.1) is specified in the domain Ω. Conjugation conditions (1.4), (1.6) and (2.1) are specified, in their turn, on γ and, on Γ, the boundary conditions are

$$y = 0, \quad \frac{\partial y}{\partial x_2} = 0, \quad x \in \Gamma_2, \tag{4.1}$$

$$\Delta y - (1 - v)\frac{\partial^2 y}{\partial x_2^2} = 0, \quad x \in \Gamma_1, \tag{4.2}$$

and

$$-\frac{\partial}{\partial x_1}D\left(\frac{\partial^2 y}{\partial x_1^2}+v\frac{\partial^2 y}{\partial x_2^2}\right)-2\frac{\partial}{\partial x_2}(1-v)D\frac{\partial^2 y}{\partial x_1 \partial x_2}=0,\ x\in\Gamma_1, \quad (4.3)$$

where $\Gamma_1=\{x\colon x_1=a_1,a_2;\ x_2\in[0,b]\}$, $\Gamma_2=\{x\colon x_1\in[a_1,a_2];\ x_2=0,b\}$.

Specify observation like (1.8):

$$Z(u)\equiv y(u),\ x\in\Gamma_1.$$

Specify the cost functional for the control $u\in\mathcal{U}=L_2(\gamma)$ by the expression

$$J(u)=\int_{\Gamma_1}\left(y(u)-z_g\right)^2 d\Gamma_1+\int_\gamma \bar{a}\,u^2 d\gamma, \quad (4.4)$$

where z_g is a known element of the space $L_2(\Gamma_1)$, $\bar{a}=\bar{a}(x)\in C(\gamma)$, $0<a_0\leq\bar{a}\leq a_1<\infty$, $a_0,a_1=\mathrm{const}$.

The following statement is valid.

Lemma 4.1. *A unique state, namely, a function* $y(u)\in V=$

$$=\left\{v\colon v\big|_{\Omega_l}\in W_2^2(\Omega_l),\ l=1,2;\ v\big|_{\Gamma_2}=0,\ \frac{\partial v}{\partial x_2}\bigg|_{\Gamma_2}=0,\ [v]_\gamma=0\right\}\ \ corresponds$$

to every control $u\in\mathcal{U}$, *delivers a minimum to functional (2.3) on* V, *and it is a unique solution in* V *to weakly stated problem (2.4), where the bilinear form* $a(\cdot,\cdot)$ *is specified by expression (1.13'), and the linear functional is*

$$l(v)=(q,v)-\int_\gamma uv\,d\gamma. \quad (4.5)$$

Proof. Estimates (1.14) are true for the bilinear form $a(\cdot,\cdot)$: $\bar{H}_2^2\times\bar{H}_2^2\to R^1$ and linear functional $l(\cdot)$: $\bar{H}_2^2\to R^1$. Illustrate the V-ellipticity of the bilinear form $a(\cdot,\cdot)$ on the subspace $V\subset\bar{H}_2^2$. In this case,

$$a(v,v)\geq D_0\left(1-v_1\right)\|v\|_{2,\Omega,0}^2\geq$$

$$\geq D_0 \mu (1-v_1) \iint\limits_{\Omega} \left(\frac{\partial v}{\partial x_2}\right)^2 dx \geq D_0 (1-v_1) \mu^2 \|v\|_{0,\Omega}^2, \tag{4.6}$$

where μ is the positive constant in the Friedrichs inequality, and

$$\|v\|_{2,\Omega,0}^2 = \iint\limits_{\Omega} \left\{ \sum_{i=1}^{2} \left(\frac{\partial^2 v}{\partial x_i^2}\right)^2 + 2\left(\frac{\partial^2 v}{\partial x_1 \partial x_2}\right)^2 \right\} dx.$$

Take inequality (4.6) and the following one, i.e. [78]

$$\|v\|_{1,\Omega_i}^2 \leq c_i \left\{\|v\|_{2,\Omega_i,0}^2 + \|v\|_{0,\Omega_i}^2\right\}, \quad \forall v \in \bar{H}_2^2, \quad c_i = \text{const} > 0, \quad i = 1,2,$$

into account, and the inequality

$$a(v,v) \geq c_0 \|v\|_{2,\Omega}^2, \quad \forall v \in V, \tag{4.7}$$

is obtained. The V-ellipticity of the bilinear form $a(\cdot,\cdot)$ on V is, therefore, provided.

According to the Lax-Milgramm lemma, a unique state, namely, a function $y(u) \in V$ corresponds to every control $u \in \mathcal{U}$, minimizes functional (2.3) on V, and it is the unique solution on V to problem like (2.4), where the bilinear form $a(\cdot,\cdot)$ is specified by expression (1.13′) and linear functional $l(\cdot)$ has the form of expression (4.5). Lemma is proved.

Rewrite cost functional (4.4) as

$$J(u) = \pi(u,u) - 2L(u) + \|z_g - y(0)\|_{L_2(\Gamma_1)}^2,$$

where

$$\pi(u,v) = \left(y(u) - y(0), y(v) - y(0)\right)_{L_2(\Gamma_1)} + \left(\bar{a} u, v\right)_{L_2(\gamma)}$$

and

$$L(v) = \left(z_g - y(0), y(v) - y(0)\right)_{L_2(\Gamma_1)}.$$

Let $\tilde{y}' = \tilde{y}(u')$ and $\tilde{y}'' = \tilde{y}(u'')$ be solutions from V to problem like (2.4) that corresponds to boundary-value problem (1.1), (1.4), (1.6), (2.1),

(4.1)–(4.3) under $q = 0$ and under a function u that is equal, respectively, to u' and u''. Then:

$$c_1 \|\tilde{y}' - \tilde{y}''\|^2_{L_2(\Gamma_1)} \le c_2 \|\tilde{y}' - \tilde{y}''\|^2_V \le c_3 a(\tilde{y}' - \tilde{y}'', \tilde{y}' - \tilde{y}'') \le$$

$$\le c_3 \|u' - u''\|_{L_2(\gamma)} \|\tilde{y}' - \tilde{y}''\|_{L_2(\gamma)} \le c_4 \|u' - u''\|_{L_2(\gamma)} \|\tilde{y}' - \tilde{y}''\|_V.$$

Therefore, the inequality

$$\|\tilde{y}' - \tilde{y}''\|_{L_2(\Gamma_1)} \le c_5 \|u' - u''\|_{L_2(\gamma)}$$

is derived that provides the continuity of the linear functional $L(\cdot)$ and bilinear form $\pi(\cdot, \cdot)$ on $\mathcal{U}$.

On the basis of [58, Chapter 1, Theorem 1.1], the validity of the following statement is proved.

Theorem 4.1. *Let a system state be determined as a generalized solution to boundary-value problem (1.1), (1.4), (1.6), (2.1), (4.1)–(4.3). Then, there exists a unique element u of a convex set $\mathcal{U}_\partial$ that is closed in $\mathcal{U}$, and relation like (1.20) takes place for u, where the cost functional is specified by expression (4.4).*

For the considered problem, inequality (1.21) has the form

$$\left(y(u) - z_g, y(v) - y(u) \right)_{L_2(\Gamma_1)} + \left(\bar{a} u, v - u \right)_{L_2(\gamma)} \ge 0, \ \forall v \in \mathcal{U}_\partial. \quad (4.8)$$

As for the control $v \in \mathcal{U}$, the conjugate state $p(v) \in V$ is specified as a generalized solution to the boundary-value problem specified, in its turn, by all the equalities of system (3.4), except the second, third and fifth ones, and by the conditions

$$p = 0, \ \frac{\partial p}{\partial x_2} = 0, \ x \in \Gamma_2, \quad (4.9)$$

$$\Delta p - (1 - v) \frac{\partial^2 p}{\partial x_2^2} = 0, \ x \in \Gamma_1, \quad (4.10)$$

$$-\left(\frac{\partial}{\partial x_1}D\left(\frac{\partial^2 p}{\partial x_1^2}+v\frac{\partial^2 p}{\partial x_2^2}\right)+2\frac{\partial}{\partial x_2}(1-v)D\frac{\partial^2 p}{\partial x_1\partial x_2}\right)\cos(n,x_1)=$$

$$=y(v)-z_g, \quad x\in\Gamma_1, \tag{4.11}$$

and

$$\left[Q_p\right]=\beta\, p, \quad x\in\gamma. \tag{4.12}$$

The generalized solution $p\in V$ to this boundary-value problem exists, it is unique and meets equation like (1.24), where the bilinear form $a(\cdot,\cdot)$ is specified by expression (1.13'), and

$$l_1(y,v)=\left(y-z_g,v\right)_{L_2(\Gamma_1)}. \tag{4.13}$$

Use the difference $y(v)-y(u)$ instead of z in equality like (1.24), where the bilinear form $a(\cdot,\cdot)$ and linear functional $l_1(\cdot,\cdot)$ are specified, respectively, by expressions (1.13') and (4.13), and the equality

$$a\big(p,y(v)-y(u)\big)=\big(y(u)-z_g,y(v)-y(u)\big)_{L_2(\Gamma_1)}$$

is obtained. Take equation (2.4), where $l(\cdot)$ has the form of expression (4.5), into account, and the following equality is derived:

$$\big(y(u)-z_g,y(v)-y(u)\big)_{L_2(\Gamma_1)}=-\big(v-u,\,p(u)\big)_{L_2(\gamma)}. \tag{4.14}$$

Use it, and inequality (4.8) has the form

$$(-p+\bar{a}\,u,v-u)_{L_2(\gamma)}\geq 0, \quad \forall v\in\mathcal{U}_\partial. \tag{4.15}$$

If the constraints are absent, i.e. when $\mathcal{U}_\partial=\mathcal{U}$, then equality (2.10) follows from condition (4.15). Therefore, when the constraints are absent and equality (2.10) is used, then conjugation condition (2.1) can be written as

$$\left[Q_y\right]=\beta\, y+p/\bar{a}, \quad x\in\gamma, \tag{4.16}$$

and problem (2.4) is transformed into equality like (2.12). The space V is specified in point 4.4.

Hence, the problem of finding the vector-function $(y,p)^{\mathrm{T}}$ $(y,p \in V)$, that satisfies equalities (1.24) and (2.12), where the bilinear form $a(\cdot,\cdot)$ and linear functional $l_1(\cdot,\cdot)$ are specified, respectively, by expressions (1.13') and (4.13) and the optimal control is specified, in its turn, by equality (2.13), is derived.

Let the considered vector solution $(y,p)^{\mathrm{T}}$ be smooth enough on $\bar{\Omega}_l$, $l = 1, 2$. Then, the equivalent differential problem of finding the vector-function $(y,p)^{\mathrm{T}}$, that satisfies equalities (1.1), (1.4), (1.6), (4.1)–(4.3), (4.16), (4.9)–(4.11) and (4.12) and the equalities of system (3.4), except the second, third and fifth ones, corresponds to the present weakly stated problem.

Therefore, in the absence of the constraints ($\mathcal{U}_{\partial} = \mathcal{U}$), the following problem is obtained: Find the vector-function $(y,p)^{\mathrm{T}}$ $(y,p \in V)$ that satisfies the equality system

$$a(y,v) + \left(p/\bar{a},v\right)_{L_2(\gamma)} = (q,v), \ \forall v \in V, \tag{4.17}$$

$$a(p,v) - (y,v)_{L_2(\Gamma_1)} = -\left(z_g,v\right)_{L_2(\Gamma_1)}, \ \forall v \in V, \tag{4.18}$$

from which the vector solution $(y,p)^{\mathrm{T}}$ is found and the optimal control is

$$u = p/\bar{a}, \ x \in \gamma.$$

The bilinear form $a(\cdot,\cdot)$ is specified by expression (1.13').

The equivalent problem of finding the vector-function $(y,p)^{\mathrm{T}} \in H = \left\{v = (v_1,v_2)^{\mathrm{T}} : v_1,v_2 \in V\right\}$, that satisfies the equality

$$a(U, z) = l(z), \ \forall z \in H, \tag{4.19}$$

where

$$U = (u_1,u_2)^{\mathrm{T}}, \ u_1 = y, \ u_2 = p,$$

$$a(U,z) = a_0(U,z) + \int_\gamma u_2 \, z_1/\bar{a} \, d\gamma - \int_{\Gamma_1} u_1 z_2 \, d\Gamma_1$$

and

$$l(z) = \iint_\Omega q z_1 \, dx - \int_{\Gamma_1} z_g z_2 \, d\Gamma_1,$$

corresponds to problem (4.17), (4.18).

If the inequality $\min\{\beta - 1/2a_0, \ c_0 - 1/2c_0'\} > 0$, where c_0 and c_0' are the positive constants, respectively, from inequality (4.7) and the embedding theorem [55], is met, then, by virtue of the Lax-Milgramm lemma, problem (4.19) has the unique solution $U \in H$. Therefore, estimate like (2.16) takes place for the finite-element approximation $\tilde{u}^N = p^N/\bar{a}$ of the optimal control $u = p/\bar{a}$.

4.5   CONTROL ON Γ_1 WITH OBSERVATION ON A SECTION γ: CASE 1

Assume that equation (1.1) is specified in the domain Ω. Conditions (1.4)–(1.6) are specified, in their turn, on γ. On Γ, the boundary conditions are conditions (4.1) and (4.2) and the specified constraint is

$$-\left(\frac{\partial}{\partial x_1} D\left(\frac{\partial^2 y}{\partial x_1^2} + v \frac{\partial^2 y}{\partial x_2^2} \right) + 2 \frac{\partial}{\partial x_2} D(1-v) \frac{\partial^2 y}{\partial x_1 \partial x_2} \right) \cos(n, x_1) =$$

$$= u, \ x \in \Gamma_1. \tag{5.1}$$

Specify observation like (1.8) for the control $u \in \mathcal{U} = L_2(\Gamma_1)$:

$$Z(u) \equiv y(u), \ x \in \gamma. \tag{5.2}$$

Specify the cost functional by the expression

$$J(u) = \int_\gamma \left(y(u) - z_g \right)^2 d\gamma + \int_{\Gamma_1} \bar{a} \, u^2 d\Gamma_1, \tag{5.3}$$

where z_g is a known element of the space $L_2(\gamma)$, $\bar{a} = \bar{a}(x) \in L_2(\Gamma_1)$, $0 < a_0 \leq \bar{a}(x) \leq a_1 < \infty$, $a_0, a_1 = \text{const}$.

The following statement is valid.

Lemma 5.1. *A unique state, namely, a function $y = y(u) \in V$ corresponds to every control $u \in \mathcal{U} = L_2(\Gamma_1)$, delivers a minimum to functional (2.3) on V, and it is a unique solution in V to weakly stated problem (2.4); in this case, the space V is specified in point 4.4, the bilinear form $a(\cdot, \cdot)$ is specified by expression (1.13′), and the linear functional $l(\cdot)$ is*

$$l(v) = (q, v) + \int_{\Gamma_1} uv \, d\Gamma_1.$$

The validity of Lemma 5.1 is stated according to the Lax-Milgramm lemma.

Rewrite cost functional (5.3) as

$$J(u) = \pi(u, u) - 2L(u) + \left\| z_g - y(0) \right\|^2_{L_2(\gamma)}, \tag{5.4}$$

where

$$\pi(u, v) = \left(y(u) - y(0), \, y(v) - y(0) \right)_{L_2(\gamma)} + \left(\bar{a} u, v \right)_{L_2(\Gamma_1)}$$

and

$$L(v) = \left(z_g - y(0), \, y(v) - y(0) \right)_{L_2(\gamma)}.$$

Let $\tilde{y}' = \tilde{y}(u')$ and $\tilde{y}'' = \tilde{y}(u'')$ be solutions from V to problem (2.4) that corresponds to boundary-value problem (1.1), (1.4)–(1.6), (4.1), (4.2), (5.1) under $q = 0$ and under a function u that is equal, respectively, to u' and u''. Then:

$$c_1 \left\| \tilde{y}' - \tilde{y}'' \right\|^2_{L_2(\gamma)} \leq c_2 \left\| \tilde{y}' - \tilde{y}'' \right\|^2_V \leq c_3 \, a\left(\tilde{y}' - \tilde{y}'', \tilde{y}' - \tilde{y}'' \right) \leq$$

$$\leq c_3 \left\| u' - u'' \right\|_{L_2(\Gamma_1)} \left\| \tilde{y}' - \tilde{y}'' \right\|_{L_2(\Gamma_1)} \leq c_4 \left\| u' - u'' \right\|_{L_2(\Gamma_1)} \left\| \tilde{y}' - \tilde{y}'' \right\|_V.$$

Therefore, the inequality $\|\tilde{y}' - \tilde{y}''\|_{L_2(\gamma)} \le c_5 \|u' - u''\|_{L_2(\Gamma_1)}$ is derived that provides the continuity of the linear functional $L(\cdot)$ and bilinear form $\pi(\cdot,\cdot)$ on $\mathcal{U}$.

On the basis of [58, Chapter 1, Theorem 1.1], the validity of the following statement is proved.

Theorem 5.1. *Let a system state be determined as a generalized solution to boundary-value problem (1.1), (1.4)–(1.6), (4.1), (4.2), (5.1). Then, there exists a unique element u of a convex set $\mathcal{U}_\partial$ that is closed in $\mathcal{U}$, and relation like (1.20) takes place for u, where the cost functional is specified by expression (5.3).*

For the considered optimal control problem, inequality (1.21) has the form

$$\left(y(u) - z_g, \, y(v) - y(u)\right)_{L_2(\gamma)} + \left(\bar{a}\,u, v - u\right)_{L_2(\Gamma_1)} \ge 0, \ \forall v \in \mathcal{U}_\partial. \quad (5.5)$$

As for the control $v \in \mathcal{U}$, the conjugate state $p(v) \in V$ is specified as a generalized solution to the boundary-value problem specified, in its turn, by the equalities of system (3.4), except the second and third ones, and by the following constraints:

$$p = 0, \ \frac{\partial p}{\partial x_2} = 0, \ x \in \Gamma_2,$$

$$\Delta p - (1 - v)\,\frac{\partial^2 p}{\partial x_2^2} = 0, \ x \in \Gamma_1,$$

$$-\frac{\partial}{\partial x_1} D\left(\frac{\partial^2 p}{\partial x_1^2} + v\frac{\partial^2 p}{\partial x_2^2}\right) - 2\frac{\partial}{\partial x_2}(1 - v)D\frac{\partial^2 p}{\partial x_1 \partial x_2} = 0, \ x \in \Gamma_1. \quad (5.6)$$

For such boundary-value problem, the generalized problem means to find a function $p \in V$ that meets the equation

$$a(p, z) = l(z), \ \forall z \in V; \quad (5.7)$$

in this case, the bilinear form $a(\cdot,\cdot)$ is specified by expression (1.13′), and the linear functional is

$$l(z) = \int_\gamma \left(-y + z_g\right) z \, d\gamma .$$ (5.8)

The generalized solution $p \in V$ to the considered problem, i.e. the one to problem (5.7) exists and it is unique in V. Use the difference $y(v) - y(u)$ in equation (5.7) instead of z, and the equality

$$a\left(p, y(v) - y(u)\right) = -\int_\gamma \left(y(u) - z_g\right)\left(y(v) - y(u)\right) d\gamma$$

or

$$-\int_\gamma \left(y(u) - z_g\right)\left(y(v) - y(u)\right) d\gamma = \int_{\Gamma_1} p(u)(v - u) d\Gamma_1$$

is obtained. Use it, and inequality (5.5) has the form

$$\left(-p(u) + \bar{a}\, u, v - u\right)_{L_2(\Gamma_1)} \geq 0, \ \forall v \in \mathcal{U}_\partial .$$

If the constraints are absent, i.e. when $\mathcal{U}_\partial = \mathcal{U}$, then:

$$-p(u) + \bar{a}\, u = 0 \quad x \in \Gamma_1 .$$ (5.9)

Therefore, when the constraints are absent and equality (5.9) is used, then boundary condition (5.1) can be written as

$$-\left(\frac{\partial}{\partial x_1} D\left(\frac{\partial^2 y}{\partial x_1^2} + v \frac{\partial^2 y}{\partial x_2^2}\right) + 2 \frac{\partial}{\partial x_2} D(1 - v) \frac{\partial^2 y}{\partial x_1 \partial x_2}\right) \cos(n, x_1) =$$

$$= p/\bar{a}, \ x \in \Gamma_1 .$$ (5.10)

Hence, in the absence of the constraints $(\mathcal{U}_\partial = \mathcal{U})$, the following problem is obtained: Find the vector-function $(y, p)^{\mathrm{T}}$ $(y, p \in V)$ that satisfies the equality system

$$a(y, v) - \int_{\Gamma_1} p v / \bar{a} \, d\Gamma_1 = (q, v), \ \forall v \in V ,$$ (5.11)

$$a(p, v) + \int_\gamma y v \, d\gamma = \int_\gamma z_g v \, d\gamma, \ \forall v \in V ,$$ (5.12)

from which the vector solution $(y, p)^{\mathrm{T}}$ is found along with the optimal control

$$u = p/\bar{a}, \quad x \in \Gamma_1.$$

The equivalent problem of finding the vector-function $U \in H = = \left\{ v = (v_1, v_2)^{\mathrm{T}} : v_1, v_2 \in V \right\}$, that satisfies equality like (4.19), where

$$U = (u_1, u_2)^{\mathrm{T}}, \quad u_1 = y, \quad u_2 = p,$$
$$a(U, z) = a_0(U, z) - \int_{\Gamma_1} u_2 z_1/\bar{a}\, d\Gamma_1 + \int_{\gamma} u_1 z_2 d\gamma$$

and

$$l(z) = (q, z_1) + \int_{\gamma} z_g z_2 d\gamma,$$

corresponds to problem (5.11), (5.12).

If the inequality $\min\left\{ \beta - \dfrac{1}{2}, \ c_0 - \dfrac{c_0'}{2a_0} \right\} > 0$, where c_0 and c_0' are the positive constants, respectively, from inequality (4.7) and the embedding theorem [55], is met, then, by virtue of the Lax-Milgramm lemma, problem like (4.19) has the unique solution $(y, p)^{\mathrm{T}} \in H$. Therefore, the estimate

$$\left\| u - u^N \right\|_{L_2(\Gamma_1)} \leq ch^{k-1}, \quad c = \mathrm{const} > 0, \tag{5.13}$$

takes place for the finite-element approximation $u^N = p^N/\bar{a}$ of the optimal control $u = p/\bar{a}$.

4.6   CONTROL ON Γ_1 WITH OBSERVATION ON A SECTION γ: CASE 2

Assume that equation (1.1) is specified in the domain Ω. Conditions (1.4)–(1.6) are specified, in their turn, on γ. On Γ, the boundary conditions are conditions (4.1) and (4.3) and the specified constraint is

$$D\left(\Delta y - (1-v)\frac{\partial^2 y}{\partial x_2^2}\right)\cos(n, x_1) = u, \quad x \in \Gamma_1. \tag{6.1}$$

Observation (5.2) is specified for the control $u \in \mathcal{U} = L_2(\Gamma_1)$. Specify the cost functional by expression (5.3), where z_g is a known element of the space $L_2(\gamma)$; $\bar{a} = \bar{a}(x) \in L_2(\Gamma_1)$, $0 < a_0 \leq \bar{a}(x) \leq a_1 < \infty$, $a_0, a_1 = \text{const}$.

The following statement is valid.

Lemma 6.1. *A unique state, namely, a function $y = y(u) \in V$ corresponds to every control $u \in \mathcal{U} = L_2(\Gamma_1)$, delivers a minimum to functional (2.3) on V, and it is a unique solution in V to weakly stated problem (2.4); in this case, the space V is specified in point 4.4, the bilinear form $a(\cdot, \cdot)$ is specified by expression (1.13′), and the linear functional $l(\cdot)$ is*

$$l(v) = (q, v) + \int_{\Gamma_1} u \frac{\partial v}{\partial x_1} d\Gamma_1. \tag{6.2}$$

The validity of Lemma 6.1 is stated according to the Lax-Milgramm lemma.

Let $\tilde{y}' = \tilde{y}(u')$ and $\tilde{y}'' = \tilde{y}(u'')$ be solutions from V to problem (2.4) that corresponds to boundary-value problem (1.1), (1.4)–(1.6), (4.1), (4.3), (6.1) under $q = 0$ and under a function u that is equal, respectively, to u' and u''. Then:

$$c_1 \|\tilde{y}' - \tilde{y}''\|_{L_2(\gamma)}^2 \leq c_2 \|\tilde{y}' - \tilde{y}''\|_V^2 \leq c_3 \, a\left(\tilde{y}' - \tilde{y}'', \tilde{y}' - \tilde{y}''\right) \leq$$

$$\leq c_3 \|u' - u''\|_{L_2(\Gamma_1)} \left\|\frac{\partial(\tilde{y}' - \tilde{y}'')}{\partial x_1}\right\|_{L_2(\Gamma_1)} \leq c_4 \|u' - u''\|_{L_2(\Gamma_1)} \|\tilde{y}' - \tilde{y}''\|_V.$$

Therefore, the inequality

$$\|\tilde{y}' - \tilde{y}''\|_{L_2(\gamma)} \leq c_5 \|u' - u''\|_{L_2(\Gamma_1)}$$

is derived that provides the continuity of the linear functional $L(\cdot)$ and bilinear form $\pi(\cdot, \cdot)$ of representation (5.4) and cost functional (5.3) on $\mathcal{U}$.

On the basis of [58, Chapter 1, Theorem 1.1], the validity of the following statement is proved.

Theorem 6.1. *Let a system state be determined as a generalized solution to boundary-value problem (1.1), (1.4)–(1.6), (4.1), (4.3), (6.1). Then, there exists a unique element u of a convex set $\mathcal{U}_\partial$ that is closed in $\mathcal{U}$, and relation like (1.20) takes place for u, where the cost functional is specified by expression (5.3).*

For the considered optimal control problem, inequality (1.21) has the form of inequality (5.5). As for the control $v \in \mathcal{U}$, the conjugate state $p(v) \in V$ is specified as a generalized solution to the boundary-value problem specified, in its turn, by the equalities of system (3.4), except the second and third ones, and by constraints (5.6).

For such boundary-value problem, the generalized problem means to find a function $p \in V$ that meets equation (5.7), where the bilinear form $a(\cdot,\cdot)$ and linear functional are specified, respectively, by expressions (1.13') and (5.8).

The generalized solution $p \in V$ to the considered problem, i.e. to problem (5.7) exists and it is unique in V. Use the difference $y(v) - y(u)$ in equation (5.7) instead of z, and the equality

$$a\left(p, y(v) - y(u)\right) = -\int_\gamma \left(y(u) - z_g\right)\left(y(v) - y(u)\right) d\gamma$$

is obtained. Therefore:

$$-\int_\gamma \left(y(u) - z_g\right)\left(y(v) - y(u)\right) d\gamma = \int_{\Gamma_1} (v - u)\frac{\partial p}{\partial x_1} d\Gamma_1. \tag{6.3}$$

Use the obtained equality, and inequality (5.5) has the form

$$\left(-\frac{\partial p}{\partial x_1} + \bar{a}\,u, v - u\right)_{L_2(\Gamma_1)} \geq 0, \quad \forall v \in \mathcal{U}_\partial. \tag{6.4}$$

If the constraints are absent, i.e. when $\mathcal{U}_\partial = \mathcal{U}$, then the equality

$$-\frac{\partial p}{\partial x_1} + \bar{a}\,u = 0, \quad x \in \Gamma_1, \tag{6.5}$$

follows from condition (6.4). Therefore, when the constraints are absent and equality (6.5) is used, then boundary condition (6.1) can be written as

$$D\left(\Delta y - (1-v)\frac{\partial^2 y}{\partial x_2^2}\right)\cos(n, x_1) = \frac{1}{a}\frac{\partial p}{\partial x_1}, \quad x \in \Gamma_1. \tag{6.6}$$

Hence, in the absence of the constraints ($\mathcal{U}_\partial = \mathcal{U}$), the following problem is obtained: Find the vector-function $(y, p)^{\mathrm{T}}$ ($y, p \in V$) that satisfies the equality system

$$a(y, v) - \int_{\Gamma_1} \frac{1}{a}\frac{\partial p}{\partial x_1}\frac{\partial v}{\partial x_1} d\Gamma_1 = (q, v), \quad \forall v \in V, \tag{6.7}$$

$$a(p, v) + \int_\gamma yv\, d\gamma = \int_\gamma z_g v\, d\gamma, \quad \forall v \in V, \tag{6.8}$$

from which the vector solution $(y, p)^{\mathrm{T}}$ is found along with the optimal control

$$u = \frac{1}{a}\frac{\partial p}{\partial x_1}, \quad x \in \Gamma_1.$$

The equivalent problem of finding the vector-function $U \in H = = \left\{ v = (v_1, v_2)^{\mathrm{T}} : v_1, v_2 \in V \right\}$, that satisfies equality like (4.19), where

$$U = (u_1, u_2)^{\mathrm{T}}, \quad u_1 = y, \quad u_2 = p,$$

$$a(U, z) = a_0(U, z) - \int_{\Gamma_1} \frac{\partial u_2}{\partial x_1}\frac{\partial z_1}{\partial x_1} \Big/ \overline{a}\, d\Gamma_1 + \int_\gamma u_1 z_2\, d\gamma$$

and

$$l(z) = (q, z_1) + \int_\gamma z_g z_2\, d\gamma,$$

corresponds to problem (6.7), (6.8).

If the inequality $\min\left\{\beta - \dfrac{1}{2},\ c_0 - \dfrac{c_0'}{2a_0}\right\} > 0$, where c_0 and c_0' are the positive constants, respectively, from inequalities (4.7) and the embedding theorems [55], is met, then, by virtue of the Lax-Milgramm lemma, problem like (4.19) has the unique solution $(y,p)^{\mathrm{T}} \in H$. Therefore, estimate (5.13) takes place for the finite-element approximation

$$\tilde{u}^N = \frac{1}{\bar{a}} \frac{\partial p^N}{\partial x_1} \text{ of the optimal control } u = \frac{1}{\bar{a}} \frac{\partial p}{\partial x_1}.$$

4.7 CONTROL ON Γ_1 WITH OBSERVATION ON Γ_1

Assume that equation (1.1) is specified in the domain Ω. Conditions (1.4)–(1.6) are specified, in their turn, on the section γ. On Γ, the boundary conditions are conditions (4.1), (4.2) and (5.1).

Specify the observation

$$Z(u) \equiv y(u), \quad x \in \Gamma_1,$$

for the control $u \in \mathcal{U} = L_2(\Gamma_1)$.

Specify the cost functional by the expression

$$J(u) = \int_{\Gamma_1} \left(y(u) - z_g\right)^2 d\Gamma_1 + \int_{\Gamma_1} \bar{a}\, u^2 d\Gamma_1, \tag{7.1}$$

where z_g is a known element of the space $L_2(\Gamma_1)$, and the function $\bar{a} = \bar{a}(x) \in L_2(\Gamma_1)$ is specified, in its turn, in point 4.5.

According to Lemma 5.1, the unique state, namely, the function $y = y(u)$ corresponds to every control $u \in \mathcal{U} = L_2(\Gamma_1)$, and it is the generalized solution to problem (1.1), (1.4)–(1.6), (4.1), (4.2), (5.1).

Rewrite cost functional (7.1) as

$$J(u) = \pi(u,u) - 2L(u) + \left\| z_g - y(0) \right\|_{L_2(\Gamma_1)}^2, \tag{7.2}$$

where

$$\pi(u,v) = \left(y(u) - y(0),\, y(v) - y(0)\right)_{L_2(\Gamma_1)} + \left(\overline{a}\,u, v\right)_{L_2(\Gamma_1)}$$

and

$$L(v) = \left(z_g - y(0),\, y(v) - y(0)\right)_{L_2(\Gamma_1)}.$$

Let $\tilde{y}' = \tilde{y}(u')$ and $\tilde{y}'' = \tilde{y}(u'')$ be solutions from V to problem like (2.4) that corresponds to boundary-value problem (1.1), (1.4)–(1.6), (4.1), (4.2), (5.1) under $q = 0$ and under a function u that is equal, respectively, to u' and u''. Then:

$$c_1 \|\tilde{y}' - \tilde{y}''\|^2_{L_2(\Gamma_1)} \le c_2 \|\tilde{y}' - \tilde{y}''\|^2_V \le a\left(\tilde{y}' - \tilde{y}'', \tilde{y}' - \tilde{y}''\right) \le$$

$$\le \|u' - u''\|_{L_2(\Gamma_1)} \|\tilde{y}' - \tilde{y}''\|_{L_2(\Gamma_1)} \le c_3 \|u' - u''\|_{L_2(\Gamma_1)} \|\tilde{y}' - \tilde{y}''\|_V.$$

Therefore, the inequality

$$\|\tilde{y}' - \tilde{y}''\|_{L_2(\Gamma_1)} \le c\|u' - u''\|_{L_2(\Gamma_1)}$$

is derived that provides the continuity of the functional $L(\cdot)$ and bilinear form $\pi(\cdot,\cdot)$ on $\mathcal{U}$.

On the basis of [58, Chapter 1, Theorem 1.1], the validity of the following statement is proved.

Theorem 7.1. *Let a system state be determined as a generalized solution to boundary-value problem (1.1), (1.4)–(1.6), (4.1), (4.2), (5.1). Then, there exists a unique element u of a convex set $\mathcal{U}_{\partial}$ that is closed in $\mathcal{U}$, and relation like (1.20) takes place for u, where the cost functional is specified by expression (7.1).*

For the considered optimal control problem, inequality (1.21) has the form

$$\left(y(u) - z_g,\, y(v) - y(u)\right)_{L_2(\Gamma_1)} + \left(\overline{a}\,u, v - u\right)_{L_2(\Gamma_1)} \ge 0, \quad \forall v \in \mathcal{U}_{\partial}. \quad (7.3)$$

As for the control $v \in \mathcal{U}$, the conjugate state $p(v) \in V$ is specified as a generalized solution to the boundary-value problem specified, in its turn, by the equalities of system (3.4), except the second, third and fifth ones, by constraints (4.9), (4.10) and (4.12) and by the equality

$$-\left(\frac{\partial}{\partial x_1}D\left(\frac{\partial^2 p}{\partial x_1^2}+v\frac{\partial^2 p}{\partial x_2^2}\right)+2\frac{\partial}{\partial x_2}D(1-v)\frac{\partial^2 p}{\partial x_1\partial x_2}\right)\cos(n,x_1)=$$

$$=-y+z_g,\ x\in\Gamma_1.$$

For such boundary-value problem, the generalized problem means to find a function $p\in V$ that meets equation like (5.7), where he bilinear form $a(\cdot,\cdot)$ is specified by expression (1.13'), and the linear functional is

$$l(z)=-\int_{\Gamma_1}\left(y-z_g\right)z\,d\Gamma_1. \tag{7.4}$$

The solution $p\in V$ to this generalized problem exists and it is unique in V. Use the difference $y(v)-y(u)$ instead of z in equation like (5.7), and the equality

$$a\left(p,y(v)-y(u)\right)=-\int_{\Gamma_1}\left(y(u)-z_g\right)\left(y(v)-y(u)\right)d\Gamma_1$$

is obtained. Use it , and inequality (7.3) has the form

$$\left(-p(u)+\bar{a}\,u,v-u\right)_{L_2(\Gamma_1)}\geq 0,\ \forall v\in\mathcal{U}_\partial. \tag{7.5}$$

If the constraints are absent, i.e. when $\mathcal{U}_\partial=\mathcal{U}$, equality (5.9) follows from condition (7.5). Therefore, when the constraints are absent and equality (5.9) is used, boundary condition (5.1) can be written in the form of boundary condition (5.10).

Hence, in the absence of the constraints ($\mathcal{U}_\partial=\mathcal{U}$), the following problem is obtained: Find the vector-function $(y,p)^{\mathrm{T}}$ $(y,p\in V)$ that satisfies the equality system

$$a(y,v)-\int_{\Gamma_1}pv/\bar{a}\,d\Gamma_1=(q,v),\ \forall v\in V, \tag{7.6}$$

$$a(p,v)+\int_{\Gamma_1}yv\,d\Gamma_1=\int_{\Gamma_1}z_gv\,d\Gamma_1,\ \forall v\in V, \tag{7.7}$$

from which the vector solution $(y,p)^{\mathrm{T}}$ is found along with the optimal control

$$u = p/\bar{a}, \quad x \in \Gamma_1.$$

The equivalent problem of finding the vector-function $U \in H = \left\{ v = (v_1, v_2)^{\mathrm{T}} : v_1, v_2 \in V \right\}$, that satisfies inequality like (4.19), where

$$U = (u_1, u_2)^{\mathrm{T}}, \ u_1 = y, \ u_2 = p,$$

$$a(U, z) = a_0(U, z) - \int_{\Gamma_1} u_2 \, z_1 / \bar{a} \, d\Gamma_1 + \int_{\Gamma_1} u_1 z_2 \, d\Gamma_1$$

and

$$l(z) = (q, z_1) + \int_{\Gamma_1} z_g z_2 \, d\Gamma_1,$$

corresponds to problem (7.6), (7.7).

If the inequality $c_0 - \dfrac{c_0'}{2a_0} - \dfrac{c_0'}{2} > 0$, where c_0 and c_0' are the positive constants, respectively, from inequalities (4.7) and the embedding theorem [58], is met, then, by virtue of the Lax-Milgramm lemma, problem like (4.19) has the unique solution $(y, p)^{\mathrm{T}} \in H$. Therefore, estimate like (5.13) takes place for the finite-element approximation $u^N = p^N / \bar{a}$ of the optimal control $u = p/\bar{a}$.

CONTROL OF A SYSTEM DESCRIBED BY A PARABOLIC EQUATION UNDER CONJUGATION CONDITIONS

Introduce the following denotations: Ω is a domain that consists of two open, non-intersecting and strictly Lipschitz domains Ω_1 and Ω_2 from an n-dimensional real linear space R^n; $\Gamma = (\partial\Omega_1 \cup \partial\Omega_2)\backslash\gamma$ $(\gamma = \partial\Omega_1 \cap \cap\partial\Omega_2 \neq \varnothing)$ is a boundary of a domain $\bar{\Omega}$, $\partial\Omega_i$ is a boundary of a domain Ω_i, $i = 1, 2$; $\Omega_T = \Omega \times (0, T)$ is a complicated cylinder, $\Gamma_T = \Gamma \times (0, T)$ is the lateral surface of a cylinder $\Omega_T \cup \gamma_T$, $\gamma_T = \gamma \times (0, T)$.

Let V be some Hilbert space and assume that V' is a space dual with respect to V. By analogy [58], introduce a space $L^2(0, T; V)$ of functions $t \to f(t)$ that map an interval $(0, T)$ into the space V of measurable functions, namely, of such ones that

$$\left(\int_0^T \|f(t)\|_V^2 \, dt\right)^{1/2} < \infty.$$

Also by analogy, specify the space $L^2(0, T; V')$. Introduce a space $W(0, T) = \left\{ f \in L^2(0, T; V) : df/dt \in L^2(0, T; V') \right\}$ that is supplied with the norm

$$\|f\|_{W(0,T)} = \left(\int_0^T \|f(t)\|_V^2 \, dt + \int_0^T \left\| \frac{df}{dt} \right\|_{V'}^2 dt \right)^{1/2}$$

and becomes the Hilbert one.

5.1   DISTRIBUTED CONTROL

Assume that the parabolic equation

$$\frac{\partial y}{\partial t} = \sum_{i,j=1}^n \frac{\partial}{\partial x_i} \left(k_{ij}(x) \frac{\partial y}{\partial x_j} \right) + f(x,t) \qquad (1.1)$$

is specified in the domain Ω_T, where

$$k_{ij}\big|_{\bar{\Omega}_l} = k_{ji}\big|_{\bar{\Omega}_l} \in C(\bar{\Omega}_l) \cap C^1(\Omega_l), \ i,j = \overline{1,n}; \ \sum_{i,j=1}^n k_{ij}\xi_i\xi_j \ge \alpha_0 \sum_{i=1}^n \xi_i^2,$$

$$\forall \xi_i, \xi_j \in R^1, \ \forall x \in \Omega, \ \alpha_0 = \text{const} > 0; \ f\big|_{\Omega_{lT}} \in C(\Omega_{lT}), \ l = 1,2,$$

$$|f| < \infty, \quad \Omega_{lT} = \Omega_l \times (0,T). \qquad (1.1')$$

The third boundary condition

$$\sum_{i,j=1}^n k_{ij} \frac{\partial y}{\partial x_j} \cos(\nu, x_i) = -\alpha\, y + \beta \qquad (1.2)$$

is specified, in its turn, on the boundary Γ_T, where $\alpha = \alpha(x) \ge \alpha^0 > 0$, $\alpha, \beta \in L_2(\Gamma)$, $\alpha^0 = \text{const}$.

On γ_T, the conjugation conditions are

$$\left[\sum_{i,j=1}^n k_{ij} \frac{\partial y}{\partial x_j} \cos(\nu, x_i) \right] = 0 \qquad (1.3)$$

and

$$\left\{\sum_{i,j=1}^{n} k_{ij}\frac{\partial y}{\partial x_j}\cos(v,x_i)\right\}^{\pm} = r[y], \tag{1.4}$$

where $0 \le r = r(x) \le r_1 < \infty$, $r_1 = \text{const}$, $[\varphi] = \varphi^+ - \varphi^-$; $\varphi^+ = \{\varphi\}^+ = \varphi(x,t)$ under $(x,t) \in \gamma_T^+ = (\partial\Omega_2 \cap \gamma) \times (0,T)$; $\varphi^- = \{\varphi\}^- = \varphi(x,t)$ under $(x,t) \in \gamma_T^- = (\partial\Omega_1 \cap \gamma) \times (0,T)$, v is an ort of a normal to γ that is called simply a normal to γ and it is directed into the domain Ω_2.

The initial condition

$$y(x,0) = y_0, \tag{1.5}$$

where $y_0 \in H = L_2(\Omega)$, is specified under $t = 0$.

Let there be a control Hilbert space $\mathcal{U}$ and operator $B \in \mathcal{L}\left(\mathcal{U}; L^2(0,T;V')\right)$.

For every control $u \in \mathcal{U}$, determine a system state $y = y(u) = y(x,t;u)$ as a generalized solution to the problem specified by the equation

$$\frac{\partial y}{\partial t} = \sum_{i,j=1}^{n}\frac{\partial}{\partial x_i}\left(k_{ij}\frac{\partial y}{\partial x_j}\right) + f + Bu \tag{1.6}$$

and by conditions (1.2)–(1.5). Without loss of generality, assume $Bu \equiv u$.

Specify the observation by the following expression:

$$Z(u) = C\,y(u), \quad C \in \mathcal{L}\left(W(0,T); \mathcal{H}\right). \tag{1.7}$$

Specify the operator

$$\mathcal{N} \in \mathcal{L}(\mathcal{U}; \mathcal{U}); \quad (\mathcal{N}u,u)_{\mathcal{U}} \ge v_0\|u\|_{\mathcal{U}}^2, \quad v_0 = \text{const} > 0. \tag{1.8}$$

Assume the following: $\mathcal{N}u = \bar{a}\,u$; in this case, $\bar{a}\big|_{\Omega_l} \in C(\Omega_l)$, $l = 1,2$; $0 < a_0 \le \bar{a} \le a_1 < \infty$, a_0, $a_1 = \text{const}$. The cost functional is

$$J(u) = \|Cy(u) - z_g\|_{\mathcal{H}}^2 + (\mathcal{N}u,u)_{\mathcal{U}}, \tag{1.9}$$

where z_g is a known element of the space $\mathcal{H}$.

The optimal control problem is: Find such an element $u \in \mathcal{U}$ that the condition

$$J(u) = \inf_{v \in \mathcal{U}_\partial} J(v) \tag{1.10}$$

is met, where $\mathcal{U}_\partial$ is some convex closed subset in $\mathcal{U}$.

Definition 1.1. If an element $u \in \mathcal{U}_\partial$ meets condition (1.10), it is called an optimal control.

The generalized problem corresponds to initial boundary-value problem (1.6), (1.2)–(1.5) and means to find a vector-function $y(x,t;u) \in$ $\in W(0,T)$ that satisfies the following equations $\forall w(x) \in V_0 =$ $= \left\{ v : \; v|_{\Omega_i} \in W_2^1(\Omega_i), \; i = 1,2 \right\}$:

$$\int_\Omega \frac{dy}{dt} w \, dx + \int_\Omega \sum_{i,j=1}^n k_{ij} \frac{\partial y}{\partial x_j} \frac{\partial w}{\partial x_i} \, dx + \int_\gamma r[y][w] \, d\gamma + \int_\Gamma \alpha y w \, d\Gamma =$$

$$= (f,w) + (Bu,w) + \int_\Gamma \beta w \, d\Gamma \tag{1.11}$$

and

$$\int_\Omega y(x,0;u)w(x) \, dx = \int_\Omega y_0(x)w(x) \, dx ; \tag{1.12}$$

in this case, $(\varphi, \psi) = \int_\Omega \varphi(x,t) \, \psi(x,t) \, dx$, $V = \left\{ v(x,t) : v|_{\Omega_i} \in W_2^1(\Omega_i), \right.$

$\left. i = 1, 2; \; \forall t \in (0,T) \right\}$.

Consider the existence and uniqueness of the solution to problem (1.11), (1.12). Since Bu and $f \in L^2(0,T;V')$, then, without loss of generality, assume the following: $Bu \equiv 0$.

The space V_0 is complete, separable and reflexive since $W_2^1(\Omega_i)$ $(i = 1,2)$ is complete, separable [41, 55, p. 69] and reflexive [32], and $V_0 \subset L_2(\Omega)$. Any element $u(x) \in W_m^l(\Omega_i)$ can be approximated by the functions $u_k(x)$ from $C^{\infty}(\Omega_i)$ within the norms of the space $W_m^l(\Omega_i')$ $\forall \bar{\Omega}_i' \subset \Omega_i$ $(i=1,2)$ [55, p. 69].

Choose [41, 54] an arbitrary fundamental system of linearly independent functions $w_k(x)$, $k = 1, 2, ...$, in V_0 and suppose that this system is orthonormal in $L_2(\Omega)$ so that $(w_k, w_l) = \delta_k^l$ ($\delta_k^k = 1$, $\delta_k^l = 0$ under $l \neq k$; $l, k = 1, 2, ...$). Under $y_0 \in L_2(\Omega)$ [49]:

$$y_0 = \sum_{i=1}^{\infty} \xi_i w_i(x), \tag{1.13}$$

where $\xi_i = (y_0, w_i)$, $i = 1, 2, ...$.

Find the approximate solution to problem (1.11), (1.12) as

$$y_m(x,t) = \sum_{i=1}^{m} g_{im}(t) w_i(x), \tag{1.14}$$

where the functions $g_{im}(t)$ are chosen in such a way that the relations

$$\left(\frac{\partial y_m}{\partial t}, w_j \right) + a\left(y_m, w_j \right) = \left(f, w_j \right) + \int_{\Gamma} \beta w_j d\Gamma, \ j = \overline{1, m}, \tag{1.15}$$

and

$$\left(y_m(x,0), w_j \right) = (y_m, w_j), \ j = \overline{1, m}, \tag{1.16}$$

where

$$a\left(y_m, w_j \right) = \int_{\Omega} \sum_{l,s=1}^{n} k_{ls} \frac{\partial y_m}{\partial x_s} \frac{\partial w_j}{\partial x_l} dx + \int_{\gamma} r[y_m][w_j] d\gamma + \int_{\Gamma} \alpha y_m w_j d\Gamma,$$

are met. Equalities (1.15) and (1.16) specify the Cauchy problem for the system of m first-order linear ordinary differential equations as for $g_{im}(t)$:

$$M_m \frac{dg_m}{dt} + K_m g_m = F_m(t), \tag{1.17}$$

$$M_m^0 g_m(0) = F_m^0; \tag{1.18}$$

in this case: $M_m = \left\{ M_{ij}^m \right\}_{i,j=1}^m$, $M_{ij}^m = (w_i, w_j)$, $K_m = \left\{ k_{ij}^m \right\}_{i,j=1}^m$,

$$k_{ij}^m = a(w_i, w_j), \quad F_m(t) = \left\{ f_i^m(t) \right\}_{i=1}^m, \quad f_i^m(t) = (f, w_i) + \int_\Gamma \beta w_i \, d\Gamma,$$

$$F_m^0 = \left\{ f_{0i}^m \right\}_{i=1}^n, \quad f_{0i}^m = (y_0, w_i), \quad M_m^0 = \left\{ M_{0ij}^m \right\}_{i,j=1}^m, \quad M_{0ij}^m = (w_i, w_j).$$

It is easy to see that the problem solution $y_m(x,t)$ exists and that such solution is unique. The following statement must be proved: $y_m \to y$ under $m \to \infty$, where $y = y(x,t)$ is the solution to problem (1.11), (1.12).

Multiply equality (1.15) by $g_{jm}(t)$ and find the sum over j. Then:

$$\left(\frac{\partial y_m}{\partial t}, y_m \right) + a(y_m, y_m) = (f, y_m) + \int_\Gamma \beta y_m \, d\Gamma,$$

i.e.:

$$\frac{1}{2} \frac{d}{dt} \| y_m \|^2 + a(y_m, y_m) = (f, y_m) + \int_\Gamma \beta y_m \, d\Gamma; \qquad (1.19)$$

in this case, $\| \varphi \| = (\varphi, \varphi)^{1/2} = \| \varphi \|(t)$.

Take the ellipticity condition and generalized Friedrichs inequality [21] into account, and the inequality

$$a(y_m, y_m) \geq \alpha_1 \| y_m \|_V^2, \quad \alpha_1 = \text{const} > 0, \qquad (1.20)$$

is derived, where $\| \varphi \|_V = \left\{ \sum_{i=1}^2 \| \varphi \|_{W_2^1(\Omega_i)}^2 \right\}^{1/2}$, $\| \cdot \|_{W_2^1(\Omega_i)}$ is the norm of the

Sobolev space $W_2^1(\Omega_i)$, $\| \varphi \|_V = \| \varphi \|_V(t)$. Consider inequality (1.20), the ε- and Cauchy-Bunyakovsky inequalities and embedding theorems [55], and the inequality

$$\|y_m\|^2(T) + 2\alpha_1 \int_0^T \|y_m\|_V^2 \, dt \le \|y_m\|^2(0) + 2 \int_0^T \left|(f, y_m)\right| dt +$$

$$+ \int_0^T \left| \int_\Gamma \beta \, y_m \, d\Gamma \right| dt \le \|y_m\|^2(0) + \varepsilon \int_0^T \|y_m\|_V^2 \, dt + \frac{1}{4\varepsilon} \int_0^T \|f\|^2 dt +$$

$$+ \varepsilon_1 c_1' \int_0^T \|y_m\|_V^2 \, dt + \frac{1}{4\varepsilon_1} \int_0^T \|\beta\|_{L_2(\Gamma)}^2 \, dt \qquad (1.21)$$

follows from equality (1.19); in this case, $\|\varphi\|_{L_2(\Gamma)}^2 = \int_\Gamma \varphi^2 d\Gamma$, $c_1' = \max_{l=1,2} c_l$,

and the constant c_l is obtained from the inequality proved within the embedding theorem applied for the domain Ω_l. Proceed from equality (1.16), and

$$(y_m(\cdot,0), \ y_m(\cdot,0)) = (y_0, \ y_m(\cdot,0)).$$

Proceed also from Cauchy-Bunyakovsky inequality, and the inequality

$$\|y_m\|^2(0) \le \|y_0\| \cdot \|y_m\|(0)$$

is derived, i.e.:

$$\|y_m\|(0) \le \|y_0\|. \qquad (1.22)$$

Take it into account, and the inequality

$$\int_0^T \|y_m\|_V^2 \, dt \le c \left(\|y_0\|^2 + \int_0^T \|f\|^2 dt + \int_0^T \|\beta\|_{L_2(\Gamma)}^2 \, dt \right)$$

follows from inequality (1.21).

Therefore, the elements y_m are in some bounded subset of the space $L^2(0,T;V)$. Hence, there exists some subsequence $\{y_\chi\}$ that weakly converges to the element z in $L^2(0,T;V)$ $\left(y_\chi \to z \in L^2(0,T;V) \right)$. Without loss of generality, assume that the whole sequence $\{y_m\}$ weakly converges to z.

Rewrite equality (1.15) as

$$\frac{d}{dt}\left(y_m, w_j\right) + a\left(y_m, w_j\right) = \left(f, w_j\right) + \int_\Gamma \beta w_j \, d\Gamma,$$

multiply its both sides by the function

$$\varphi(t) \in C^1\left([0,T]\right), \quad \varphi(T) = 0, \tag{1.22'}$$

and find the integral from 0 to T of the result:

$$\int_0^T \left\{ -\left(y_m(\cdot,t), \varphi'_j(\cdot,t)\right) + a\left(y_m(\cdot,t), \varphi_j(\cdot,t)\right) \right\} dt =$$

$$= \int_0^T \left(f, \varphi_j\right) dt + \int_0^T \int_\Gamma \beta \varphi_j \, d\Gamma dt + \left(y_m, \varphi_j\right)(0); \tag{1.23}$$

in this case: $\varphi_j(x,t) = \varphi(t) w_j(x)$, $\varphi'_j(x,t) = \dfrac{d\varphi(t)}{dt} w_j(x)$.

By virtue of the aforesaid weak convergence, it is possible to pass in equality (1.23) to the limit under $m \to \infty$, and here is the equality

$$\int_0^T \left\{ -\left(z, \varphi'_j\right) + a\left(z, \varphi_j\right) \right\} dt = \int_0^T \left(f, \varphi_j\right) dt + \int_0^T \int_\Gamma \beta \varphi_j \, d\Gamma dt + \left(z, \varphi_j\right)(0). \tag{1.24}$$

Consider the assumptions as for $\{w_j\}$, and it can be seen that the matrix M_m^0 from condition (1.18) is diagonal $\left(M_{0ij}^m = \left(w_i, w_j\right), \ M_{0ij}^m = 0\right.$ under $i \neq j$, $i,j = \overline{1,m}\big)$. The equality $g_{im}(0) = \left(y_0, w_i\right) / \left(w_i, w_i\right)$ follows from condition (1.18), i.e. $g_{im}(0)$ $(i = \overline{1,m})$ are the Fourier coefficients for the function y_0. By virtue of [49]:

$$y_m(x,0) = \sum_{i=1}^m g_{im}(0) w_i(x) \to y_0(x) \text{ under } m \to \infty.$$

Therefore: $z(x,0) = y_0(x)$.

Equality (1.24) is true for the arbitrary function φ that meets conditions (1.22'). Thus, the following may be assumed: $\varphi \in D((0,T))$ [58]. Then, the equality

$$\int_0^T \{-(z,w_j)\varphi' + a(z,\varphi_j)\} dt = \int_0^T (f,\varphi_j) dt + \int_0^T \int_\Gamma \beta\varphi_j \, d\Gamma dt$$

follows from equality (1.24). Hence:

$$\int_0^T \left[\frac{d}{dt}(z,w_j) + a(z,w_j) - (f,w_j) - \int_\Gamma \beta w_j d\Gamma \right] \varphi(t) dt = 0,$$

i.e.

$$\left(\frac{d}{dt} z, w_j \right) + a(z,w_j) = (f,w_j) + \int_\Gamma \beta w_j \, d\Gamma . \tag{1.25}$$

Take the space V_0 and assumptions as for the functions w_j into account, proceed from equality (1.25), and the following equality is true $\forall w \in V_0$:

$$\left(\frac{d}{dt} z, w \right) + a(z,w) = (f,w) + \int_\Gamma \beta w d\Gamma . \tag{1.26}$$

The forthcoming equality is derived from relation (1.16):

$$(z(\cdot,0), w(\cdot)) = (y_0(\cdot), w(\cdot)), \quad \forall w \in V_0 . \tag{1.27}$$

Therefore, the function $z \in L^2(0,T;V)$ is the solution to problem (1.11), (1.12) $\forall f \in L^2(0,T;V')$ and under $Bu = 0$, i.e. to problem (1.26), (1.27).

Illustrate the uniqueness of the solution to problem (1.26), (1.27) by contradiction. Let there exist two solutions: $z'(x,t)$ and $z''(x,t) \in L^2(0,T;V)$. Then, on the basis of equality (1.26), the equality

$$\frac{d}{dt}\|\bar{z}\|^2 + a(\bar{z},\bar{z}) = 0 \tag{1.28}$$

is obtained, where $\bar{z} = z' - z'' \neq 0$.

Consider equality (1.27), and the contradiction

$$0 < \|\bar{z}\|^2 (T) + \alpha_0 \int_0^T \|\bar{z}\|_V^2 \, dt \le 0, \quad \alpha_0 = \text{const} > 0, \tag{1.29}$$

follows from equality (1.28).

Therefore, the validity of the following statement is proved.

Theorem 1.1. *Initial boundary-value problem (1.1)–(1.5) has a unique generalized solution* $y(x,t) \in L^2(0,T;V)$.

Proceed from equality (1.28) and contradiction (1.29), and it is easy to see that $y(u_1) \neq y(u_2)$ under $u_1 \neq u_2$, i.e. under $Bu_1 \neq Bu_2$. Assume the following: $Bu \equiv u$.

Let $\tilde{y}' = \tilde{y}(u')$ and $\tilde{y}'' = \tilde{y}(u'')$ be solutions from $L^2(0,T;V)$ to problem (1.11), (1.12) under $f = 0$ and $\beta = 0$ and under a function $u = u(x,t)$ that is equal, respectively, to u' and u''. Then:

$$\frac{d}{dt}\|\tilde{y}' - \tilde{y}''\|^2 + \alpha_0 \|\tilde{y}' - \tilde{y}''\|_V^2 \le \|u' - u''\| \cdot \|\tilde{y}' - \tilde{y}''\|_V. \tag{1.30}$$

Therefore, the inequality

$$\|\tilde{y}' - \tilde{y}''\|_{V \times L_2} \le \frac{1}{\alpha_0} \|u' - u''\|_{L_2 \times L_2} \tag{1.31}$$

is obtained, where

$$\|\varphi\|_{L_2 \times L_2}^2 = \int_0^T \|\varphi\|_{L_2}^2 \, dt, \ \|\varphi\|_{L_2}^2 = \|\varphi\|^2 = \int_\Omega \varphi^2(x,t) dx, \ \|\varphi\|_{V \times L_2}^2 = \int_0^T \|\varphi\|_V^2 \, dt.$$

Rewrite expression (1.9) as

$$J(u) = \pi(u,u) - 2L(u) + \int_0^T \|z_g - y(0)\|^2 \, dt, \tag{1.32}$$

where

$$\pi(u,v) = \big(y(u) - y(0), y(v) - y(0)\big)_{\mathscr{H}} + (\bar{a}u, v)_{\mathscr{U}},$$

$$L(v) = \big(z_g - y(0), y(v) - y(0)\big)_{\mathscr{H}}; \tag{1.33}$$

in this case, $(z,v)_{\mathscr{H}} = \int\limits_0^T (z,v)dt$, $(z,v)_{\mathscr{U}} = \int\limits_0^T (z,v)dt$, $(z,v) = \int\limits_\Omega zvdx$.

Inequality (1.31) provides the continuity of the linear functional $L(\cdot)$ and bilinear form $\pi(\cdot,\cdot)$ on $\mathscr{U}$.

On the basis of [58, Chapter 1, Theorem 1.1], the validity of the following statement is proved.

Theorem 1.2. *Let a system state be determined as a solution to problem (1.11), (1.12). Then, there exists a unique element u of a convex set $\mathscr{U}_\partial$ that is closed in $\mathscr{U}$, and*

$$J(u) = \inf_{v \in \mathscr{U}_\partial} J(v) \tag{1.34}$$

takes place for u.

A control $u \in \mathscr{U}_\partial$ is optimal if and only if the inequality

$$\langle J'(u), v - u \rangle \geq 0, \quad \forall v \in \mathscr{U}_\partial,$$

is true, i.e. under

$$\big(y(u) - z_g, y(v) - y(u)\big)_{\mathscr{H}} + (\mathscr{N}u, v - u)_{\mathscr{U}} \geq 0. \tag{1.35}$$

As for the control $v \in \mathscr{U}$, the conjugate state is specified by the relations

$$-\frac{\partial p}{\partial t} - \sum_{i,j=1}^n \frac{\partial}{\partial x_i}\left(k_{ij}\frac{\partial p}{\partial x_j}\right) = y(v) - z_g, \ (x,t) \in \Omega_T,$$

$$\sum_{i,j=1}^n k_{ij}\frac{\partial p}{\partial x_j}\cos(v, x_i) = -\alpha\, p, \ (x,t) \in \Gamma_T,$$

$$\left[\sum_{i,j=1}^n k_{ij}\frac{\partial p}{\partial x_j}\cos(v, x_i)\right] = 0, \ (x,t) \in \gamma_T, \tag{1.36}$$

$$\left\{ \sum_{i,j=1}^{n} k_{ij} \frac{\partial p}{\partial x_j} \cos(\nu, x_i) \right\}^{\pm} = r[p], \ (x,t) \in \gamma_T,$$

$$p(x,T) = 0, \ x \in \bar{\Omega}_1 \cup \bar{\Omega}_2 .$$

Substitute a time $T - t$ for the time t, proceed from Theorem 1.1, and it is concluded that boundary-value problem (1.36) has the unique generalized solution $p(v) \in L^2(0,T;V)$ as the unique one to the equality system

$$-\left(\frac{d}{dt} p(v), w \right) + a(p,w) = \left(y(v) - z_g, w \right), \ \forall w \in V_0 , \tag{1.37}$$

$$(p,w) = 0, \ t = T . \tag{1.38}$$

Multiply the first equality of relations (1.36) (under $v = u$) in a scalar way by $y(v) - y(u)$ and find the integral from 0 to T of the result.

Consider the equality

$$\int_0^T \left(-\frac{dp(u)}{dt}, y(v) - y(u) \right) dt = \int_0^T \left(p(u), \frac{d}{dt} (y(v) - y(u)) \right) dt \tag{1.39}$$

and obtain the equality

$$\int_0^T \left(y(u) - z_g, y(v) - y(u) \right) dt = \int_0^T \left(p(u), \frac{d}{dt} (y(v) - y(u)) \right) dt +$$

$$+ \int_0^T a \left(p(u), y(v) - y(u) \right) dt . \tag{1.40}$$

Note, that when equality (1.39) is taken into account, equality (1.40) can be derived from equality (1.37). Take equation (1.11) into consideration, and the equality

$$\int_0^T \left(y(u) - z_g, y(v) - y(u) \right) dt = \int_0^T (p(u), v - u) \, dt \tag{1.41}$$

is found from equality (1.40). Therefore, inequality (1.35) has the form

$$\int_0^T \left(p(u) + \bar{a}\,u, v - u \right) dt \geq 0, \quad \forall v \in \mathcal{U}_\partial. \tag{1.42}$$

Thus, the optimal control $u \in \mathcal{U}_\partial$ is specified by relations (1.11), (1.12), (1.37), (1.38) and (1.42).

If the constraints are absent, i.e. when $\mathcal{U}_\partial = \mathcal{U}$, then the equality

$$p(u) + \bar{a}\,u = 0$$

is obtained from inequality (1.42).

The control

$$u = -p/\bar{a}, \quad (x,t) \in \Omega_T, \tag{1.43}$$

is found from the latter equality.

If the solution $(y, p)^{\mathrm{T}}$ to problem (1.11), (1.12), (1.37), (1.38), (1.43) is smooth enough on $\bar{\Omega}_{lT}$, viz., $y,\ p\big|_{\bar{\Omega}_{lT}} \in C^{1,0}(\bar{\Omega}_{lT}) \cap C^{2,0}(\Omega_{lT}) \cap$
$\cap C^{0,1}(\Omega_{lT})$, $l = 1, 2$, then the differential problem of finding the vector-function $(y, p)^{\mathrm{T}}$, that satisfies equality (1.43) and the equalities

$$\frac{\partial y}{\partial t} - \sum_{i,j=1}^n \frac{\partial}{\partial x_i}\left(k_{ij} \frac{\partial y}{\partial x_j} \right) + p/\bar{a} = f, \quad (x,t) \in \Omega_T,$$

$$-\frac{\partial p}{\partial t} - \sum_{i,j=1}^n \frac{\partial}{\partial x_i}\left(k_{ij} \frac{\partial p}{\partial x_j} \right) - y = -z_g, \quad (x,t) \in \Omega_T,$$

$$\sum_{i,j=1}^n k_{ij} \frac{\partial y}{\partial x_j} \cos(\nu, x_i) = -\alpha y + \beta, \quad (x,t) \in \Gamma_T,$$

$$\sum_{i,j=1}^n k_{ij} \frac{\partial p}{\partial x_j} \cos(\nu, x_i) = -\alpha\,p, \quad (x,t) \in \Gamma_T,$$

$$\left[\sum_{i,j=1}^{n} k_{ij} \frac{\partial y}{\partial x_j} \cos(v, x_i)\right] = 0, \quad (x,t) \in \gamma_T,$$

$$\left[\sum_{i,j=1}^{n} k_{ij} \frac{\partial p}{\partial x_j} \cos(v, x_i)\right] = 0, \quad (x,t) \in \gamma_T, \qquad (1.44)$$

$$\left\{\sum_{i,j=1}^{n} k_{ij} \frac{\partial y}{\partial x_j} \cos(v, x_i)\right\}^{\pm} = r[y], \quad (x,t) \in \gamma_T,$$

$$\left\{\sum_{i,j=1}^{n} k_{ij} \frac{\partial p}{\partial x_j} \cos(v, x_i)\right\}^{\pm} = r[p], \quad (x,t) \in \gamma_T,$$

$$y(x,0) = y_0, \quad p(x,T) = 0, \quad x \in \bar{\Omega}_1 \cup \bar{\Omega}_2,$$

corresponds to problem (1.11), (1.12), (1.37), (1.38), (1.43).

5.2 CONTROL UNDER CONJUGATION CONDITION WITH OBSERVATION THROUGHOUT A WHOLE DOMAIN

Assume that equation (1.1) is specified in the domain Ω_T. On the boundary Γ_T, the boundary condition has the form of expression (1.2).

For every control $u \in \mathcal{U} = L_2(\gamma_T)$, determine a state $y = y(u)$ as a generalized solution to the boundary-value problem specified, in its turn, by equation (1.1), boundary condition (1.2), initial condition (1.5) and the conjugation conditions

$$[y] = 0, \quad (x,t) \in \gamma_T, \qquad (2.1)$$

and

$$\left[\sum_{i,j=1}^{n} k_{ij} \frac{\partial y}{\partial x_j} \cos(\nu, x_i)\right] = \omega + u, \quad (x,t) \in \gamma_T, \tag{2.2}$$

where $\omega = \omega(x,t) \in L_2(\gamma_T)$.

Since there exists the generalized solution $y(u) \in W(0,T)$ to boundary-value problem (1.1), (1.2), (1.5), (2.1), (2.2), then such solution is reasonable on $\bar{\Omega}_{lT}$ $(l=1,2)$.

The generalized problem corresponds to initial boundary-value problem (1.1), (1.2), (1.5), (2.1), (2.2) and means to find a function $y(x,t;u) \in W(0,T)$ that satisfies the following equations $\forall w(x) \in V_0 = \{v: v|_{\Omega_i} \in W_2^1(\Omega_i), i=1,2; [v]=0\}$:

$$\int_{\Omega} \frac{dy}{dt} w \, dx + \int_{\Omega} \sum_{i,j=1}^{n} k_{ij} \frac{\partial y}{\partial x_j} \frac{\partial w}{\partial x_i} dx + \int_{\Gamma} \alpha y w \, d\Gamma =$$

$$= (f,w) - \int_{\gamma} \omega w \, d\gamma - \int_{\gamma} u w \, d\gamma + \int_{\Gamma} \beta w \, d\Gamma \tag{2.3}$$

and

$$\int_{\Omega} y(x,0;u) w(x) \, dx = \int_{\Omega} y_0(x) w(x) \, dx; \tag{2.4}$$

in this case, $V = \left\{v(x,t): v|_{\Omega_i} \in W_2^1(\Omega_i), i=1,2; [v]=0, \forall t \in [0,T]\right\}$.

The following statement takes place.

Theorem 2.1. *Initial boundary-value problem (1.1), (1.2), (1.5), (2.1), (2.2) has a unique generalized solution* $y(x,t;u) \in L^2(0,T;V)$ $\forall u \in \mathcal{U}$.

The validity of Theorem 2.1 is stated by analogy with the proof of Theorem 1.1.

Proceed from equations (2.3) and (2.4), and it is easy to see that $y(u_1) \neq y(u_2)$ under $u_1 \neq u_2$. If $\tilde{y}' = \tilde{y}(u')$ and $\tilde{y}'' = \tilde{y}(u'')$ are generalized solutions from $L^2(0,T;V)$ to problem (2.3), (2.4) under $f=0$, $\beta=0$

and $\omega = 0$ and under a function u that is equal, respectively, to u' and u'', then the inequality

$$\frac{d}{dt}\|\tilde{y}' - \tilde{y}''\|^2 + \alpha_0\|\tilde{y}' - \tilde{y}''\|_V^2 \le$$

$$\le \|u' - u''\|_{L_2(\gamma)}\|\tilde{y}' - \tilde{y}''\|_{L_2(\gamma)} \le c_0\|u' - u''\|_{L_2(\gamma)}\|\tilde{y}' - \tilde{y}''\|_V$$

is derived, where c_0 is the constant in the inequality of the embedding theorem [55].

Therefore, the inequality

$$\|\tilde{y}' - \tilde{y}''\|_{V \times L_2} \le c_1\|u' - u''\|_{L_2(\gamma) \times L_2} \tag{2.5}$$

is obtained, where $\|\varphi\|_{L_2(\gamma) \times L_2}^2 = \int_0^T \|\varphi\|_{L_2(\gamma)}^2 dt$, $\|\varphi\|_{L_2(\gamma)}^2 = \int_\gamma \varphi^2 d\gamma$ and from

which the inequality

$$\|\tilde{y}' - \tilde{y}''\|_{L_2 \times L_2} \le c_1\|u' - u''\|_{L_2(\gamma) \times L_2}, \tag{2.6}$$

where $c_1 = \text{const} > 0$, follows that provides the continuity of the linear functional $L(\cdot)$ and bilinear form $\pi(\cdot,\cdot)$ on $\mathcal{U}$. In this case, the linear functional $L(\cdot)$ and bilinear form $\pi(\cdot,\cdot)$ are specified by expressions (1.33), where $(\bar{a}u, v)_{\mathcal{U}} = \int_0^T \int_\gamma \bar{a}\, uv\, d\gamma\, dt$.

Specify the observation in the form of expression (1.7), where $Cy(u) \equiv y(u)$. Bring a value of cost functional (1.9) in correspondence with every control $u \in \mathcal{U}$; in this case, z_g is a known element from $L^2(0,T;V)$ and the cost functional is

$$J(u) = \int_0^T \int_\Omega \left(y(u) - z_g\right)^2 dx\, dt + \int_0^T \int_\gamma \bar{a}u^2 d\gamma\, dt, \tag{2.7}$$

where $0 < a_0 \le \bar{a} \le a_1 < \infty$, $a_0, a_1 = \text{const}$, $\bar{a} \in L_2(\gamma)$.

On the basis of [58, Chapter 1, Theorem 1.1], the validity of the following statement is proved.

Theorem 2.2. *If a system state is determined as a solution to problem (2.3), (2.4), then there exists a unique element u of a convex set $\mathcal{U}_\partial$ that is closed in $\mathcal{U}$, and relation (1.34) takes place for u, where the cost functional has the form of expression (2.7).*

As for the control $v \in \mathcal{U}$, the conjugate state $p(v)$ is specified by the relations

$$-\frac{\partial p}{\partial t} - \sum_{i,j=1}^{n} \frac{\partial}{\partial x_i}\left(k_{ij}\frac{\partial p}{\partial x_j}\right) = y(v) - z_g, \ (x,t) \in \Omega_T,$$

$$\sum_{i,j=1}^{n} k_{ij}\frac{\partial p}{\partial x_j}\cos(v, x_i) = -\alpha\, p, \ (x,t) \in \Gamma_T,$$

$$[p] = 0, \ (x,t) \in \gamma_T, \tag{2.8}$$

$$\left[\sum_{i,j=1}^{n} k_{ij}\frac{\partial p}{\partial x_j}\cos(v, x_i)\right] = 0, \ (x,t) \in \gamma_T,$$

$$p(x,T) = 0, \ x \in \bar{\Omega}_1 \cup \bar{\Omega}_2.$$

Problem (2.8) has the unique generalized solution $p(v) \in L^2(0,T;V)$ as the unique one to the following equality system:

$$-\int_{\Omega}\frac{dp}{dt}w\,dx + \int_{\Omega}\sum_{i,j=1}^{n} k_{ij}\frac{\partial p}{\partial x_j}\frac{\partial w}{\partial x_i}\,dx + \int_{\Gamma}\alpha p w\,d\Gamma = \int_{\Omega}\left(y - z_g\right)w\,dx, \ \forall w \in V_0,$$

$$\int_{\Omega} p(x,T;\cdot)w(x)\,dx = 0. \tag{2.9}$$

Multiply the first equality of relations (2.8) (under $v = u$) in a scalar way by the difference $y(v) - y(u)$ and find the integral of the result over Ω_T. Consider equality (1.39), and equality like (1.40) is obtained, where

$$a(z,w) = \int\limits_{\Omega} \sum_{i,j=1}^{n} k_{ij} \frac{\partial z}{\partial x_j} \frac{\partial w}{\partial x_i} dx + \int\limits_{\Gamma} \alpha z w d\Gamma. \qquad (2.10)$$

Proceed from equation (2.3), equality (1.39), and the equality

$$\int\limits_0^T \left(y(u) - z_g, y(v) - y(u) \right) dt = -\int\limits_0^T \int\limits_\gamma p(u) \, (v-u) d\gamma \, dt$$

is found from system (2.9).

Therefore, the control $u \in \mathcal{U}_\partial$ is optimal if and only if the inequality

$$\int\limits_0^T \int\limits_\gamma (-p(u) + \bar{a}u) \, (v-u) d\gamma \, dt \geq 0, \ \forall v \in \mathcal{U}_\partial, \qquad (2.11)$$

takes place.

Thus, the optimal control $u \in \mathcal{U}_\partial$ is specified by relations (2.3), (2.4), (2.9) and (2.11). If the constraints are absent, i.e. when $\mathcal{U}_\partial = \mathcal{U}$, then the equality

$$-p(u) + \bar{a}u = 0, \ (x,t) \in \gamma_T,$$

is obtained from inequality (2.11).

The control

$$u = p/\bar{a}, \ (x,t) \in \gamma_T, \qquad (2.12)$$

is found from the latter equality.

If the solution $(y,p)^T$ to problem (2.3), (2.4), (2.9), (2.11) is smooth enough on $\bar{\Omega}_{lT}$, viz. $y, p|_{\bar{\Omega}_{lT}} \in C^{1,0}(\bar{\Omega}_{lT}) \cap C^{2,0}(\Omega_{lT}) \cap$

$\cap C^{0,1}(\Omega_{lT})$, $l = 1, 2$, then the differential problem of finding the vector-function $(y,p)^T$, that satisfies the relations

$$\frac{\partial y}{\partial t} - \sum_{i,j=1}^{n} \frac{\partial}{\partial x_i} \left(k_{ij} \frac{\partial y}{\partial x_j} \right) = f, \ (x,t) \in \Omega_T,$$

$$-\frac{\partial p}{\partial t} - \sum_{i,j=1}^{n} \frac{\partial}{\partial x_i}\left(k_{ij}\frac{\partial p}{\partial x_j}\right) - y = -z_g, \quad (x,t)\in\Omega_T,$$

$$\sum_{i,j=1}^{n} k_{ij}\frac{\partial p}{\partial x_j}\cos(\nu,x_i) = -\alpha\, p, \quad (x,t)\in\Gamma_T,$$

$$[y]=0,\ [p]=0,\ (x,t)\in\gamma_T,$$

$$\left[\sum_{i,j=1}^{n} k_{ij}\frac{\partial y}{\partial x_j}\cos(\nu,x_i)\right] = \omega + p/\bar{a},\ (x,t)\in\gamma_T,$$

$$\left[\sum_{i,j=1}^{n} k_{ij}\frac{\partial p}{\partial x_j}\cos(\nu,x_i)\right] = 0,\ (x,t)\in\gamma_T,$$

$$y(x,0)=y_0,\ p(x,T)=0,\ x\in\bar{\Omega}_1\cup\bar{\Omega}_2,$$

$$u = p/\bar{a},\ (x,t)\in\gamma_T,$$

corresponds to problem (2.3), (2.4), (2.9), (2.11).

5.3 CONTROL UNDER CONJUGATION CONDITION WITH BOUNDARY OBSERVATION

Assume that equation (1.1) is specified in the domain Ω_T. On the boundary Γ_T, the boundary condition has the form of expression (1.2). For every control $u\in\mathcal{U}=L_2(\gamma_T)$, determine a state $y(u)$ as a generalized solution to the boundary-value problem specified, in its turn, by equation (1.1), boundary condition (1.2), initial condition (1.5) and the conjugation conditions

$$[y]=0,\ (x,t)\in\gamma_T, \tag{3.1}$$

and

$$\left[\sum_{i,j=1}^{n} k_{ij} \frac{\partial y}{\partial x_j} \cos(\nu, x_i)\right] = \omega + u, \quad (x,t) \in \gamma_T, \tag{3.2}$$

where $\omega = \omega(x,t) \in L_2(\gamma_T)$.

The cost functional is

$$J(u) = \int_0^T \int_\Gamma \left(y(u) - z_g\right)^2 d\Gamma \, dt + \int_0^T \int_\gamma \bar{a} u^2 d\gamma \, dt . \tag{3.3}$$

The generalized problem corresponds to initial boundary-value problem (1.1), (1.2), (1.5), (3.1), (3.2) and means to find a function $y(x,t;u) \in W(0,T)$ that satisfies equations (2.3) and (2.4) $\forall w(x) \in V_0$; in this case, the spaces V of $W(0,T)$ and V_0 are specified in point 5.2.

By virtue of Theorem 2.1, initial boundary-value problem (1.1), (1.2), (1.5), (3.1), (3.2) has the unique generalized solution $y(x,t) \in L^2(0,T;V)$ $\forall u \in \mathcal{U}$.

Proceed from equations (2.3) and (2.4), and it is easy to see that $y(u_1) \neq y(u_2)$ under $u_1 \neq u_2$. Let $\tilde{y}' = \tilde{y}(u')$ and $\tilde{y}'' = \tilde{y}(u'')$ be solutions from $L^2(0,T;V)$ to problem (2.3), (2.4) under $f = 0$, $\beta = 0$ and $\omega = 0$ and under a function u that is equal, respectively, to u' and u''. Then, the inequality (2.5) is true. Consider the embedding theorems [55], and the inequality

$$\left\|\tilde{y}' - \tilde{y}''\right\|_{L_2(\Gamma) \times L_2} \leq c_2 \left\|u' - u''\right\|_{L_2(\gamma) \times L_2}$$

is obtained from inequality (2.5). The obtained inequality provides the continuity of the linear functional $L(\cdot)$ and bilinear form $\pi(\cdot,\cdot)$ of representation like (1.32) for cost functional (3.3) on $\mathcal{U}$; in this case:

$$\pi(u,v) = \left(y(u) - y(0), y(v) - y(0)\right)_{L_2(\Gamma) \times L_2} + \left(\bar{a} u, v\right)_{L_2(\gamma) \times L_2}$$

and

$$L(v) = \left(z_g - y(0), y(v) - y(0)\right)_{L_2(\Gamma) \times L_2} .$$

On the basis of [58, Chapter 1, Theorem 1.1], the validity of the following statement is proved.

Theorem 3.1. *Let a system state be determined as a solution to problem (2.3), (2.4). Then, there exists a unique element u of a convex set $\mathcal{U}_{\partial}$ that is closed in $\mathcal{U}$, and relation (1.34) takes place for u, where the cost functional has the form of expression (3.3).*

As for the control $v \in \mathcal{U}$, the conjugate state is specified by the relations

$$-\frac{\partial p}{\partial t} - \sum_{i,j=1}^{n} \frac{\partial}{\partial x_i}\left(k_{ij} \frac{\partial p}{\partial x_j} \right) = 0, \ (x,t) \in \Omega_T,$$

$$\sum_{i,j=1}^{n} k_{ij} \frac{\partial p}{\partial x_j} \cos(v, x_i) = -\alpha p + y(v) - z_g, \ (x,t) \in \Gamma_T, \tag{3.4}$$

$$[p] = 0, \ (x,t) \in \gamma_T,$$

$$\left[\sum_{i,j=1}^{n} k_{ij} \frac{\partial p}{\partial x_j} \cos(v, x_i) \right] = 0, \ (x,t) \in \gamma_T,$$

$$p(x,T) = 0, \ x \in \bar{\Omega}_1 \cup \bar{\Omega}_2.$$

Problem (3.4) has the unique generalized solution $p(v) \in L^2(0,T;V)$ as the unique one to the following equality system:

$$-\int_{\Omega} \frac{dp}{dt} w\, dx + \int_{\Omega} \sum_{i,j=1}^{n} k_{ij} \frac{\partial p}{\partial x_j} \frac{\partial w}{\partial x_i}\, dx + \int_{\Gamma} \alpha pw\, d\Gamma = \int_{\Gamma} \left(y(v) - z_g \right) w\, d\Gamma, \forall w \in V_0,$$

$$\int_{\Omega} p(x,T;\cdot) w(x)\, dx = 0. \tag{3.5}$$

Multiply the first equality of relations (3.4) (under $v = u$) in a scalar way by the difference $y(v) - y(u)$ and find the integral of the result from 0 to T. Consider equality (1.39) and obtain the equality

$$0 = \int_0^T \left(p(u), \frac{d}{dt}(y(v) - y(u)) \right) dt + \int_0^T a\left(p(u), y(v) - y(u) \right) dt -$$

$$-\int\limits_0^T\int\limits_\Gamma\big(y(u)-z_g\big)\big(y(v)-y(u)\big)d\Gamma dt, \tag{3.6}$$

where the bilinear form $a(\cdot,\cdot)$ is specified by expression (2.10).

Proceed from equation (2.3), and the equality

$$0=-\int\limits_0^T\int\limits_\gamma p(u)(v-u)d\gamma\,dt-\int\limits_0^T\int\limits_\Gamma\big(y(u)-z_g\big)\big(y(v)-y(u)\big)d\Gamma dt,$$

i.e.

$$\int\limits_0^T\int\limits_\Gamma\big(y(u)-z_g\big)\big(y(v)-y(u)\big)d\Gamma dt=-\int\limits_0^T\int\limits_\gamma p(u)\,(v-u)d\gamma\,dt$$

follows from equality (3.6).

Therefore, inequality (1.35) has the form

$$\int\limits_0^T\int\limits_\gamma\big(-p(u)+\bar a\,u\big)\,(v-u)d\gamma\,dt\ge 0 \tag{3.7}$$

for the optimization problem considered in the present point.

Thus, the optimal control $u\in\mathscr{U}_\partial$ is specified by relations (2.3), (2.4), (3.5) and (3.7). If the constraints are absent, i.e. $\mathscr{U}_\partial=\mathscr{U}$, then the following equality is obtained from inequality (3.7):

$$u=p/\bar a,\ (x,t)\in\gamma_T. \tag{3.8}$$

If the solution $(y,p)^{\mathrm{T}}$ to problem (2.3), (2.4), (3.5), (3.8) is smooth enough on $\bar\Omega_{lT}$, $l=1,2$, then the differential problem of finding the vector-function $(y,p)^{\mathrm{T}}$, that satisfies the equalities

$$\frac{\partial y}{\partial t}-\sum\limits_{i,j=1}^n\frac{\partial}{\partial x_i}\left(k_{ij}\frac{\partial y}{\partial x_j}\right)=f,\ (x,t)\in\Omega_T,$$

$$-\frac{\partial p}{\partial t}-\sum\limits_{i,j=1}^n\frac{\partial}{\partial x_i}\left(k_{ij}\frac{\partial p}{\partial x_j}\right)=0,\ (x,t)\in\Omega_T,$$

$$\sum_{i,j=1}^{n} k_{ij} \frac{\partial y}{\partial x_j} \cos(v, x_i) = -\alpha\, y + \beta, \quad (x,t) \in \Gamma_T,$$

$$\sum_{i,j=1}^{n} k_{ij} \frac{\partial p}{\partial x_j} \cos(v, x_i) = -\alpha\, p + y - z_g, \quad (x,t) \in \Gamma_T, \tag{3.9}$$

$$[y] = 0, \quad [p] = 0, \quad (x,t) \in \gamma_T,$$

$$\left[\sum_{i,j=1}^{n} k_{ij} \frac{\partial y}{\partial x_j} \cos(v, x_i) \right] = \omega + p/\bar{a}, \quad (x,t) \in \gamma_T,$$

$$\left[\sum_{i,j=1}^{n} k_{ij} \frac{\partial p}{\partial x_j} \cos(v, x_i) \right] = 0, \quad (x,t) \in \gamma_T,$$

$$y(x,0) = y_0, \quad p(x,T) = 0, \quad x \in \bar{\Omega}_1 \cup \bar{\Omega}_2,$$

corresponds to problem (2.3), (2.4), (3.5), (3.8).

5.4 CONTROL UNDER CONJUGATION CONDITION WITH FINAL OBSERVATION

Assume that equation (1.1) is specified in the domain Ω_T. On the boundary Γ_T, the boundary condition has the form of expression (1.2). For every control $u \in \mathcal{U} = L_2(\gamma_T)$, determine a state $y(u)$ as a generalized solution to the boundary-value problem specified, in its turn, by equation (1.1), boundary condition (1.2), initial condition (1.5) and conjugation conditions (3.1) and (3.2).

The cost functional is

$$J(u) = \int_{\Omega} \left(y(x,T;u) - z_g \right)^2 dx + \int_0^T \int_\gamma \bar{a} u^2 \, d\gamma\, dt, \tag{4.1}$$

where $0 < a_0 \leq \bar{a} \leq a_1 < \infty$; $a_0, a_1 = \text{const}$, and it may be rewritten as

$$J(u) = \pi(u,u) - 2L(u) + \int_{\Omega} \left(z_g(x) - y(x,T;0)\right)^2 dx \, ;$$

in this case,

$$\pi(u,v) = \left(y(\cdot,T;u) - y(\cdot,T;0), y(\cdot,T;v) - y(\cdot,T;0)\right) + \int_0^T \int_\gamma \bar{a} uv \, d\gamma \, dt$$

and

$$L(v) = \left(z_g(\cdot) - y(\cdot,T;0), y(\cdot,T;v) - y(\cdot,T;0)\right).$$

The generalized problem corresponds to initial boundary-value problem (1.1), (1.2), (1.5), (3.1), (3.2) and means to find a function $y(x,t;u) \in$ $\in W(0,T)$ that satisfies equations (2.3) and (2.4) $\forall w(x) \in V_0$; the spaces $W(0,T)$ and V_0 are specified in point 5.2.

Theorem 2.1 takes place. It is stated in point 5.3 that $y(u_1) \neq y(u_2)$ under $u_1 \neq u_2$. If $\tilde{y}' = \tilde{y}(u')$ and $\tilde{y}'' = \tilde{y}(u'')$ are solutions from $L^2(0,T;V)$ to problem (2.3), (2.4) under $f = 0$, $\beta = 0$ and $\omega = 0$ and under a function u that is equal, respectively, to u' and u'', then inequality (2.5) is true.

Proceed from equation (2.3) and obtain the equality

$$\|\tilde{y}' - \tilde{y}''\|_{L_2}^2 (T) + \int_0^T a\left(\tilde{y}' - \tilde{y}'', \tilde{y}' - \tilde{y}''\right) dt = -\int_0^T \int_\gamma (u' - u'')(\tilde{y}' - \tilde{y}'') d\gamma \, dt,$$

where the bilinear form $a(\cdot,\cdot)$ is specified by expression (2.10).

Consider inequality (2.5), the Cauchy-Bunyakovsky inequality and embedding theorems, and the inequality

$$\|\tilde{y}' - \tilde{y}''\|^2 (T) \leq \|u' - u''\|_{L_2(\gamma)\times L_2} \|\tilde{y}' - \tilde{y}''\|_{L_2(\gamma)\times L_2} \leq$$

$$\leq c_0 \|u' - u''\|_{L_2(\gamma)\times L_2} \|\tilde{y}' - \tilde{y}''\|_{V\times L_2} \leq c_0 \, c_1 \|u' - u''\|_{L_2(\gamma)\times L_2}^2$$

or

$$\|\tilde{y}' - \tilde{y}''\|(T) \le c_2 \|u' - u''\|_{L_2(\gamma) \times L_2}$$

is derived that provides the continuity of the linear functional $L(\cdot)$ and bilinear form $\pi(\cdot,\cdot)$ on $\mathcal{U}$.

On the basis of [58, Chapter 1, Theorem 1.1], the validity of the following statement is proved.

Theorem 4.1. *If a system state is determined as a solution to problem (2.3), (2.4), then there exists a unique element u of convex set $\mathcal{U}_\partial$ that is closed in $\mathcal{U}$, and relation (1.34) takes place for u, where the cost functional has the form of expression (4.1).*

As for the control $v \in \mathcal{U}$, the conjugate state $p(v)$ is specified by the relations

$$-\frac{\partial p}{\partial t} - \sum_{i,j=1}^{n} \frac{\partial}{\partial x_i}\left(k_{ij}\frac{\partial p}{\partial x_j}\right) = 0, \quad (x,t) \in \Omega_T,$$

$$\sum_{i,j=1}^{n} k_{ij}\frac{\partial p}{\partial x_j}\cos(v,x_i) = -\alpha\, p, \quad (x,t) \in \Gamma_T,$$

$$[p] = 0, \quad (x,t) \in \gamma_T, \tag{4.2}$$

$$\left[\sum_{i,j=1}^{n} k_{ij}\frac{\partial p}{\partial x_j}\cos(v,x_i)\right] = 0, \quad (x,t) \in \gamma_T,$$

$$p(x,T;v) = y(x,T;v) - z_g, \quad x \in \bar{\Omega}_1 \cup \bar{\Omega}_2.$$

Problem (4.2) has the unique generalized solution $p(v) \in L^2(0,T;V)$ as the unique one to the equality system

$$-\int_\Omega \frac{dp}{dt}w\,dx + \int_\Omega \sum_{i,j=1}^{n} k_{ij}\frac{\partial p}{\partial x_j}\frac{\partial w}{\partial x_i}\,dx + \int_\Gamma \alpha pw\,d\Gamma = 0, \quad \forall w \in V_0,$$

$$\int_\Omega p(x,T;v)\,w(x)\,dx = \int_\Omega \left(y(x,T;v) - z_g\right)w\,dx. \tag{4.3}$$

Multiply the first equality of relations (4.2) (under $v = u$) in a scalar way by the difference $y(v) - y(u)$ and find the integral from 0 to T of the result. Then, consider the equality

$$\int_0^T \left(-\frac{dp(u)}{dt}, y(v) - y(u) \right) dt = \int_0^T \left(p(u), \frac{d}{dt}(y(v) - y(u)) \right) dt -$$

$$- \int_\Omega \left(y(x, T; u) - z_g \right) (y(x, T; v) - y(x, T; u)) dx$$

and obtain the equality

$$0 = \int_0^T \left(p(u), \frac{d}{dt}(y(v) - y(u)) \right) dt + a(p, y(v) - y(u)) -$$

$$- \int_\Omega \left(y(x, T; u) - z_g \right) (y(x, T; v) - y(x, T; u)) dx, \qquad (4.4)$$

where the bilinear form $a(\cdot, \cdot)$ is specified by expression (2.10).

Proceed from equation (2.3), and the equality

$$0 = -\int_0^T \int_\gamma p(u)(v - u) d\gamma dt - \int_\Omega \left(y(x, T; u) - z_g \right) (y(x, T; v) - y(x, T; u)) dx$$

follows from equality (4.4), i.e.:

$$\int_\Omega \left(y(x, T; u) - z_g \right) (y(x, T; v) - y(x, T; u)) dx = -\int_0^T \int_\gamma p(u)(v - u) d\gamma \, dt.$$

Then, inequality (1.35) has the form

$$\int_0^T \int_\gamma (-p + \bar{a}u)(v - u) d\gamma \, dt \geq 0 \qquad (4.5)$$

for the optimization problem considered in the present point.

Thus, the optimal control $u \in \mathcal{U}_{\partial}$ is specified by relations (2.3), (2.4), (4.3) and (4.5). If the constraints are absent, i.e. when $\mathcal{U}_{\partial} = \mathcal{U}$, then equality (3.8) is obtained from inequality (4.5). If the solution $(y, p)^{\mathrm{T}}$ to problem (2.3), (2.4), (3.8), (4.3) is smooth enough on $\overline{\Omega}_{lT}$, $l = 1, 2$, then the differential problem of finding the vector-function $(y, p)^{\mathrm{T}}$, that satisfies the equalities

$$\frac{\partial y}{\partial t} - \sum_{i,j=1}^{n} \frac{\partial}{\partial x_i}\left(k_{ij}\frac{\partial y}{\partial x_j}\right) = f, \ (x,t) \in \Omega_T,$$

$$-\frac{\partial p}{\partial t} - \sum_{i,j=1}^{n} \frac{\partial}{\partial x_i}\left(k_{ij}\frac{\partial p}{\partial x_j}\right) = 0, \ (x,t) \in \Omega_T,$$

$$\sum_{i,j=1}^{n} k_{ij}\frac{\partial y}{\partial x_j}\cos(\nu, x_i) = -\alpha\, y + \beta, \ (x,t) \in \Gamma_T,$$

$$\sum_{i,j=1}^{n} k_{ij}\frac{\partial p}{\partial x_j}\cos(\nu, x_i) = -\alpha\, p, \ (x,t) \in \Gamma_T,$$

$$[y] = 0, \ [p] = 0, \ (x,t) \in \gamma_T,$$

$$\left[\sum_{i,j=1}^{n} k_{ij}\frac{\partial y}{\partial x_j}\cos(\nu, x_i)\right] = \omega + p/\overline{a}, \ (x,t) \in \gamma_T,$$

$$\left[\sum_{i,j=1}^{n} k_{ij}\frac{\partial p}{\partial x_j}\cos(\nu, x_i)\right] = 0, \ (x,t) \in \gamma_T,$$

and

$$y(x,0) = y_0, \quad p(x,T) = y(x,T) - z_g, \quad x \in \bar{\Omega}_1 \cup \bar{\Omega}_2,$$

corresponds to problem (2.3), (2.4), (3.8), (4.3).

5.5   CONTROL UNDER BOUNDARY CONDITION WITH OBSERVATION UNDER CONJUGATION CONDITION

Assume that equation (1.1) is specified in the domain Ω_T. The boundary condition

$$\sum_{i,j=1}^{n} k_{ij} \frac{\partial y}{\partial x_j} \cos(\nu, x_i) = -\alpha\, y + \beta + u \tag{5.1}$$

is specified, in its turn, on the boundary Γ_T, where $u \in L_2(\Gamma_T)$. On γ_T, the conjugation conditions are

$$[y] = 0,$$

$$\left[\sum_{i,j=1}^{n} k_{ij} \frac{\partial y}{\partial x_j} \cos(\nu, x_i) \right] = \omega \tag{5.2}$$

and the initial condition is specified by equality (1.5).

For every control $u \in \mathcal{U} = L_2(\Gamma_T)$, determine a state $y(x,t;u)$ as a generalized solution to initial boundary-value problem (1.1), (1.5), (5.1), (5.2). The cost functional is

$$J(u) = \int_0^T \int_\gamma \left(y(\cdot,t;u) - z_g(\cdot) \right)^2 d\gamma\, dt + \int_0^T \int_\Gamma \bar{a} u^2 d\Gamma dt, \tag{5.3}$$

where $0 < a_0 \le \bar{a} \le a_1 < \infty$, $a_0, a_1 = \text{const}$, and it may be rewritten as

$$J(u) = \pi(u,u) - 2L(u) + \int_0^T \int_\gamma \left(z_g(\cdot) - y(\cdot,t;0) \right)^2 d\gamma\, dt;$$

in this case,

$$\pi(u,v) = \int\limits_0^T \int\limits_\gamma (y(u) - y(0))(y(v) - y(0)) \, d\gamma \, dt + \int\limits_0^T \int\limits_\Gamma \bar{a} u v \, d\Gamma \, dt$$

and

$$L(v) = \int\limits_0^T \int\limits_\gamma (z_g - y(0))(y(v) - y(0)) \, d\gamma \, dt .$$

The generalized problem corresponds to initial boundary-value problem (1.1), (1.5), (5.1), (5.2) and means to find a function $y(x,t;u) \in W(0,T)$ that satisfies the following equations $\forall w(x) \in V_0$:

$$\int\limits_\Omega \frac{dy}{dt} w \, dx + a(y,w) = (f,w) + \int\limits_\Gamma \beta w \, d\Gamma + \int\limits_\Gamma u w \, d\Gamma - \int\limits_\gamma \omega w \, d\gamma \qquad (5.4)$$

and

$$\int\limits_\Omega y(x,0;u) w(x) \, dx = \int\limits_\Omega y_0(x) w(x) \, dx ; \qquad (5.5)$$

in this case, the spaces $W(0,T)$ and V_0 are specified in point 5.2 and the bilinear form $a(\cdot,\cdot)$ is specified, in its turn, by expression (2.10).

Theorem 5.1. *Initial boundary-value problem (1.1), (1.5), (5.1), (5.2) has a unique generalized solution* $y(x,t;u) \in L^2(0,T;V)$ $\forall u \in \mathcal{U}$.

Proceed from equation (5.4) and the embedding theorems, and it is easy to see that $y(u_1) \neq y(u_2)$ under $u_1 \neq u_2$. Let $\tilde{y}' = \tilde{y}(u')$ and $\tilde{y}'' = \tilde{y}(u'')$ be solutions from $L^2(0,T;V)$ to problem (5.4), (5.5) under f, ω and $\beta = 0$ and under a function u that is equal, respectively, to u' and u'' Then, the inequality

$$\|\tilde{y}' - \tilde{y}''\|_{V \times L_2} \leq c_1 \|u' - u''\|_{L_2(\Gamma) \times L_2} \qquad (5.6)$$

is obtained from equation (5.4). Consider the embedding theorems, and

$$\|\tilde{y}' - \tilde{y}''\|_{L_2(\gamma) \times L_2} \leq c_2 \|u' - u''\|_{L_2(\Gamma) \times L_2}$$

is derived from inequality (5.6). The derived inequality provides the continuity of the linear functional $L(\cdot)$ and bilinear form $\pi(\cdot,\cdot)$ on $\mathcal{U}$.

On the basis of [58, Chapter 1, Theorem 1.1], the validity of the following statement is proved.

Theorem 5.2. *Let a system state be determined as a solution to problem (5.4), (5.5). Then, there exists a unique element u of a convex set $\mathcal{U}_{\partial}$ that is closed in $\mathcal{U}$, and relation (1.34) takes place for u, where the cost functional has the form of expression (5.3).*

As for the control $v \in \mathcal{U}$, the conjugate state $p(v)$ is specified by the relations

$$-\frac{\partial p}{\partial t} - \sum_{i,j=1}^{n} \frac{\partial}{\partial x_i}\left(k_{ij}\frac{\partial p}{\partial x_j}\right) = 0, \ (x,t) \in \Omega_T,$$

$$\sum_{i,j=1}^{n} k_{ij}\frac{\partial p}{\partial x_j}\cos(v,x_i) = -\alpha\, p, \ (x,t) \in \Gamma_T,$$

$$[p] = 0, \ (x,t) \in \gamma_T, \tag{5.7}$$

$$\left[\sum_{i,j=1}^{n} k_{ij}\frac{\partial p}{\partial x_j}\cos(v,x_i)\right] = y(v) - z_g, \ (x,t) \in \gamma_T,$$

$$p(x,T) = 0, \ x \in \bar{\Omega}_1 \cup \bar{\Omega}_2.$$

Problem (5.7) has the unique generalized solution $p(v) \in L^2(0,T;V)$ as the unique one to the following equality system:

$$-\int_{\Omega}\frac{dp}{dt}w dx + \int_{\Omega}\sum_{i,j=1}^{n} k_{ij}\frac{\partial p}{\partial x_j}\frac{\partial w}{\partial x_i}dx + \int_{\Gamma}\alpha pwd\Gamma = -\int_{\gamma}\left(y(v)-z_g\right)wd\gamma, \tag{5.8}$$

$$\int_{\Omega} p(x,T;v)\, w(x)dx = 0.$$

Multiply the first equality of relations (5.7) (under $v = u$) in a scalar way by the difference $y(v) - y(u)$ and take the integral from 0 to T of the result. Consider equality (1.39) and obtain the equality

$$0 = \int_0^T \left(p(u), \frac{d}{dt}(y(v) - y(u)) \right) dt + \int_0^T a(p(u), y(v) - y(u)) dt +$$

$$+ \int_0^T \int_\gamma (y(u) - z_g)(y(v) - y(u)) d\gamma dt, \tag{5.9}$$

where the bilinear form is specified by expression (2.10).

Proceed from equation (5.4), and the equality

$$0 = \int_0^T \int_\Gamma p(u) (v - u) d\Gamma dt + \int_0^T \int_\gamma (y(u) - z_g)(y(v) - y(u)) d\gamma dt,$$

i.e.

$$\int_0^T \int_\gamma (y(u) - z_g)(y(v) - y(u)) d\gamma dt = - \int_0^T \int_\Gamma p(u) (v - u) d\Gamma dt$$

follows from equality (5.9). Then, inequality (1.35) has the form

$$\int_0^T \int_\Gamma (-p(u) + \bar{a}u)(v - u) d\Gamma \geq 0, \quad \forall v \in \mathcal{U}_\partial, \tag{5.10}$$

for the optimization problem considered in the present point.

Thus, the optimal control $u \in \mathcal{U}_\partial$ is specified by relations (5.4), (5.5), (5.8) and (5.10). If the constraints are absent, i.e. when $\mathcal{U}_\partial = \mathcal{U}$, then the following equality is obtained from inequality (5.10):

$$u = p/\bar{a}, \quad (x,t) \in \Gamma_T. \tag{5.11}$$

If the solution $(y, p)^{\mathrm{T}}$ to problem (5.4), (5.5), (5.8), (5.11) is smooth enough on $\bar{\Omega}_{lT}$, $l = 1,2$, then the differential problem of finding the vector-function $(y, p)^T$, that satisfies the second and fourth equalities of system (5.7), system (3.9), except its third, fourth, seventh and eighth equalities, and the conditions

$$\sum_{i,j=1}^{n} k_{ij} \frac{\partial y}{\partial x_j} \cos(\nu, x_i) = -\alpha y + \beta + p/\bar{a}, \ (x,t) \in \Gamma_T,$$

and

$$\left[\sum_{i,j=1}^{n} k_{ij} \frac{\partial y}{\partial x_j} \cos(\nu, x_i) \right] = \omega, \ (x,t) \in \gamma_T,$$

corresponds to problem (5.4), (5.5), (5.8), (5.11).

Remark. In Chapter 5 and in the nextcoming ones, the following is assumed everywhere:

$$L_2(D_T) = \left\{ v(x,t) : (v,v)_{L_2(D)}^{1/2} < \infty, \forall t \in [0,T], \left(\int_0^T (v,v)_{L_2(D)} dt \right)^{1/2} < \infty \right\};$$

in this case, $D_T = D \times (0, T)$.

CONTROL OF A SYSTEM DESCRIBED BY A PARABOLIC EQUATION IN THE PRESENCE OF CONCENTRATED HEAT CAPACITY

Introduce the following denotations: Ω is a domain that consists of two open, non-intersecting and strictly Lipschitz domains Ω_1 and Ω_2 from an n-dimensional real linear space R^n, $\Gamma = (\partial\Omega_1 \cup \partial\Omega_2)\backslash\gamma$ $(\gamma = \partial\Omega_1 \cap \cap\partial\Omega_2 \neq \varnothing)$ is a boundary of a domain $\bar{\Omega}$, $\partial\Omega_i$ is a boundary of a domain Ω_i, $i = 1,2$; $\Omega_T = \Omega \times (0,T)$ is a complicated cylinder; $\Gamma_T = \Gamma \times (0,T)$ is the lateral surface of a cylinder $\Omega_T \cup \gamma_T$, $\gamma_T = \gamma \times (0,T)$.

Let V be some Hilbert space and assume that V' is a space dual with respect to V. By analogy [58], introduce a space $L^2(0,T;V)$ of functions $t \to f(t)$ that map an interval $(0,T)$ into the space V of measurable functions, namely, of such ones that

$$\left(\int_0^T \|f(t)\|_V^2 \, dt\right)^{1/2} < \infty.$$

Also by analogy, specify the space $L^2(0,T;V')$. Introduce a space $W(0,T) = \left\{ f \in L^2(0,T;V) : \dfrac{df}{dt} \in L^2(0,T;V') \right\}$ that is supplied with the norm

$$\|f\|_{W(0,T)} = \left(\int_0^T \|f(t)\|_V^2 \, dt + \int_0^T \left\| \frac{df}{dt} \right\|_{V'}^2 \, dt \right)^{1/2} < \infty$$

and becomes the Hilbert one.

6.1   DISTRIBUTED CONTROL

Assume that the parabolic equation

$$\frac{\partial y}{\partial t} = \sum_{i,j=1}^{n} \frac{\partial}{\partial x_i} \left(k_{ij}(x) \frac{\partial y}{\partial x_j} \right) + f(x,t) \tag{1.1}$$

is specified in the domain Ω_T, where

$$k_{ij}\big|_{\bar{\Omega}_l} = k_{ji}\big|_{\bar{\Omega}_l} \in C(\bar{\Omega}_l) \cap C^1(\Omega_l), \quad i,j = \overline{1,n};$$

$$\sum_{i,j=1}^{n} k_{ij}\,\xi_i\,\xi_j \geq \alpha_0 \sum_{i=1}^{n} \xi_i^2, \quad \forall \xi_i, \xi_j \in R^1, \ \forall x \in \Omega, \ \alpha_0 = \text{const} > 0;$$

$$f\big|_{\Omega_{lT}} \in C(\Omega_{lT}), \ l = 1,2; \ |f| < \infty, \ \Omega_{lT} = \Omega_l \times (0,T).$$

The third boundary condition

$$\sum_{i,j=1}^{n} k_{ij} \frac{\partial y}{\partial x_j} \cos(\nu, x_i) = -\alpha\, y + \beta \tag{1.2}$$

is specified, in its turn, on the boundary $\Gamma_T = \Gamma \times (0,T)$, where $0 < \alpha^0 \leq \alpha = \alpha(x)$, $\alpha, \beta \in L_2(\Gamma)$.

On γ_T, the conjugation conditions are [91, 21]

$$[y] = 0 \tag{1.3}$$

and

$$\left[q_y\right] = c\,\frac{\partial y}{\partial t}, \quad 0 < \bar{c}_0 \le c \le c_0, \tag{1.4}$$

where $c \in L_2(\gamma)$, $[\varphi] = \varphi^+ - \varphi^-$, $\varphi^+ = \{\varphi\}^+ = \varphi(x,t)$ under $(x,t) \in \gamma_T^+ =$

$= (\partial\Omega_2 \cap \gamma) \times (0,T)$, $\varphi^- = \{\varphi\}^- = \varphi(x,t)$ under $(x,t) \in \gamma_T^- = (\partial\Omega_1 \cap \gamma) \times$

$\times (0,T)$, $\gamma_T = \gamma \times (0,T)$,

$$q_y = \sum_{i,j=1}^{n} k_{ij}(x)\,\frac{\partial y}{\partial x_j}\cos(v,x_i),$$

and v is an ort of a normal to γ that is called simply a normal to γ and it is directed into the domain Ω_2.

The initial condition

$$y(x,0) = y_0(x), \quad x \in \bar{\Omega}_1 \cup \bar{\Omega}_2, \tag{1.5}$$

where $y_0 \in L_2(\Omega)$, $[y_0] = 0$, $\|y_0\|_{L_2(\gamma)} < \infty$, $\|v\|_{L_2(\gamma)} = \left\{\int_\gamma v^2 d\gamma\right\}^{1/2}$, is

specified under $t = 0$.

Let there be a control Hilbert space $\mathscr{U}$ and operator $B \in$

$\in \mathscr{L}\left(\mathscr{U}; L^2(0,T;V')\right)$.

For every control $u \in \mathscr{U}$, determine a system state $y = y(u) = y(x,t; u)$ as a generalized solution to the problem specified by the equation

$$\frac{\partial y}{\partial t} = \sum_{i,j=1}^{n}\frac{\partial}{\partial x_i}\left(k_{ij}\frac{\partial y}{\partial x_j}\right) + f + Bu, \quad (x,t) \in \Omega_T, \tag{1.6}$$

and by conditions (1.2)–(1.5).

Specify the observation by the following expression:

$$Z(u) = C\,y(u), \quad C \in \mathscr{L}(W(0,T); \mathscr{H}). \tag{1.7}$$

Specify the operator

$$\mathscr{N} \in \mathscr{L}(\mathscr{U};\mathscr{U}), \quad (\mathscr{N}u,u)_{\mathscr{U}} \ge \nu_0 \|u\|_{\mathscr{U}}^2, \quad \nu_0 = \text{const} > 0. \qquad (1.8)$$

Assume the following: $\mathscr{N}u = \bar{a}u$; in this case, $\bar{a}\big|_{\Omega_l} \in C(\Omega_l)$, $l = 1,2$; $0 < a_0 \le \bar{a} \le a_1 < \infty$, $a_0, a_1 = \text{const}$. The cost functional is

$$J(u) = \left\| C\, y(u) - z_g \right\|_{\mathscr{H}}^2 + (\mathscr{N}u,u)_{\mathscr{U}}, \qquad (1.9)$$

where z_g is a known element of the space $\mathscr{H}$.

The optimal control problem is: Find such an element $u \in \mathscr{U}_\partial$ that the condition

$$J(u) = \inf_{v \in \mathscr{U}_\partial} J(v) \qquad (1.10)$$

is met, where $\mathscr{U}_\partial$ is some convex closed subset in $\mathscr{U}$.

Definition 1.1. If an element $u \in \mathscr{U}_\partial$ meets condition (1.10), it is called an optimal control.

The generalized problem corresponds to initial boundary-value problem (1.6), (1.2)–(1.5) and means [21] to find a vector-function $y(x,t;u) \in$ $\in W(0,T)$ that satisfies the following equations $\forall w(x) \in V_0 =$ $= \left\{ v(x): \ v\big|_{\Omega_i} \in W_2^1(\Omega_i), \ i = 1,2, \ [v] = 0 \right\}$:

$$\int_\Omega \frac{dy}{dt} w\, dx + \int_\Omega \sum_{i,j=1}^n k_{ij} \frac{\partial y}{\partial x_j} \frac{\partial w}{\partial x_i}\, dx + \int_\gamma c\frac{dy}{dt} w\, d\gamma +$$

$$+ \int_\Gamma \alpha y w\, d\Gamma = (f,w) + (Bu,w) + \int_\Gamma \beta w\, d\Gamma \qquad (1.11)$$

and

$$\int_\Omega y(x,0; u)\, w\, dx + \int_\gamma c\, y(x,0; u)\, w\, d\gamma =$$

$$= \int_\Omega y_0(x)\, w\, dx + \int_\gamma c\, y_0(x)\, w\, d\gamma ; \qquad (1.12)$$

in this case, $(\varphi, \psi) = \int_{\Omega} \varphi(x,t)\psi(x,t)\,dx$, $V = \left\{ v(x,t) : v\big|_{\Omega_i} \in W_2^1(\Omega_i), \right.$

$i = 1, 2, \forall t \in [0,T], [v] = 0 \left. \right\}$. Consider the existence and uniqueness of

the solution to problem (1.11), (1.12). Since Bu and $f \in L^2(0,T; V')$,
then, without loss of generality, assume the following: $Bu \equiv 0$.

The space V_0 is complete, separable and reflexive [41, 55, 32].

Choose [41, 54] an arbitrary fundamental system of linearly
independent functions $w_k(x)$, $k = 1, 2, ...$, in V_0.

Remark. Functions $w_k(x)$ may be chosen as eigenfunctions that
correspond to eigenvalues λ_k, $k = 1, 2, ...$, of the spectral problem: Find

$$\{\lambda, u\} \in \left\{ R^1 \times V_0, \ u \neq 0 \right\} : a(u,w) = \lambda b(u,w), \ \forall w \in V_0, \qquad (1.13)$$

where

$$a(u,v) = \int_{\Omega} \sum_{i,j=1}^{n} k_{ij} \frac{\partial u}{\partial x_j} \frac{\partial v}{\partial x_i}\,dx + \int_{\Gamma} \alpha uv\,d\Gamma, \ b(u,v) = (u,v) + (cu,v)_{L_2(\gamma)}.$$

The bilinear form $b(\cdot,\cdot)$ is symmetric and positively specified on the
complete Hilbert space V_B obtained by way of completing the set V_0 as for
the norm $\|\cdot\|_B = b^{1/2}(\cdot,\cdot)$. In this case: $V_0 \subset V_B \subset L_2(\Omega)$. Therefore, spectral
problem (1.13) has a countable spectrum and the eigenfunction system
$\{w_j(x)\}$ for it is complete both in V_0 and V_B. Let this system be
orthonormal in V_B so that $b(w_k, w_l) = \delta_k^l$ ($\delta_k^k = 1$, $\delta_k^l = 0$ under
$l,k = 1, 2, ...$). As for $y_0 \in V_B$ [49]:

$$y_0 = \sum_{i=1}^{\infty} \xi_i w_i(x), \qquad (1.13')$$

where $\xi_i = b(y_0, w_i)$, $i = 1, 2, ...$.

The approximate solution to problem (1.11), (1.12) is given as

$$y_m(x,t) = \sum_{i=1}^{m} g_{im}(t) w_i(x), \tag{1.14}$$

where the functions $g_{im}(t)$ are chosen in such a way that the relations

$$\left(\frac{\partial y_m}{\partial t}, w_j\right) + \left(c\frac{\partial y_m}{\partial t}, w_j\right)_{L_2(\gamma)} + a(y_m, w_j) =$$

$$= (f, w_j) + (\beta, w_j)_{L_2(\Gamma)}, \quad j = \overline{1, m}, \tag{1.15}$$

and

$$\left(y_m(\cdot, 0), w_j\right) + \left(c\, y_m(\cdot, 0), w_j\right)_{L_2(\gamma)} =$$

$$= (y_0, w_j) + (c\, y_0, w_j)_{L_2(\gamma)}, \quad j = \overline{1, m}, \tag{1.16}$$

where

$$(\varphi, \psi)_{L_2(\gamma)} = \int_\gamma \varphi(x,t)\psi(x,t)\,d\gamma, \quad (\varphi, \psi)_{L_2(\Gamma)} = \int_\Gamma \varphi(x,t)\psi(x,t)\,d\Gamma$$

and

$$a(y_m, w_j) = \int_\Omega \sum_{l,s=1}^{n} k_{ls} \frac{\partial y_m}{\partial x_s} \frac{\partial w_j}{\partial x_l}\,dx + \int_\Gamma \alpha\, y_m w_j\,d\Gamma,$$

are met.

Equalities (1.15) and (1.16) specify the Cauchy problem for the system of m first-order linear ordinary differential equations as for $g_{im}(t)$:

$$M_m \frac{dg_m}{dt} + K_m g_m = F(t), \quad t \in (0, T), \tag{1.17}$$

$$M_0^m g_m(0) = F_0^m; \tag{1.18}$$

in this case,

$$M_m = \left\{ M_{ij}^m \right\}_{i,j=1}^{m}, \quad M_{ij}^m = (w_i, w_j) + (c\, w_i, w_j)_{L_2(\gamma)}, \quad K_m = \left\{ k_{ij}^m \right\}_{i,j=1}^{m},$$

$$k_{ij}^m = a\left(w_i, w_j\right), \quad F_m(t) = \left\{f_i^m(t)\right\}_{i=1}^m, \quad f_i^m(t) = (f, w_i) + (\beta, w_i)_{L_2(\Gamma)},$$

$$F_0^m = \left\{f_{0i}^m\right\}_{i=1}^m, \quad f_{0i}^m = (y_0, w_i) + (c\, y_0, w_i)_{L_2(\gamma)}, \quad M_0^m = \left\{M_{0ij}^m\right\}_{i,j=1}^m,$$

$$M_{0ij}^m = \left(w_i, w_j\right) + \left(c\, w_i, w_j\right)_{L_2(\gamma)}, \quad g_m(t) = \left\{g_{im}(t)\right\}_{i=1}^m.$$

It is easy to see that the solution $g_m(t)$ to problem (1.15), (1.16) exists and that such solution is unique. The following statement must be proved: $y_m \to y$ under $m \to \infty$, where $y = y(x,t)$ is the solution to problem (1.11), (1.12).

Multiply equality (1.15) by $g_{jm}(t)$ and find the sum over j for the result. Then:

$$\left(\frac{\partial y_m}{\partial t}, y_m\right) + \left(c\frac{\partial y_m}{\partial t}, y_m\right)_{L_2(\gamma)} + a\left(y_m, y_m\right) = (f, y_m) + (\beta, y_m)_{L_2(\Gamma)},$$

i.e.:

$$\frac{1}{2}\frac{d}{dt}\|y_m\|^2 + \frac{1}{2}\frac{d}{dt}\left\|\sqrt{c}\, y_m\right\|_{L_2(\gamma)}^2 + a\left(y_m, y_m\right) =$$

$$= (f, y_m) + (\beta, y_m)_{L_2(\Gamma)}; \tag{1.19}$$

in this case, $\|\varphi\| = (\varphi, \varphi)^{1/2} = \|\varphi\|(t)$, $\left\|\sqrt{c}\,\varphi\right\|_{L_2(\gamma)} = \left(\sqrt{c}\,\varphi, \sqrt{c}\,\varphi\right)_{L_2(\gamma)}^{1/2} =$

$= \left\|\sqrt{c}\,\varphi\right\|_{L_2(\gamma)}(t)$. Take the ellipticity condition and generalized Friedrichs inequality [21] into account, and the inequality

$$a\left(y_m, y_m\right) \geq \alpha_1 \|y_m\|_V^2, \quad \alpha_1 = \text{const} > 0, \tag{1.20}$$

is derived, where $\|\varphi\|_V = \left\{\sum_{i=1}^2 \|\varphi\|_{W_2^1(\Omega_i)}^2\right\}^{1/2}$, $\|\cdot\|_{W_2^1(\Omega_i)}$ is the norm of the Sobolev space $W_2^1(\Omega_i)$, $\|\varphi\|_V = \|\varphi\|_V(t)$.

Consider inequality (1.20), the ε- and Cauchy-Bunyakovsky inequalities and embedding theorems [55], and the inequality

$$\|y_m\|^2(T) + \left\|\sqrt{c}\, y_m\right\|^2_{L_2(\gamma)}(T) + 2\alpha_1 \int_0^T \|y_m\|^2_V \, dt \leq$$

$$\leq \|y_m\|^2(0) + \left\|\sqrt{c}\, y_m\right\|^2_{L_2(\gamma)}(0) + 2\int_0^T \left|(f, y_m)\right| dt + 2\int_0^T \left|\int_\Gamma \beta\, y_m d\Gamma\right| dt \leq$$

$$\leq \|y_m\|^2(0) + \left\|\sqrt{c}\, y_m\right\|^2_{L_2(\gamma)}(0) + 2\varepsilon \int_0^T \|y_m\|^2_V \, dt +$$

$$+ \frac{1}{2\varepsilon}\int_0^T \|f\|^2 dt + 2\varepsilon_1 c_1' \int_0^T \|y_m\|^2_V \, dt + \frac{1}{2\varepsilon_1}\int_0^T \|\beta\|^2_{L_2(\Gamma)} dt \qquad (1.21)$$

follows from equality (1.19); in this case, $c_1' = \max\limits_{l=1,2} c_l$ and the constant c_l

is obtained from the inequality proved in the embedding theorem applied for the domain Ω_l. Proceed from equality (1.16), and

$$\left(y_m(\cdot, 0), y_m(\cdot, 0)\right) + \left(c\, y_m(\cdot, 0), y_m(\cdot, 0)\right)_{L_2(\gamma)} =$$

$$= \left(y_0(\cdot), y_m(\cdot, 0)\right) + \left(c\, y_0(\cdot), y_m(\cdot, 0)\right)_{L_2(\gamma)}.$$

Proceed also from the ε- and Cauchy-Bunyakovsky inequalities, and the inequality

$$\|y_m\|^2(0) + \left\|\sqrt{c}\, y_m\right\|^2_{L_2(\gamma)}(0) \leq c_2\left(\|y_0\|^2 + \left\|\sqrt{c}\, y_0\right\|^2_{L_2(\gamma)}\right) \qquad (1.22)$$

is derived. Take it into account, and the inequality

$$\int_0^T \|y_m\|^2_V \, dt \leq c\left(\|y_0\|^2 + \left\|\sqrt{c}\, y_0\right\|^2_{L_2(\gamma)} + \int_0^T \|f\|^2 dt + \int_0^T \|\beta\|^2_{L_2(\Gamma)} dt\right)$$

follows from inequality (1.21). Therefore, the elements y_m are in some bounded subset of the space $L^2(0,T;V)$. Hence, there exists a subsequence $\{y_\chi\}$ that weakly converges to the element z in $L^2(0,T;V)$ $\left(y_\chi \to z \in L^2(0,T;V)\right)$. Without loss of generality, assume that the whole sequence $\{y_m\}$ weakly converges to z.

Rewrite equality (1.15) as

$$\frac{d}{dt}(y_m, w_j) + \frac{d}{dt}(c\,y_m, w_j)_{L_2(\gamma)} + a(y_m, w_j) =$$

$$= (f, w_j) + (\beta, w_j)_{L_2(\Gamma)}, \quad j = \overline{1, m},$$

multiply its both sides by the function

$$\varphi(t) \in C^1([0,T]), \quad \varphi(T) = 0, \tag{1.23}$$

and find the integral from 0 to T of the result:

$$\int_0^T \left\{ -(y_m(\cdot,t), \varphi_j'(\cdot,t)) - (c\,y_m, \varphi_j')_{L_2(\gamma)} + a(y_m, \varphi_j) \right\} dt = \int_0^T (f, \varphi_j) dt +$$

$$+ \int_0^T (\beta, \varphi_j)_{L_2(\Gamma)} dt + (y_m, \varphi_j)(0) + (c\,y_m, \varphi_j)_{L_2(\gamma)}(0); \tag{1.24}$$

in this case, $\varphi_j(x,t) = \varphi(t)\,w_j(x)$, $\varphi_j'(x,t) = \dfrac{d\varphi}{dt}\,w_j(x)$.

By virtue of the aforesaid weak convergence, it is possible to pass in equality (1.24) to the limit under $m \to \infty$, and the following equality is obtained:

$$\int_0^T \left\{ -(z, \varphi_j') - (c\,z, \varphi_j')_{L_2(\gamma)} + a(z, \varphi_j) \right\} dt = \int_0^T (f, \varphi_j) dt +$$

$$+ \int_0^T (\beta, \varphi_j)_{L_2(\Gamma)} dt + (z, \varphi_j)(0) + (cz, \varphi_j)_{L_2(\gamma)}(0) . \qquad (1.25)$$

Consider the assumptions as for $\{w_j\}$, and it can be seen that the matrix M_0^m from condition (1.18) is identity ($M_{0ii}^m = 1$, $M_{0ij}^m = 0$ under $i \neq j$, $i, j = \overline{1, m}$). The equality $g_{im}(0) = b(y_0, w_i)/b(w_i, w_i)$ follows from condition (1.18), i.e. $g_{im}(0)$ ($i = \overline{1, m}$) are the Fourier coefficients for the function y_0. By virtue of [49]:

$$y_m(x, 0) = \sum_{i=1}^m g_{im}(0) w_i(x) \to y_0(x) \text{ under } m \to \infty.$$

Therefore: $z(x, 0) = y_0(x)$.

Equality (1.25) is true for the arbitrary function φ that meets conditions (1.23). Thus, the following can be assumed: $\varphi \in D((0, T))$ [58]. Then, the equality

$$\int_0^T \left\{ -(z, w_j) \varphi' - (cz, w_j)_{L_2(\gamma)} \varphi' + a(z, w_j) \varphi \right\} dt =$$

$$= \int_0^T (f, \varphi_j) dt + \int_0^T (\beta, \varphi_j)_{L_2(\Gamma)} dt, \quad j = \overline{1, m},$$

follows from equality (1.25). Hence:

$$\int_0^T \left\{ \frac{d}{dt}(z, w_j) + \frac{d}{dt}(cz, w_j)_{L_2(\gamma)} + a(z, w_j) - (f, w_j) - (\beta, w_j)_{L_2(\Gamma)} \right\} \varphi(t) \, dt = 0,$$

i.e.

$$\left(\frac{dz}{dt}, w_j \right) + \left(c \frac{dz}{dt}, w_j \right)_{L_2(\gamma)} + a(z, w_j) =$$

$$= (f, w_j) + (\beta, w_j)_{L_2(\Gamma)}. \tag{1.26}$$

Take equality (1.26), the space V_0 and assumptions as for the functions w_j into account, and it is stated that the equality

$$\left(\frac{dz}{dt}, w\right) + \left(c\frac{dz}{dt}, w\right)_{L_2(\gamma)} + a(z, w) =$$

$$= (f, w) + (\beta, w)_{L_2(\Gamma)} \tag{1.27}$$

is true $\forall w \in V_0$. The following equality is derived from relation (1.16):

$$(z(\cdot, 0), w(\cdot)) + (c\,z(\cdot, 0), w(\cdot))_{L_2(\gamma)} =$$

$$= (y_0(\cdot), w(\cdot)) + (y_0(\cdot), w(\cdot))_{L_2(\gamma)}. \tag{1.28}$$

Therefore, the function $z \in L^2(0, T; V)$ is the solution to problem (1.11), (1.12) $\forall f \in L^2(0, T; V')$ and under $Bu = 0$, i.e. to problem (1.27), (1.28).

Illustrate the uniqueness of the solution to problem (1.27), (1.28) by contradiction. Let there exist two solutions: $z_1(x, t)$ and $z_2(x, t) \in L^2(0, T; V)$. Then, on the basis of equality (1.27), the equality

$$\frac{d}{dt}(\bar{z}, \bar{z}) + \frac{d}{dt}(c\bar{z}, \bar{z})_{L_2(\gamma)} + 2a(\bar{z}, \bar{z}) = 0 \tag{1.29}$$

is obtained, where $\bar{z} = z_1 - z_2$.

Consider equality (1.28), and the contradiction

$$0 < (\bar{z}, \bar{z})(T) + (c\bar{z}, \bar{z})_{L_2(\gamma)}(T) + \alpha_0 \int_0^T \|\bar{z}\|_V^2 \, dt \le 0, \quad \alpha_0 = \text{const} > 0,$$

follows from equality (1.29).

Therefore, the validity of the following statement is proved.

Theorem 1.1. *Initial boundary-value problem (1.1)–(1.5) has a unique generalized solution* $y(x, t) \in L^2(0, T; V)$.

Let $\tilde{y}' = \tilde{y}(u')$ and $\tilde{y}'' = \tilde{y}(u'')$ be solutions from $L^2(0,T;V)$ to problem (1.11), (1.12) under $f = 0$ and $\beta = 0$ and under a function $u = u(x,t)$ that is equal, respectively, to u' and u''. Then, the inequality

$$\frac{1}{2}\frac{d}{dt}(\tilde{y}' - \tilde{y}'', \tilde{y}' - \tilde{y}'') + \frac{1}{2}\frac{d}{dt}(c(\tilde{y}' - \tilde{y}''), \tilde{y}' - \tilde{y}'')_{L_2(\gamma)} +$$

$$+\alpha_0 \|\tilde{y}' - \tilde{y}''\|_V^2 \le \|u' - u''\| \cdot \|\tilde{y}' - \tilde{y}''\|_V \tag{1.30}$$

is obtained for $Bu \equiv u$. Therefore,

$$\|\tilde{y}' - \tilde{y}''\|_{V \times L_2} \le \frac{1}{\alpha_0}\|u' - u''\|_{L_2 \times L_2}, \tag{1.30'}$$

where

$$\|\varphi\|_{L_2 \times L_2}^2 = \int_0^T \|\varphi\|_{L_2}^2 dt, \quad \|\varphi\|_{L_2}^2 = \|\varphi\|^2 = \int_\Omega \varphi^2(x,t)dx, \quad \|\varphi\|_{V \times L_2}^2 = \int_0^T \|\varphi\|_V^2 dt.$$

Rewrite functional (1.9) as

$$J(u) = \pi(u,u) - 2L(u) + \int_0^T \|z_g - y(0)\|_{L_2}^2 dt, \tag{1.31}$$

where

$$\pi(u,v) = \left(y(u) - y(0), \ y(v) - y(0)\right)_{\mathcal{H}} + \left(\bar{a}u, v\right)_{\mathcal{U}}, \tag{1.32}$$

$$L(v) = \left(z_g - y(0), \ y(v) - y(0)\right)_{\mathcal{H}};$$

in this case,

$$(z,v)_{\mathcal{H}} = (z,v)_{\mathcal{U}} = \int_0^T (z,v)dt, \quad (z,v) = \int_\Omega zv\,dx.$$

Inequality (1.30') provides the continuity of the linear functional $L(\cdot)$ and bilinear form $\pi(\cdot,\cdot)$ on $\mathcal{U}$.

On the basis of [58, Chapter 1, Theorem 1.1], the validity of the following statement is proved.

Theorem 1.2. *Let a system state be determined as a solution to problem (1.11), (1.12). Then, there exists a unique element u of a convex set $\mathcal{U}_\partial$ that is closed in $\mathcal{U}$, and*

$$J(u) = \inf_{v \in \mathcal{U}_\partial} J(v) \tag{1.33}$$

takes place for u.

A control $u \in \mathcal{U}_\partial$ is optimal if and only if the following inequality is true:

$$\left(y(u) - z_g, \; y(v) - y(u) \right)_{\mathcal{H}} + (\mathcal{N}u, v - u)_{\mathcal{U}} \geq 0, \; \forall v \in \mathcal{U}_\partial. \tag{1.34}$$

As for the control $v \in \mathcal{U}$, the conjugate state $p(v)$ is specified by the relations

$$-\frac{\partial p}{\partial t} - \sum_{i,j=1}^{n} \frac{\partial}{\partial x_i} \left(k_{ij} \frac{\partial p}{\partial x_j} \right) = y(v) - z_g, \; (x,t) \in \Omega_T,$$

$$\sum_{i,j=1}^{n} k_{ij} \frac{\partial p}{\partial x_j} \cos(v, x_i) = -\alpha p, \; (x,t) \in \Gamma_T,$$

$$[p] = 0, \; (x,t) \in \gamma_T,$$

$$\left[\sum_{i,j=1}^{n} k_{ij} \frac{\partial p}{\partial x_j} \cos(v, x_i) \right] = -c \frac{\partial p}{\partial t}, \; (x,t) \in \gamma_T, \tag{1.34'}$$

$$p(x,T) = 0, \; x \in \bar{\Omega}_1 \cup \bar{\Omega}_2.$$

Substitute a time $T - t$ for the time t, proceed from Theorem 1.1, and it is concluded that initial boundary-value problem (1.34′) has the unique generalized solution $p(v) \in L^2(0,T;V)$ as the unique one to the following equality system:

$$-\left(\frac{d}{dt} p(v), w \right) - \int_\gamma c \frac{dp}{dt} w \, d\gamma + \int_\Omega \sum_{i,j=1}^{n} k_{ij} \frac{\partial p}{\partial x_j} \frac{\partial w}{\partial x_i} \, dx +$$

$$+ \int_{\Gamma} \alpha \, p \, w \, d\Gamma = \left(y(v) - z_g, w \right), \tag{1.35}$$

$$\int_{\Omega} p(x,T;v) \, w \, dx + \int_{\gamma} c \, p(x,T;v) \, w \, d\gamma = 0. \tag{1.36}$$

Choose the difference $y(v) - y(u)$ instead of w in equality (1.35), consider equation (1.11), and the equality

$$\int_0^T \left(y(u) - z_g, y(v) - y(u) \right) dt =$$

$$= - \int_0^T \left(\left(\frac{d}{dt} p(u), y(v) - y(u) \right) + \int_{\gamma} c \frac{dp}{dt} (y(v) - y(u)) \, d\gamma \right) dt +$$

$$+ \int_0^T a \left(p(u), (y(v) - y(u)) \right) dt =$$

$$= - \left\{ (p(u), y(v) - y(u)) + (cp(u), \, y(v) - y(u))_{L_2(\gamma)} \right\} \Big|_0^T +$$

$$+ \int_0^T \left\{ \left(p(u), \frac{d}{dt} (y(v) - y(u)) \right) + \left(c \, p(u), \frac{d}{dt} (y(v) - y(u)) \right)_{L_2(\gamma)} \right\} dt +$$

$$+ \int_0^T a \left(p(u), y(v) - y(u) \right) dt = \int_0^T (p(u), \, v - u) \, dt$$

is obtained, i.e.

$$\int_0^T \left(y(u) - z_g, y(v) - y(u) \right) dt = \int_0^T (p(u), \, v - u) \, dt. \tag{1.37}$$

Therefore, inequality (1.34) has the form

$$\int_0^T (p(u) + \bar{a} \, u, \, v - u) \, dt \geq 0, \quad \forall v \in \mathcal{U}_\partial. \tag{1.38}$$

Thus, the optimal control $u \in \mathcal{U}_\partial$ is specified by relations (1.11), (1.12), (1.35), (1.36) and (1.38).

If the constraints are absent, i.e. when $\mathcal{U}_\partial = \mathcal{U}$, then the equality

$$p(u) + \bar{a}\,u = 0, \quad (x,t) \in \Omega_T,$$

is obtained from inequality (1.38). The optimal control

$$u = -p/\bar{a}, \quad (x,t) \in \Omega_T, \tag{1.39}$$

is found from the latter equality. If the solution $(y, p)^{\mathrm{T}}$ to problem (1.11), (1.12), (1.35), (1.36), (1.39) is smooth enough on $\overline{\Omega}_{lT}$, viz., $y\big|_{\overline{\Omega}_{lT}}$,

$p\big|_{\overline{\Omega}_{lT}} \in C^{1,0}(\overline{\Omega}_{lT}) \cap C^{2,0}(\Omega_{lT}) \cap C^{0,1}(\Omega_{lT}), l = 1,2$, then the differential

problem of finding the vector-function $(y, p)^{\mathrm{T}}$, that satisfies the equality system

$$\frac{\partial y}{\partial t} - \sum_{i,j=1}^{n} \frac{\partial}{\partial x_i}\left(k_{ij}\frac{\partial y}{\partial x_j}\right) + p/\bar{a} = f, \quad (x,t) \in \Omega_T,$$

$$-\frac{\partial p}{\partial t} - \sum_{i,j=1}^{n} \frac{\partial}{\partial x_i}\left(k_{ij}\frac{\partial p}{\partial x_i}\right) - y = -z_g, \quad (x,t) \in \Omega_T,$$

$$\sum_{i,j=1}^{n} k_{ij}\frac{\partial y}{\partial x_j}\cos(v,x_i) = -\alpha y + \beta, \quad (x,t) \in \Gamma_T,$$

$$\sum_{i,j=1}^{n} k_{ij}\frac{\partial p}{\partial x_j}\cos(v,x_i) = -\alpha p, \quad (x,t) \in \Gamma_T,$$

$$[y] = 0, \quad [p] = 0, \quad (x,t) \in \gamma_T,$$

$$\left[\sum_{i,j=1}^{n} k_{ij}\frac{\partial y}{\partial x_j}\cos(v,x_i)\right] = c\frac{\partial y}{\partial t}, \quad (x,t) \in \gamma_T, \tag{1.40}$$

$$\left[\sum_{i,j=1}^{n} k_{ij} \frac{\partial p}{\partial x_j} \cos(\nu, x_i)\right] = -c \frac{\partial p}{\partial t}, \quad (x,t) \in \gamma_T,$$

$$y(x,0) = y_0, \quad p(x,T) = 0, \quad x \in \bar{\Omega}_1 \cup \bar{\Omega}_2,$$

corresponds to problem (1.11), (1.12), (1.35), (1.36), (1.39), where the control $u = u(x,t)$ is found by formula (1.39).

6.2   CONTROL UNDER CONJUGATION CONDITION WITH OBSERVATION THROUGHOUT A WHOLE DOMAIN

Assume that equation (1.1) is specified in the domain Ω_T. On the boundary Γ_T, the boundary condition has the form of expression (1.2).

For every control $u \in \mathcal{U} = L_2(\gamma_T)$, determine a state $y = y(u)$ as a generalized solution to the initial boundary-value problem specified, in its turn, by equation (1.1), boundary condition (1.2), initial condition (1.5) and the conjugation conditions

$$[y] = 0, \quad (x,t) \in \gamma_T, \tag{2.1}$$

and

$$\left[\sum_{i,j=1}^{n} k_{ij} \frac{\partial y}{\partial x_j} \cos(\nu, x_i)\right] = c \frac{\partial y}{\partial t} + u, \quad (x,t) \in \gamma_T. \tag{2.2}$$

The generalized problem corresponds to initial boundary-value problem (1.1), (1.2) (1.5), (2.1), (2.2) and means to find a function $y(x,t;u) \in W(0,T)$ that satisfies the following equations $\forall w(x) \in V_0$:

$$\int_{\Omega} \frac{dy}{dt} w dx + \int_{\gamma} c \frac{dy}{dt} w d\gamma + \int_{\Omega} \sum_{i,j=1}^{n} k_{ij} \frac{\partial y}{\partial x_j} \frac{\partial w}{\partial x_i} dx + \int_{\Gamma} \alpha y w d\Gamma =$$

$$= (f, w) - \int_{\gamma} uw \, d\gamma + \int_{\Gamma} \beta w \, d\Gamma \tag{2.3}$$

and

$$\int_{\Omega} y(x, 0; u) w \, dx + \int_{\gamma} c \, y(x, 0; u) w \, d\gamma =$$

$$= \int_{\Omega} y_0 w \, dx + \int_{\gamma} c \, y_0 w \, d\gamma. \tag{2.4}$$

The following statement takes place.

Theorem 2.1. *Initial boundary-value problem (1.1), (1.2), (1.5), (2.1), (2.2) has a unique generalized solution* $y(x, t; u) \in W(0, T) \quad \forall u \in \mathcal{U}$.

The validity of Theorem 2.1 is stated by analogy with the proof of Theorem 1.1.

If $\tilde{y}' = \tilde{y}(u')$ and $\tilde{y}'' = \tilde{y}(u'')$ are solutions to problem (2.3), (2.4) under $f = 0$ and $\beta = 0$ and under a function u that is equal, respectively, to u' and u'', then the inequality

$$\frac{1}{2} \frac{d}{dt} \|\tilde{y}' - \tilde{y}''\|^2 + \frac{1}{2} \frac{d}{dt} \left\| \sqrt{c} \left(\tilde{y}' - \tilde{y}'' \right) \right\|^2_{L_2(\gamma)} + \alpha_0 \|\tilde{y}' - \tilde{y}''\|^2_V \leq$$

$$\leq \|u' - u''\|_{L_2(\gamma)} \|\tilde{y}' - \tilde{y}''\|_{L_2(\gamma)} \leq c_0 \|u' - u''\|_{L_2(\gamma)} \|\tilde{y}' - \tilde{y}''\|_V \tag{2.4'}$$

is derived, where c_0 is the constant obtained from the inequality of the embedding theorem [55].

Therefore, here is the inequality

$$\|\tilde{y}' - \tilde{y}''\|_{V \times L_2} \leq c_1 \|u' - u''\|_{L_2(\gamma) \times L_2}, \tag{2.5}$$

where $\|\varphi\|^2_{L_2(\gamma) \times L_2} = \int_0^T \|\varphi\|^2_{L_2(\gamma)} dt$ and from which the inequality

$$\|\tilde{y}' - \tilde{y}''\|_{L_2 \times L_2} \leq c_1 \|u' - u''\|_{L_2(\gamma) \times L_2} \tag{2.6}$$

follows that provides the continuity of the linear functional $L(\cdot)$ and bilinear form $\pi(\cdot, \cdot)$ on $\mathcal{U}$. In this case, the linear functional $L(\cdot)$ and bilinear form $\pi(\cdot,\cdot)$ are specified by expressions (1.32), where $(\bar{a}u,v)_{\mathcal{U}} =$

$$= \int\limits_0^T\int\limits_\gamma \bar{a}\,uv\,d\gamma\,dt\,.$$

Specify the observation in the form of expression (1.7), where $C\,y(u) \equiv y(u)$. Bring a value of cost functional (1.9) in correspondence with every control $u \in \mathcal{U}$; in this case, z_g is a known element from $L^2(0,T;V)$ and the cost functional is

$$J(u) = \int\limits_0^T\int\limits_\Omega \big(y(u) - z_g\big)^2 dx\,dt + \int\limits_0^T\int\limits_\gamma \bar{a}u^2 d\gamma\,dt\,, \qquad (2.7)$$

where $0 < a_0 \le \bar{a} \le a_1 < \infty$; $a_0, a_1 = \text{const}$, $\bar{a} \in L_2(\gamma)$.

On the basis of [58, Chapter 1, Theorem 1.1], the validity of the following statement is proved.

Theorem 2.2. *If a system state is determined as a solution to problem (2.3), (2.4), then there exists a unique element u of a convex set $\mathcal{U}_\partial$ that is closed in $\mathcal{U}$, and relation (1.10) takes place for u, where the cost functional has the form of expression (2.7).*

As for the control $v \in \mathcal{U}$, the conjugate state $p(v)$ is specified by system (1.34'), for which the generalized problem is written by equality system (1.35), (1.36). Choose the difference $y(v) - y(u)$ instead of w in equality (1.35), consider equation (2.3), and the equality

$$\int\limits_0^T\big(y(u) - z_g, y(v) - y(u)\big)dt = -\int\limits_0^T\bigg(\frac{d}{dt}p(u),\ y(v) - y(u)\bigg)dt -$$

$$-\int\limits_0^T\int\limits_\gamma c\frac{dp}{dt}(y(v) - y(u))d\gamma\,dt + \int\limits_0^T a\big(p(u), y(v) - y(u)\big)dt =$$

$$= -\bigg\{\big(p(u), y(v) - y(u)\big) + \big(cp(u), y(v) - y(u)\big)_{L_2(\gamma)}\bigg\}\bigg|_0^T +$$

$$+ \int_0^T \left\{ \left(p(u), \frac{d}{dt}(y(v) - y(u)) \right) + \left(cp(u), \frac{d}{dt}(y(v) - y(u)) \right)_{L_2(\gamma)} \right\} dt +$$

$$+ \int_0^T a \left(p(u), y(v) - y(u) \right) dt = - \int_0^T \int_\gamma (v - u) p(u) \, d\gamma \, dt$$

is obtained, i.e.

$$\int_0^T \left(y(u) - z_g, y(v) - y(u) \right) dt = - \int_0^T (v - u, p(u))_{L_2(\gamma)} \, dt.$$

Therefore, the necessary condition for the optimality of the control u is

$$\int_0^T \int_\gamma (-p(u) + \bar{a} u)(v - u) d\gamma \, dt \geq 0, \quad \forall v \in \mathcal{U}_\partial. \tag{2.8}$$

Thus, the optimal control $u \in \mathcal{U}_\partial$ is specified by relations (1.35), (1.36), (2.3), (2.4) and (2.8). If the constraints are absent, i.e. when $\mathcal{U}_\partial = \mathcal{U}$, then the equality

$$-p(u) + \bar{a} u = 0, \quad (x, t) \in \gamma_T, \tag{2.9}$$

is obtained from condition (2.8) and the optimal control

$$u = p / \bar{a}, \quad (x, t) \in \gamma_T, \tag{2.10}$$

is found from equality (2.9).

If the solution $(y, p)^T$ to problem (1.35), (1.36), (2.3), (2.4), (2.9) is smooth enough on $\bar{\Omega}_{IT}$, viz., $y \big|_{\bar{\Omega}_{IT}}$, $p \big|_{\bar{\Omega}_{IT}} \in C^{1,0}(\bar{\Omega}_{IT}) \cap C^{2,0}(\Omega_{IT}) \cap$

$\cap C^{0,1}(\Omega_{IT})$, $l = 1, 2$, then the equivalent differential problem of finding the vector-function $(y, p)^T$, that satisfies the system specified by equalities (1.1), (1.2), (1.5), (2.1) and (1.34′) and by the constraint

$$\left[\sum_{i,j=1}^n k_{ij} \frac{\partial y}{\partial x_j} \cos(v, x_i) \right] = c \frac{\partial y}{\partial t} + p / \bar{a}, \quad (x, t) \in \gamma_T, \tag{2.11}$$

corresponds to problem (1.35), (1.36), (2.3), (2.4), (2.9), where the optimal control is found by formula (2.10).

6.3   CONTROL UNDER CONJUGATION CONDITION WITH BOUNDARY OBSERVATION

For every control $u \in \mathcal{U} = L_2(\gamma_T)$, determine a state $y(x,t;u)$ as a generalized solution to the initial boundary-value problem specified by equation (1.1), boundary condition (1.2), initial condition (1.5) and conjugation conditions (2.1) and (2.2). The cost functional is

$$J(u) = \int_0^T \int_\Gamma \left(y(u) - z_g \right)^2 d\Gamma dt + \int_0^T \int_\gamma \bar{a} u^2 d\gamma dt. \tag{3.1}$$

The generalized problem corresponds to initial boundary-value problem (1.1), (1.2), (1.5), (2.1), (2.2) and means to find a function $y(x,t;u) \in$ $\in W(0,T)$ that satisfies equations (2.3) and (2.4) $\forall w \in V_0$; in this case, the spaces $W(0,T)$ and V_0 are specified in point 6.1. Theorem 2.1 takes place. Consider the embedding theorems, and the inequality

$$c_0 \left\| \tilde{y}' - \tilde{y}'' \right\|_{L_2(\Gamma) \times L_2} \leq \left\| \tilde{y}' - \tilde{y}'' \right\|_{V \times L_2} \leq c_1 \left\| u' - u'' \right\|_{L_2(\gamma) \times L_2},$$

i.e.

$$\left\| \tilde{y}' - \tilde{y}'' \right\|_{L_2(\Gamma) \times L_2} \leq c_2 \left\| u' - u'' \right\|_{L_2(\gamma) \times L_2}, \tag{3.2}$$

where $\|\varphi\|_{L_2(\Gamma) \times L_2}^2 = \int_0^T \|\varphi\|_{L_2(\Gamma)}^2 dt$, $\|\varphi\|_{L_2(\Gamma)}^2 = \int_\Gamma \varphi^2(x,t) d\Gamma$, $c_2 = c_1/c_0$, is derived from inequality (2.5).

The derived inequality provides the continuity of the linear functional $L(\cdot)$ and bilinear form $\pi(\cdot,\cdot)$ on $\mathcal{U}$ for the representation

$$J(u) = \pi(u,u) - 2L(u) + \left\| z_g - y(0) \right\|_{L_2(\Gamma)}^2 \tag{3.3}$$

of cost functional (3.1), where

$$\pi(u,v) = \left(y(u) - y(0), \ y(v) - y(0) \right)_{L_2(\Gamma) \times L_2} + (\bar{a}\, u, v)_{L_2(\gamma) \times L_2}$$

and

$$L(v) = \left(z_g - y(0), \ y(v) - y(0) \right)_{L_2(\Gamma) \times L_2};$$

in this case,

$$(\varphi, \psi)_{L_2(\gamma) \times L_2} = \int_0^T \int_\gamma \varphi \psi \, d\gamma \, dt = \int_0^T (\varphi, \psi)_{L_2(\gamma)} dt, \quad (\varphi, \psi)_{L_2(\Gamma) \times L_2} =$$

$$= \int_0^T (\varphi, \psi)_{L_2(\Gamma)} dt.$$

On the basis of [58, Chapter 1, Theorem 1.1], the validity of the following statement is proved.

Theorem 3.1. *Let a system state be determined as a solution to problem (2.3), (2.4). Then, there exists a unique element u of a convex set $\mathcal{U}_\partial$ that is closed in $\mathcal{U}$, and relation (1.10) takes place for u, where the cost functional has the form of expression (3.1).*

As for the control $v \in \mathcal{U}$, the conjugate state $p(v)$ is specified by the equalities

$$-\frac{\partial p}{\partial t} - \sum_{i,j=1}^n \frac{\partial}{\partial x_i} \left(k_{ij} \frac{\partial p}{\partial x_j} \right) = 0, \quad (x,t) \in \Omega_T,$$

$$\sum_{i,j=1}^n k_{ij} \frac{\partial p}{\partial x_j} \cos(v, x_i) = -\alpha p + y - z_g, \quad (x,t) \in \Gamma_T,$$

$$[p] = 0, \quad (x,t) \in \gamma_T,$$

$$\left[\sum_{i,j=1}^n k_{ij} \frac{\partial p}{\partial x_j} \cos(v, x_i) \right] = -\, c\, \frac{\partial p}{\partial t}, \quad (x,t) \in \gamma_T, \tag{3.4}$$

$$p(x,T) = 0, \quad x \in \bar{\Omega}_1 \cup \bar{\Omega}_2.$$

Problem (3.4) has the unique generalized solution $p(v) \in W(0,T)$ as the unique one to the following equality system:

$$-\int_\Omega \frac{dp}{dt} w \, dx - \int_\gamma c \frac{dp}{dt} w \, d\gamma + \int_\Omega \sum_{i,j=1}^n k_{ij} \frac{\partial p}{\partial x_j} \frac{\partial w}{\partial x_i} dx + \int_\Gamma \alpha p w \, d\Gamma =$$

$$= \int_\Gamma (y - z_g) w \, d\Gamma, \quad \forall w \in V_0, \tag{3.5}$$

$$\int_\Omega p(x,T;\cdot) w \, dx + \int_\gamma c \, p(x,T;\cdot) w \, d\gamma = 0. \tag{3.6}$$

It is easy to state the existence of the unique solution $p(v) \in W(0,T)$ to problem (3.5), (3.6).

Choose the difference $y(v) - y(u)$ instead of w in equality (3.5), consider equation (2.3), and the equality

$$\int_0^T \left((y(u) - z_g), (y(v) - y(u)) \right)_{L_2(\Gamma)} dt =$$

$$= -\left\{ (p(u), y(v) - y(u)) + (c \, p(u), y(v) - y(u))_{L_2(\gamma)} \right\} \Big|_0^T +$$

$$+ \int_0^T \left\{ \left(p(u), \frac{d}{dt}(y(v) - y(u)) \right) + \left(c \, p(u), \frac{d}{dt}(y(v) - y(u)) \right)_{L_2(\gamma)} \right\} dt +$$

$$+ \int_0^T a(p(u), y(v) - y(u)) \, dt = -\int_0^T \int_\gamma (v - u) p(u) \, d\gamma \, dt$$

is obtained, i.e.

$$\int_0^T (y(u) - z_g, y(v) - y(u))_{L_2(\Gamma)} dt = -\int_0^T (v - u, p(u))_{L_2(\gamma)} dt. \tag{3.7}$$

Therefore, the necessary condition for the optimality of the control u is

$$\int_0^T \int_\gamma (-p(u) + \bar{a}\,u)(v - u)\,d\gamma\,dt \geq 0, \quad \forall v \in \mathcal{U}_\partial. \tag{3.8}$$

Thus, the optimal control $u \in \mathcal{U}_\partial$ is specified by relations (2.3), (2.4), (3.5), (3.6) and (3.8). If the constraints are absent, i.e. when $\mathcal{U}_\partial = \mathcal{U}$, then equality (2.9) is obtained from condition (3.8). The optimal control u in the form of equality (2.10) is derived from equality (2.9).

If the solution $(y, p)^{\mathrm{T}}$ to problem (2.3), (2.4), (3.5), (3.6), (2.9) is smooth enough on $\bar{\Omega}_{lT}$, $l = 1, 2$, then the equivalent differential problem of finding the vector-function $(y, p)^{\mathrm{T}}$, that satisfies the system specified by equalities (1.1), (1.2), (1.5), (2.1), (2.11) and (3.4), corresponds to problem (2.3), (2.4), (3.5), (3.6), (2.9), where the optimal control is found by formula (2.10).

6.4 CONTROL UNDER CONJUGATION CONDITION WITH FINAL OBSERVATION

For every control $u \in \mathcal{U} = L_2(\gamma_T)$, determine a state $y(u)$ as a generalized solution to the initial boundary-value problem specified by equation (1.1), boundary condition (1.2), initial condition (1.5) and conjugation conditions (2.1) and (2.2).

The cost functional is

$$J(u) = \int_\Omega \big(y(x, T; u) - z_g(x) \big)^2 dx + \int_0^T \int_\gamma \bar{a}\,u^2\,d\gamma\,dt, \tag{4.1}$$

where $0 < a_0 \leq \bar{a} \leq a_1 < \infty$; $a_0, a_1 = \text{const}$, and it may be rewritten as

$$J(u) = \pi(u, u) - 2L(u) + \int_\Omega \big(z_g(x) - y(x, T; 0) \big)^2 dx;$$

in this case,

$$\pi(u,v) = \left(y(\cdot,T;u) - y(\cdot,T;0),\ y(\cdot,T;v) - y(\cdot,T;0) \right) +$$

$$+ \int_0^T \int_\gamma \bar{a}\, u\, v\, d\gamma\, dt, \tag{4.1'}$$

$$L(v) = \left(z_g(\cdot) - y(\cdot,T;0),\ y(\cdot,T;v) - y(\cdot,T;0) \right).$$

The generalized problem corresponds to initial boundary-value problem (1.1), (1.2), (1.5), (2.1), (2.2) and means to find a function $y(x,t;u) \in$ $\in W(0,T)$ that satisfies equations (2.3) and (2.4) $\forall w(x) \in V_0$. The existence of the unique generalized solution to problem (1.1), (1.2), (1.5), (2.1), (2.2) $\forall u \in \mathcal{U}$ is provided by Theorem 2.1. If $\tilde{y}' = \tilde{y}(u')$ and $\tilde{y}'' = \tilde{y}(u'')$ are solutions from $W(0,T)$ to problem (2.3), (2.4) under $f = 0$ and $\beta = 0$ and under a function u that is equal, respectively, to u' and u'', then inequality (2.5) is true. Proceed from inequality (2.4'), consider equation (2.4) and derive the inequality

$$\left\| \tilde{y}' - \tilde{y}'' \right\|^2 (T) \leq 2 \left\| u' - u'' \right\|_{L_2(\gamma) \times L_2} \left\| \tilde{y}' - \tilde{y}'' \right\|_{L_2(\gamma) \times L_2} \leq$$

$$\leq 2c_0 \left\| u' - u'' \right\|_{L_2(\gamma) \times L_2} \left\| \tilde{y}' - \tilde{y}'' \right\|_{V \times L_2} \leq 2c_0 c_1 \left\| u' - u'' \right\|^2_{L_2(\gamma) \times L_2},$$

i.e. the inequality

$$\left\| \tilde{y}' - \tilde{y}'' \right\| (T) \leq c_2 \left\| u' - u'' \right\|_{L_2(\gamma) \times L_2} \tag{4.2}$$

is thus obtained that provides the continuity of the linear functional $L(\cdot)$ and bilinear form $\pi(\cdot,\cdot)$ on $\mathcal{U}$. In this case, the linear functional $L(\cdot)$ and bilinear form $\pi(\cdot,\cdot)$ are specified by expressions (4.1'). On the basis of [58, Chapter 1, Theorem 1.1], the validity of the following statement is proved.

Theorem 4.1. *If a system state is determined as a solution to problem (2.3), (2.4), then there exists a unique element u of a convex set $\mathcal{U}_\partial$ that is closed in $\mathcal{U}$, and relation (1.10) takes place for u, where the cost functional has the form of expression (4.1).*

As for the control $v \in \mathcal{U}$, the conjugate state $p(v)$ is specified by the equality system

$$-\frac{\partial p}{\partial t} - \sum_{i,j=1}^{n} \frac{\partial}{\partial x_i}\left(k_{ij}\frac{\partial p}{\partial x_j}\right) = 0, \quad (x,t) \in \Omega_T,$$

$$\sum_{i,j=1}^{n} k_{ij}\frac{\partial p}{\partial x_j}\cos(v,x_i) = -\alpha p, \quad (x,t) \in \Gamma_T,$$

$$[p] = 0, \quad (x,t) \in \gamma_T, \tag{4.3}$$

$$\left[\sum_{i,j=1}^{n} k_{ij}\frac{\partial p}{\partial x_j}\cos(v,x_i)\right] = -c\frac{\partial p}{\partial t}, \quad (x,t) \in \gamma_T,$$

$$p(x,T;\cdot) + \delta\left(x - x^*\right)c\,p(x,T;\cdot) = y(v) - z_g, \quad x \in \bar{\Omega}_1 \cup \bar{\Omega}_2, \quad x^* \in \gamma,$$

where δ is the Dirac delta-function. Problem (4.3) has the unique generalized solution as the unique one to the following equality system:

$$-\left(\frac{d}{dt}p(v),w\right) - \int_{\gamma} c\frac{dp}{dt}w\,d\gamma + \int_{\Omega}\sum_{i,j=1}^{n} k_{ij}\frac{\partial p}{\partial x_j}\frac{\partial w}{\partial x_i}\,dx +$$

$$+ \int_{\Gamma}\alpha\,p\,w\,d\Gamma = 0, \quad \forall w \in V_0, \quad t \in (0,T), \tag{4.4}$$

$$\int_{\Omega}p(x,T;v)w\,dx + \int_{\gamma}cp(x,T;v)w\,dx =$$

$$= \int_{\Omega}\left(y(v) - z_g\right)w\,dx, \quad t = T. \tag{4.5}$$

Choose the difference $y(v) - y(u)$ instead of w in system (4.4), consider equations (2.3) and (2.4), equality (4.5), and the equality

$$-\left\{\left(p(u), y(v)-y(u)\right)+\left(c\,p,\ y(v)-y(u)\right)_{L_2(\gamma)}\right\}\Big|_0^T +$$

$$+\int_0^T\left(p(u),\left(\frac{dy(v)}{dt}-\frac{dy(u)}{dt}\right)\right)dt+\int_0^T\left(c\,p,\ \frac{dy(v)}{dt}-\frac{dy(u)}{dt}\right)_{L_2(\gamma)}dt +$$

$$+\int_\Omega\sum_{i,j=1}^n k_{ij}\frac{\partial p}{\partial x_j}\frac{\partial}{\partial x_i}\left(y(v)-y(u)\right)dx+\int_\Gamma\alpha\,p\left(y(v)-y(u)\right)d\Gamma =$$

$$=-\left(y(u)-z_g,\ y(v)-y(u)\right)(T)-\int_0^T\int_\gamma(v-u)p(u)\,d\gamma\,dt = 0$$

is obtained, i.e.

$$\left(y(u)-z_g,\ y(v)-y(u)\right)(T) = -\int_0^T\left(p(u),v-u\right)_{L_2(\gamma)}dt. \qquad (4.6)$$

Therefore, the necessary condition for the optimality of the control u is

$$\left(-p(u)+\bar a\,u,\ v-u\right)_{L_2(\gamma)\times L_2}\geq 0,\quad \forall v\in \mathcal{U}_\partial. \qquad (4.7)$$

Thus, the optimal control $u\in\mathcal{U}_\partial$ is specified by relations (2.3), (2.4), (4.4), (4.5) and (4.7). If the constraints are absent, i.e. when $\mathcal{U}_\partial=\mathcal{U}$, then the equality

$$-p+\bar a\,u = 0,\quad (x,t)\in\gamma_T, \qquad (4.8)$$

is obtained from condition (4.7). If the solution $(y,p)^{\mathrm T}$ to problem (2.3), (2.4), (4.4), (4.5), (4.8) is smooth enough on $\overline\Omega_{lT}$, $l=1,2$, then the equivalent differential problem of finding the vector-function $(y,p)^{\mathrm T}$, that satisfies the system specified by equalities (1.1), (1.2), (1.5), (2.1), (2.11) and (4.3), corresponds to problem (2.3), (2.4), (4.4), (4.5), (4.8), where the optimal control is found by formula (2.10).

6.5 CONTROL UNDER BOUNDARY CONDITION WITH OBSERVATION UNDER CONJUGATION CONDITION

Assume that equation (1.1) is specified in the domain Ω_T. The boundary condition

$$\sum_{i,j=1}^{n} k_{ij} \frac{\partial y}{\partial x_j} \cos(v, x_i) = -\alpha y + \beta + u \qquad (5.1)$$

is specified, in its turn, on the boundary Γ_T, where $u \in L_2(\Gamma_T)$. On γ_T, the conjugation conditions are

$$[y] = 0 \qquad (5.2)$$

and

$$\left[\sum_{i,j=1}^{n} k_{ij} \frac{\partial y}{\partial x_j} \cos(v, x_i) \right] = c \frac{\partial y}{\partial t} \qquad (5.3)$$

and the initial condition is specified by equality (1.5).

For every control $u \in \mathcal{U} = L_2(\Gamma_T)$, determine a state $y(x,t;u)$ as a generalized solution to initial boundary-value problem (1.1), (1.5), (5.1)–(5.3). The cost functional is

$$J(u) = \int_0^T \int_\gamma \left(y(\cdot,t;u) - z_g(\cdot,t) \right)^2 d\gamma \, dt + \int_0^T \int_\Gamma \bar{a} \, u^2 d\Gamma dt, \qquad (5.4)$$

where $0 < a_0 \leq \bar{a} \leq a_1 < \infty$; a_0, $a_1 = \text{const}$, and it may be rewritten as

$$J(u) = \pi(u,u) - 2L(u) + \int_0^T \int_\gamma \left(z_g(\cdot,t) - y(\cdot,t;0) \right)^2 d\gamma \, dt ;$$

in this case,

$$\pi(u,v) = \int_0^T \int_\gamma \left(y(u) - y(0) \right) \left(y(v) - y(0) \right) d\gamma \, dt + \int_0^T \int_\Gamma \bar{a} \, u v \, d\Gamma dt$$

and

$$L(v) = \int_0^T \int_\gamma \left(z_g - y(0)\right)\left(y(v) - y(0)\right) d\gamma \, dt.$$

The generalized problem corresponds to initial boundary-value problem (1.1), (1.5), (5.1)–(5.3) and means to find a vector-function $y(x,t;u) \in W(0,T)$ that satisfies the following equations $\forall w(x) \in V_0$:

$$\int_\Omega \frac{dy}{dt} w \, dx + \int_\gamma c \frac{dy}{dt} w \, d\gamma + a(y,w) =$$

$$= (f,w) + (\beta,w)_{L_2(\Gamma)} + (u,w)_{L_2(\Gamma)} \qquad (5.5)$$

and

$$\int_\Omega y(x,0;\cdot) w \, dx + \int_\gamma c \, y(x,0;\cdot) w \, d\gamma = \int_\Omega y_0(x) w \, dx + \int_\gamma c \, y_0 w \, d\gamma. \qquad (5.6)$$

Theorem 5.1. *Initial boundary-value problem (1.1), (1.5), (5.1)–(5.3) has a unique generalized solution* $y(x,t;u) \in W(0,T)$ $\forall u \in \mathcal{U}$.

Let $\tilde{y}' = \tilde{y}(u')$ and $\tilde{y}'' = \tilde{y}(u'')$ be solutions from $W(0,T)$ to problem (5.5), (5.6) under $f = 0$ and $\beta = 0$ and under a function u that is equal, respectively, to u' and u''. Then, the inequality

$$\|\tilde{y}' - \tilde{y}''\|_{V \times L_2} \le c_1 \|u' - u''\|_{L_2(\Gamma) \times L_2} \qquad (5.7)$$

is obtained from equation (5.5). Consider the embedding theorems, and the inequality

$$\|\tilde{y}' - \tilde{y}''\|_{L_2(\gamma) \times L_2} \le c_2 \|u' - u''\|_{L_2(\Gamma) \times L_2} \qquad (5.8)$$

is derived from inequality (5.7).

The derived inequality provides the continuity of the linear functional $L(\cdot)$ and bilinear form $\pi(\cdot,\cdot)$ on $\mathcal{U}$.

On the basis of [58, Chapter 1, Theorem 1.1], the validity of the following statement is proved.

Theorem 5.2. *Let a system state be determined as a solution to problem (5.5), (5.6). Then, there exists a unique element u of a convex set $\mathcal{U}_\partial$ that is closed in $\mathcal{U}$, and relation (1.10) takes place for u, where the cost functional has the form of expression (5.4).*

As for the control $v \in \mathcal{U}$, the conjugate state $p(v)$ is specified by the equality system

$$-\frac{\partial p}{\partial t} - \sum_{i,j=1}^{n} \frac{\partial}{\partial x_i}\left(k_{ij}\frac{\partial p}{\partial x_j}\right) = 0, \quad (x,t) \in \Omega_T,$$

$$\sum_{i,j=1}^{n} k_{ij}\frac{\partial p}{\partial x_j}\cos(v,x_i) = -\alpha p, \quad (x,t) \in \Gamma_T,$$

$$[p] = 0, \quad (x,t) \in \gamma_T, \tag{5.9}$$

$$\left[\sum_{i,j=1}^{n} k_{ij}\frac{\partial p}{\partial x_j}\cos(v,x_i)\right] = -c\frac{\partial p}{\partial t} + y(v) - z_g, \quad (x,t) \in \gamma_T,$$

$$p(x,T; \cdot) = 0, \quad x \in \bar{\Omega}_1 \bigcup \bar{\Omega}_2.$$

Problem (5.9) has the unique generalized solution $p(v) \in W(0,T)$ as the unique one to the following equality system:

$$-\int_\Omega \frac{dp}{dt}w\,dx - \int_\gamma c\frac{dp}{dt}w\,d\gamma + a(p,w) = -\int_\gamma \left(y(v) - z_g\right)w\,d\gamma, \tag{5.10}$$

$$\int_\Omega p(x,T;v)w\,dx + \int_\gamma c\,p(\cdot,T;v)\,w\,d\gamma = 0. \tag{5.11}$$

Choose the difference $y(v) - y(u)$ instead of w in equality (5.10), consider equations (5.5) and (5.6), equality (5.11), and the equality

$$-\left\{ (p(u), y(v)-y(u))+(cp,\ y(v)-y(u))_{L_2(\gamma)} \right\}\Big|_0^T +$$

$$+\int_0^T \left(p(u),\ \frac{d}{dt}(y(v)-y(u)) \right) dt + \int_0^T \left(cp,\ \frac{d}{dt}(y(v)-y(u)) \right)_{L_2(\gamma)} dt +$$

$$+\int_0^T a\left(p, y(v)-y(u) \right) dt = -\int_0^T \left(y(u)-z_g,\ y(v)-y(u) \right)_{L_2(\gamma)} dt$$

is obtained, or

$$-\int_0^T \left(y(u)-z_g, y(v)-y(u) \right)_{L_2(\gamma)} dt = \int_0^T ((v-u), p(u))_{L_2(\Gamma)} dt\ .$$

Therefore, the necessary condition for the optimality of the control u is

$$(-p+\bar{a}\, u, v-u)_{L_2(\Gamma)\times L_2} \geq 0 \quad \forall v \in \mathcal{U}_\partial. \tag{5.12}$$

Thus, the optimal control $u \in \mathcal{U}_\partial$ is specified by relations (5.5), (5.6), (5.10), (5.11) and (5.12). If the constraints are absent, i.e. when $\mathcal{U}_\partial = \mathcal{U}$, then the equality

$$-p+\bar{a}\, u = 0, \quad (x,t) \in \Gamma_T, \tag{5.13}$$

is obtained from condition (5.12). If the solution $(y, p)^T$ to problem (5.5), (5.6), (5.10), (5.11), (5.13) is smooth enough on $\overline{\Omega}_{lT}$, $l=1,2$, then the equivalent differential problem of finding the vector-function $(y, p)^T$, that satisfies the system specified by equalities (1.1), (1.5), (5.2), (5.3) and (5.9) and by the constraint

$$\sum_{i,j=1}^n k_{ij} \frac{\partial y}{\partial x_j} \cos(v, x_i) = -\alpha y + \beta + p/\bar{a}, \quad (x,\ t) \in \Gamma_T, \tag{5.14}$$

corresponds to problem (5.5), (5.6), (5.10), (5.11), (5.13), where the optimal control is found by the formula

$$u = p/\bar{a}, \quad (x,t) \in \Gamma_T. \tag{5.15}$$

6.6   BOUNDARY CONTROL WITH FINAL OBSERVATION

Assume that equation (1.1) is specified in the domain Ω_T. On γ_T, the conjugation conditions are specified by constraints (1.3) and (1.4). The initial condition has the form of expression (1.5), and the boundary condition for concentrated heat capacity [91]

$$\sum_{i,j=1}^{n} k_{ij} \frac{\partial y}{\partial x_j} \cos(\nu, x_i) = -\alpha y - c_0 \frac{\partial y}{\partial t} + \beta + u, \quad (x,t) \in \Gamma_T, \tag{6.1}$$

is specified on the boundary Γ_T, where the coefficient $\alpha = \alpha(x)$ is specified, in its turn, in point 6.1, $0 \le c_0' < c_0 = c_0(x)$, $c_0 \in L_2(\Gamma)$, $u = u(x,t) \in \mathcal{U} = L_2(\Gamma_T)$.

For every control $u \in \mathcal{U}$, determine a system state $y = y(u) = y(x,t;u)$ as a generalized solution to the initial boundary-value problem specified by controls (1.1), (1.3)–(1.5) and (6.1). Specify the observation by the expression

$$Z(u) = C y(u), \quad C y(u) = y(x,T;u), \quad x \in \bar{\Omega}_1 \cup \bar{\Omega}_2. \tag{6.2}$$

The cost functional is

$$J(u) = \int_\Omega \left(y(x,T;u) - z_g(x) \right)^2 dx + \int_0^T \int_\Gamma \bar{a} u^2 d\Gamma dt, \tag{6.3}$$

where $0 < a_0 \le \bar{a} \le a_1 < \infty$; $a_0, a_1 = \text{const}$, and it may be rewritten as

$$J(u) = \pi(u,u) - 2L(u) + \int_\Omega \left(z_g(x) - y(x,T;0) \right)^2 dx; \tag{6.3'}$$

in this case,

$$\pi(u,v) = \big(y(\cdot,T;u) - y(\cdot,T;0), y(\cdot,T;v) - y(\cdot,T;0)\big) + \int_0^T \int_\Gamma \bar{a}\, u\, v\, d\Gamma\, dt$$

and

$$L(v) = \big(z_g(\cdot) - y(\cdot,T;0),\ y(\cdot,T;v) - y(\cdot,T;0)\big).$$

The generalized problem corresponds to initial boundary-value problem (1.1), (1.3)–(1.5), (6.1) and means to find a vector-function $y(x,t;u) \in W(0,T)$ that satisfies the following equations $\forall w(x) \in V_0$:

$$\int_\Omega \frac{dy}{dt} w\, dx + \int_\Omega \sum_{i,j=1}^n k_{ij} \frac{\partial y}{\partial x_j} \frac{\partial w}{\partial x_i} dx + \int_\Gamma c_0 \frac{dy}{dt} w\, d\Gamma +$$

$$+ \int_\gamma c \frac{dy}{dt} w\, d\gamma + \int_\Gamma \alpha y w\, d\Gamma = (f,w) + \int_\Gamma \beta w\, d\Gamma + \int_\Gamma u w\, d\Gamma \qquad (6.4)$$

and

$$\int_\Omega y(x,0;u) w\, dx + \int_\gamma c\, y(x,0;u) w\, d\gamma + \int_\Gamma c_0 y(x,0;u) w\, d\Gamma =$$

$$= \int_\Omega y_0(x) w\, dx + \int_\gamma c\, y_0 w\, d\gamma + \int_\Gamma c_0 y_0 w\, d\Gamma. \qquad (6.5)$$

The forthcoming statement takes place.

Theorem 6.1. *Initial boundary-value problem (1.1), (1.3)–(1.5), (6.1) has a unique generalized solution $y(x,t;u) \in L^2(0,T;V)$.*

The validity of Theorem 6.1 is stated by analogy with the proof of Theorem 1.1.

Remark. When Theorem 6.1 is proved, functions $w_j(x)$ may be chosen as eigenfunctions that correspond to eigenvalues λ_j, $j = 1,2,\dots$, of the spectral problem: Find

$$\{\lambda, u\} \in \{R^1 \times V_0,\ u \neq 0\}:\quad a(u,w) = \lambda b(u,w),\quad \forall w \in V_0,$$

where

$$a(u,v) = \int_{\Omega} \sum_{i,j=1}^{n} k_{ij} \frac{\partial u}{\partial x_j} \frac{\partial v}{\partial x_i} dx + \int_{\Gamma} \alpha u v \, d\Gamma$$

and

$$b(u,v) = (u,v) + (cu,v)_{L_2(\gamma)} + (c_0 u, v)_{L_2(\Gamma)}.$$

Let $\tilde{y}' = \tilde{y}(u')$ and $\tilde{y}'' = \tilde{y}(u'')$ be solutions from $L^2(0,T;V)$ to problem (6.4), (6.5) under $f = 0$ and $\beta = 0$ and under a function u that is equal, respectively, to u' and u''. Then:

$$\frac{1}{2}\frac{d}{dt}(\tilde{y}' - \tilde{y}'', \tilde{y}' - \tilde{y}'') + \frac{1}{2}\frac{d}{dt}(c(\tilde{y}' - \tilde{y}''), \tilde{y}' - \tilde{y}'')_{L_2(\gamma)} +$$

$$+ \frac{1}{2}\frac{d}{dt}(c_0(\tilde{y}' - \tilde{y}''), \tilde{y}' - \tilde{y}'')_{L_2(\Gamma)} + \alpha_0 \|\tilde{y}' - \tilde{y}''\|_V^2 \le$$

$$\le \|u' - u''\|_{L_2(\Gamma)} \|\tilde{y}' - \tilde{y}''\|_{L_2(\Gamma)} \le \bar{c}_0 \|u' - u''\|_{L_2(\Gamma)} \|\tilde{y}' - \tilde{y}''\|_V. \quad (6.6)$$

Find the integral of inequality (6.6) over the interval $(0, T)$, and the inequality

$$\|\tilde{y}' - \tilde{y}''\|^2 (T) + \|\sqrt{c}(\tilde{y}' - \tilde{y}'')\|_{L_2(\gamma)}^2 (T) + \|\sqrt{c_0}(\tilde{y}' - \tilde{y}'')\|_{L_2(\Gamma)}^2 (T) +$$

$$+ \alpha_0 \|\tilde{y}' - \tilde{y}''\|_{V \times L_2}^2 \le c_0' \|u' - u''\|_{L_2(\Gamma) \times L_2} \|\tilde{y}' - \tilde{y}''\|_{V \times L_2} \quad (6.7)$$

is obtained from inequality (6.6). Since the inequality $\|\tilde{y}' - \tilde{y}''\|_{V \times L_2} \le$

$$\le \frac{c_0'}{\alpha_0} \|u' - u''\|_{L_2(\Gamma) \times L_2}$$ is true, then the one, i.e.

$$\|\tilde{y}' - \tilde{y}''\|_{L_2} (T) \le c_1 \|u' - u''\|_{L_2(\Gamma) \times L_2} \quad (6.8)$$

is derived from inequality (6.7).

The derived inequality provides the continuity of the linear functional $L(\cdot)$ and bilinear form $\pi(\cdot, \cdot)$ on $\mathcal{U}$ for representation (6.3') of cost functional (6.3). On the basis of [58, Chapter 1, Theorem 1.1], the validity of the following statement is proved.

Theorem 6.2. *If a system state is determined as a solution to problem (6.4), (6.5), then there exists a unique element u of a convex set $\mathcal{U}_\partial$ that is closed in $\mathcal{U}$, and relation (1.10) takes place for u, where the cost functional has the form of expression (6.3).*

As for the control $v \in \mathcal{U}$, the conjugate state $p(v)$ is specified by the equality system

$$-\frac{\partial p}{\partial t} - \sum_{i,j=1}^{n} \frac{\partial}{\partial x_i}\left(k_{ij}\frac{\partial p}{\partial x_j}\right) = 0, \quad (x,t) \in \Omega_T,$$

$$\sum_{i,j=1}^{n} k_{ij}\frac{\partial p}{\partial x_j}\cos(v,x_i) = -\alpha p + c_0\frac{\partial p}{\partial t}, \quad (x,t) \in \Gamma_T,$$

$$[p] = 0, \quad (x,t) \in \gamma_T, \tag{6.9}$$

$$\left[\sum_{i,j=1}^{n} k_{ij}\frac{\partial p}{\partial x_j}\cos(v,x_i)\right] = -c\frac{\partial p}{\partial t}, \quad (x,t) \in \gamma_T,$$

$$p(x,T; \cdot) + \delta(x-x')cp(x,T; \cdot) + \delta(x-x'')c_0\,p(x,T; \cdot) = y(v) - z_g,$$

$$x \in \bar{\Omega}_1 \cup \bar{\Omega}_2, \; x' \in \gamma, \; x'' \in \Gamma,$$

where δ is the Dirac delta-function.

Problem (6.9) has the unique generalized solution as the unique one to the following equality system:

$$-\left(\frac{dp}{dt},w\right) - \int_\gamma c\frac{dp}{dt}w\,d\gamma - \int_\Gamma c_0\frac{dp}{dt}w\,d\Gamma + a(p,w) = 0, \tag{6.10}$$

$$\int_\Omega p(x,T;v)w\,dx + \int_\gamma cp(x,T;v)w\,d\gamma +$$

$$+ \int_\Gamma c_0 p(x,T;v)w\,d\Gamma = \int_\Omega \left(y(x,T;v) - z_g\right)w\,dx. \tag{6.11}$$

Choose the difference $y(v) - y(u)$ instead of w in equality (6.10), consider equations (6.4) and (6.5), equality (6.11), and the equality

$$-\big(p(u),\, y(v)-y(u)\big)+\big(cp,\, y(v)-y(u)\big)_{L_2(\gamma)}+$$

$$+\big(c_0 p,\, y(v)-y(u)\big)_{L_2(\Gamma)}\Big|_0^T+$$

$$+\int_0^T\!\!\left(p(u),\,\frac{d}{dt}\big(y(v)-y(u)\big)\right)dt+\int_0^T\!\!\left(cp,\,\frac{d}{dt}\big(y(v)-y(u)\big)\right)_{L_2(\gamma)}dt+$$

$$+\int_0^T\!\!\left(c_0 p,\,\frac{d}{dt}\big(y(v)-y(u)\big)\right)_{L_2(\Gamma)}dt+\int_0^T a\big(p,\, y(v)-y(u)\big)dt=$$

$$=-\big(y(u)-z_g,\, y(v)-y(u)\big)(T)+\int_0^T\big((v-u),\,p(u)\big)_{L_2(\Gamma)}\,dt=0$$

is obtained, i.e.

$$\big(y(u)-z_g,\, y(v)-y(u)\big)(T)=\big(p(u),\,v-u\big)_{L_2(\Gamma)\times L_2},\quad \forall v\in\mathcal{U}_\partial.$$

Therefore, the necessary condition for the optimality of the control u may be written as follows:

$$\big(p(u)+\bar{a}u,\,v-u\big)_{L_2(\Gamma)\times L_2}\geq 0\quad \forall v\in\mathcal{U}_\partial. \tag{6.12}$$

Thus, the optimal control $u\in\mathcal{U}_\partial$ is specified by relations (6.4), (6.5), (6.10), (6.11) and (6.12). If the constraints are absent, i.e. when $\mathcal{U}_\partial=\mathcal{U}$, then the equality

$$p+\bar{a}u=0,\quad (x,t)\in\Gamma_T, \tag{6.13}$$

is obtained from condition (6.12). If the solution $(y,p)^{\mathrm{T}}$ to problem (6.4), (6.5), (6.10)–(6.12) is smooth enough on $\overline{\Omega}_{lT}$, $l=1,2$, then the equivalent differential problem of finding the vector-function $(y,p)^{\mathrm{T}}$, that satisfies the system specified by equalities (1.1), (1.3)–(1.5) and (6.9) and by the constraint

$$\sum_{i,j=1}^{n}k_{ij}\frac{\partial y}{\partial x_j}\cos(\nu,x_i)=-\alpha y-c_0\frac{\partial y}{\partial t}+\beta-p/\bar{a},\quad (x,t)\in\gamma_T,$$

corresponds to problem (6.4), (6.5), (6.10)–(6.12), where the equality
$$u = -p/\bar{a}, \quad (x,t) \in \Gamma_T,$$
specifies the optimal control u for the considered system.

CONTROL OF A SYSTEM DESCRIBED BY A PSEUDOPARABOLIC EQUATION UNDER CONJUGATION CONDITIONS

Let there be the following denotations: Ω is a domain that consists of two open, non-intersecting and strictly Lipschitz domains Ω_1 and Ω_2 from an n-dimensional real linear space R^n; $\Gamma=(\partial\Omega_1 \cup \partial\Omega_2)\backslash\gamma$ $(\gamma = \partial\Omega_1 \cap \cap \partial\Omega_2 \neq \varnothing)$ is a boundary of a domain $\bar{\Omega}$, $\partial\Omega_i$ is a boundary of a domain Ω_i, $i=1,2$, $\Omega_T = \Omega \times (0,T)$ is a complicated cylinder, $\Gamma_T = \Gamma \times (0,T)$ is the lateral surface of a cylinder $\Omega_T \cup \gamma_T$, $\gamma_T = \gamma \times (0,T)$.

Let V be some Hilbert space and assume that V' is a space dual with respect to V. By analogy [58], introduce a space $L^2(0,T;V)$ of functions $t \to f(t)$ that map an interval $(0,T)$ into the space V of measurable functions, namely, of such ones that

$$\left(\int_0^T \|f\|_V^2 \, dt\right)^{1/2} < \infty.$$

Also by analogy, specify the space $L^2(0,T;V')$. Introduce a space

$$W(0,T) = \left\{f : f, \frac{df}{dt} \in L^2(0,T;V)\right\}.$$

7.1   DISTRIBUTED CONTROL

Assume that the pseudoparabolic equation [67]

$$-\sum_{i,j=1}^{n}\frac{\partial}{\partial x_i}\left(a_{ij}\frac{\partial^2 y}{\partial x_j \partial t}\right)+a(x)\frac{\partial y}{\partial t}-\sum_{i,j=1}^{n}\frac{\partial}{\partial x_i}\left(k_{ij}\frac{\partial y}{\partial x_j}\right)=f(x,t) \quad (1.1)$$

is specified in the domain Ω_T, where

$$a_{ij}\big|_{\bar\Omega_l}=a_{ji}\big|_{\bar\Omega_l}\in C(\bar\Omega_l)\cap C^1(\Omega_l),\ \ k_{ij}\big|_{\bar\Omega_l}=k_{ji}\big|_{\bar\Omega_l}\in C(\bar\Omega_l)\cap C^1(\Omega_l),$$

$$a\big|_{\Omega_l}\in C(\Omega_l),\ \ f\big|_{\Omega_{lT}}\in C(\Omega_{lT}),\ \ l=1,2;\ \ |f|<\infty,\ \ 0<\bar a_0\le a<\infty,$$

$$\alpha_0\sum_{i=1}^{n}\xi_i^2\le\sum_{i,j=1}^{n}a_{ij}\xi_i\xi_j\le\alpha_0'\sum_{i=1}^{n}\xi_i^2,$$

$$\alpha_1\sum_{i=1}^{n}\xi_i^2\le\sum_{i,j=1}^{n}k_{ij}\xi_i\xi_j\le\alpha_1'\sum_{i=1}^{n}\xi_i^2,\ \ \forall x\in\Omega,\ \ \forall \xi_i,\xi_j\in R^1; \quad (1.1')$$

$$\bar a_0,\ \alpha_0,\ \alpha_1,\ \alpha_0',\ \alpha_1'=\text{const}>0.$$

The boundary condition

$$\sum_{i,j=1}^{n}\left(a_{ij}\frac{\partial^2 y}{\partial x_j \partial t}+k_{ij}\frac{\partial y}{\partial x_j}\right)\cos(\nu,x_i)=-\alpha y+\beta \quad (1.2)$$

is specified, in its turn, on the boundary Γ_T, where $\alpha=\alpha(x)\ge$ $\ge\alpha^0>0$; $\alpha,\beta\in L_2(\Gamma)$; $\alpha^0=\text{const}$; and ν is an outer normal.

On γ_T, the conjugation conditions are

$$\left[\sum_{i,j=1}^{n}\left(a_{ij}\frac{\partial^2 y}{\partial x_j \partial t}+k_{ij}\frac{\partial y}{\partial x_j}\right)\cos(\nu,x_i)\right]=0 \quad (1.3)$$

and

$$\left\{ \sum_{i,j=1}^{n} \left(a_{ij} \frac{\partial^2 y}{\partial x_j \partial t} + k_{ij} \frac{\partial y}{\partial x_j} \right) \cos(v, x_i) \right\}^{\pm} = r[y], \qquad (1.4)$$

where $0 \le r = r(x) \le r_1 < \infty$; $r_1 = \text{const}$, $[\varphi] = \varphi^+ - \varphi^-$; $\varphi^+ = \{\varphi\}^+ = \varphi(x, t)$
under $(x, t) \in \gamma_T^+ = (\partial\Omega_2 \cap \gamma) \times (0, T)$; $\varphi^- = \{\varphi\}^- = \varphi(x, t)$ under $(x, t) \in$
$\in \gamma_T^- = (\partial\Omega_1 \cap \gamma) \times (0, T)$, v is a normal to γ (an ort of a normal to γ) and
such normal is directed into the domain Ω_2.

The initial condition

$$y(x, 0) = y_0(x), \quad x \in \bar{\Omega}_1 \cup \bar{\Omega}_2, \qquad (1.5)$$

where $y_0 \in V_0 = \left\{ v(x) : v|_{\Omega_i} \in W_2^1(\Omega_i), \ i = 1, 2 \right\}$, is specified under $t = 0$.

Let there be a control Hilbert space $\mathcal{U}$ and operator $B \in$
$\in \mathcal{L}(\mathcal{U}; L^2(0, T; V'))$. For every control $u \in \mathcal{U}$, determine a system state
$y = y(u) = y(x, t; u)$ as a generalized solution to the problem specified by
the equation

$$A\left(\frac{\partial y}{\partial t} \right) + K(y) = f + Bu, \qquad (1.6)$$

where

$$A(z) = -\sum_{i,j=1}^{n} \frac{\partial}{\partial x_i} \left(a_{ij} \frac{\partial z}{\partial x_j} \right) + az, \quad K(z) = -\sum_{i,j=1}^{n} \frac{\partial}{\partial x_i} \left(k_{ij} \frac{\partial z}{\partial x_j} \right), \quad (1.7)$$

and by conditions (1.2)–(1.5). Further on, without loss of generality,
assume $Bu \equiv u$ and, for the sake of simplicity, use the denotation
$y = y(x, t)$.

Specify the observation by the following expression:

$$Z(u) = Cy(u), \quad C \in \mathcal{L}(W(0, T); \mathcal{H}). \qquad (1.8)$$

Specify the operator

$$\mathcal{N} \in \mathcal{L}(\mathcal{U}; \mathcal{U}); \ (\mathcal{N}u, u)_{\mathcal{U}} \ge v_0 \|u\|_{\mathcal{U}}^2, \ v_0 = \text{const} > 0. \qquad (1.9)$$

Assume the following: $\mathcal{N}u = \bar{a}u$; in this case, $\bar{a}\big|_{\Omega_l} \in C(\Omega_l)$, $l = 1,2$; $0 < \bar{a}_0 \leq \bar{a} \leq \bar{a}_1 < \infty$, $\bar{a}_0, \bar{a}_1 = \text{const}$. The cost functional is

$$J(u) = \left\| Cy(u) - z_g \right\|_{\mathcal{H}}^2 + (\mathcal{N}u, u)_{\mathcal{U}}, \tag{1.10}$$

where z_g is a known element of the space $\mathcal{H}$.

The optimal control problem is: Find such an element $u \in \mathcal{U}$ that the condition

$$J(u) = \inf_{v \in \mathcal{U}_\partial} J(v) \tag{1.11}$$

is met, where $\mathcal{U}_\partial$ is some convex closed subset in $\mathcal{U}$.

The generalized problem corresponds to initial boundary-value problem (1.6), (1.2)–(1.5) and means to find a function $y(x,t;u) \in W(0,T)$ that satisfies the following equations $\forall w(x) \in V_0 = \left\{ v : v\big|_{\Omega_i} \in W_2^1(\Omega_i), \ i = 1,2 \right\}$:

$$a_0(y', w) + a(y, w) = (f, w) + (u, w) + \int_\Gamma \beta w \, d\Gamma, \ t \in (0,T), \tag{1.12}$$

and

$$a_0\big(y(\cdot, 0; u), w(\cdot) \big) = a_0\big(y_0(\cdot), w(\cdot) \big); \tag{1.13}$$

in this case,

$$y' = \frac{dy}{dt}, \ a_0(y', w) = \int_\Omega \left(\sum_{i,j=1}^{n} a_{ij} \frac{\partial y'}{\partial x_j} \frac{\partial w}{\partial x_i} + ay'w \right) dx, \tag{1.13'}$$

$$a(y, w) = \int_\Omega \sum_{i,j=1}^{n} k_{ij} \frac{\partial y}{\partial x_j} \frac{\partial w}{\partial x_i} dx + \int_\gamma r[y][w] d\gamma + \int_\Gamma \alpha y w \, d\Gamma, \tag{1.13''}$$

$W_2^1(\Omega_i)$ is the space of the Sobolev functions that are specified on the domain Ω_i, $i = 1,2$, and, when the space $W(0,T)$ is specified, then the

space V is $V = \{v(x,t):v|_{\Omega_i} \in W_2^1(\Omega_i), \ i=1,2; \ \forall t \in [0,T]\}, \quad (\varphi,\psi) =$
$$= \int_\Omega \varphi(x,t)\psi(x,t)\,dx.$$

Consider the existence and uniqueness of the solution to problem (1.12), (1.13). Since Bu and $f \in L^2(0,T;V')$ and $V' = \{v(x,t):(v,v) < \infty, \forall t \in (0,T)\}$, then, without loss of generality, assume the following: $Bu \equiv 0$.

The space V_0 is complete, separable and reflexive [41, 49, 55]. Choose an arbitrary fundamental system of linearly independent functions $w_k(x)$, $k=1,2,\ldots$, in V_0. Let this system be orthonormal in $L_2(\Omega)$ so that $(w_k,w_l) = \delta_k^l$ ($\delta_k^k = 1$, $\delta_k^l = 0$ under $l \neq k$; $l,k=1,2,\ldots$). As for $y_0 \in L_2(\Omega)$ [49]:

$$y_0 = \sum_{i=1}^\infty \xi_i w_i(x), \qquad (1.14)$$

where $\xi_i = (y_0,w_i)$, $i=1, 2, \ldots$.

Remark. Functions $w_j(x)$ may be chosen as eigenfunctions that correspond to eigenvalues λ_j, $j=1,2,\ldots$, of the spectral problem: Find

$$(\lambda,u) \in \{R^1 \times V_0, u \neq 0\} : a_0(u,w) = \lambda(u,w), \ \forall w \in V_0.$$

The approximate solution to problem (1.12), (1.13) is given as

$$y_m(x,t) = \sum_{i=1}^m g_{im}(t) w_i(x), \qquad (1.15)$$

where the functions $g_{im}(t)$ are chosen in such a way that the relations

$$a_0\left(\frac{\partial y_m}{\partial t}, w_j\right) + a(y_m,w_j) = (f,w_j) + \int_\Gamma \beta w_j\,d\Gamma, \ j=\overline{1,m}, \quad (1.16)$$

and

$$a_0\left(y_m(\cdot,0),w_j(\cdot)\right)=a_0\left(y_0(\cdot),w_j(\cdot)\right),\ j=\overline{1,m}, \qquad (1.17)$$

are met. Equalities (1.16) and (1.17) specify the Cauchy problem for the system of m first-order linear ordinary differential equations as for $g_{im}(t)$:

$$M_m\frac{dg_m}{dt}+K_mg_m=F_m(t), \qquad (1.18)$$

$$M_m^0g_m(0)=F_m^0; \qquad (1.19)$$

in this case,

$$M_m=\left\{M_{ij}^m\right\}_{i,j=1}^m,\ M_{ij}^m=a_0(w_i,w_j),\ K_m=\left\{k_{ij}^m\right\}_{i,j=1}^m,$$

$$k_{ij}^m=a(w_i,w_j),\ F_m(t)=\left\{f_i^m(t)\right\}_{i=1}^m,\ f_i^m(t)=(f,w_i)+\int_\Gamma\beta w_id\Gamma,$$

$$F_m^0=\left\{f_{0i}^m\right\}_{i=1}^m,\ f_{0i}^m=(y_0,w_i),\ M_m^0=\left\{M_{mij}^0\right\}_{i,j=1}^m,\ M_{mij}^0=a_0(w_i,w_j).$$

Since the bilinear form $a_0(\cdot,\cdot)$ is V-elliptic on V_0, then the symmetric matrix M_m is positively specified. In this case: $\|v\|_V=\left\{\sum_{i=1}^2\|v\|_{W_2^1(\Omega_i)}^2\right\}^{1/2}$.

Therefore, the solution to Cauchy problem (1.18), (1.19) exists and such solution is unique. The following statement must be proved: $y_m\to y$ under $m\to\infty$, where $y=y(x,t)$ is the solution to problem (1.12), (1.13).

Multiply equality (1.16) by $g_{im}(t)$ and find the sum over j for the result. Then:

$$a_0\left(\frac{\partial y_m}{\partial t},y_m\right)+a\left(y_m,y_m\right)=(f,y_m)+\int_\Gamma\beta y_md\Gamma,$$

i.e.:

$$\frac{1}{2}\frac{d}{dt}a_0\left(y_m,y_m\right)+a\left(y_m,y_m\right)=(f,y_m)+\int_\Gamma\beta y_md\Gamma. \qquad (1.20)$$

Take conditions (1.1′) and the generalized Friedrichs inequality [21] into account, and the inequalities

$$a_0(y_m, y_m) \geq c_0 \|y_m\|_V^2 \geq c_0 \|y_m\|^2 \text{ and } a(y_m, y_m) \geq c_1 \|y_m\|_V^2 \quad (1.21)$$

are derived, where $c_0 = \min\{\alpha_0, \overline{a}_0\}$, $c_1 = \frac{1}{2}\alpha_1 \min(1, \mu)$, μ is the positive constant from the generalized Friedrichs inequality and $\|\varphi\| = (\varphi, \varphi)^{1/2} =$

$$= \left(\int_\Omega \varphi^2(x,t)dx \right)^{1/2}.$$

Consider inequalities (1.21), the ε- and Cauchy-Bunyakovsky inequalities and embedding theorems [55], and the inequality

$$c_0 \|y_m\|^2(T) + 2c_1 \int_0^T \|y_m\|_V^2 \, dt \leq a_0(y_m, y_m)(0) + 2\int_0^T |(f, y_m)| dt +$$

$$+2\int_0^T \left| \int_\Gamma \beta \, y_m \, d\Gamma \right| dt \leq c_0' \|y_m\|_V^2(0) + 2\varepsilon \int_0^T \|y_m\|_V^2 \, dt + \frac{1}{2\varepsilon} \int_0^T \|f\|^2 \, dt +$$

$$+2\varepsilon_1 c_1 \int_0^T \|y_m\|_V^2 \, dt + \frac{1}{2\varepsilon_1} \int_0^T \|\beta\|_{L_2(\Gamma)}^2 \, dt \quad (1.22)$$

follows from equality (1.20); in this case, $\|\varphi\|_{L_2(\Gamma)}^2 =$

$$= \int_\Gamma \varphi^2(x,t) d\Gamma, \; c_1 = \max_{l=1,2} \overline{c}_l \text{ and the constant } \overline{c}_l \text{ is obtained from the}$$

inequality proved in the embedding theorem applied for the domain Ω_l.

Proceed from equality (1.17), consider the first condition from assumptions (1.1′), boundedness of the functions a_{ij} and a on Ω_l, $l = 1, 2$, and the Cauchy-Bunyakovsky inequality, and here is the conclusion:

$$c_0 \|y_m\|_V^0 (0) \le c_0' \|y_0\|_V \|y_m\|_V (0),$$

i.e.:

$$\|y_m\|_V (0) \le \frac{c_0'}{c_0} \|y_0\|_V .$$ (1.23)

Take inequality (1.23) into account, and the inequality

$$\int_0^T \|y_m\|_V^2 (t)dt \le c \left(\|y_0\|_V^2 + \int_0^T \|f\|^2 dt + \int_0^T \|\beta\|_{L_2(\Gamma)}^2 dt \right)$$

follows from inequality (1.22).

Therefore, the elements y_m are in some bounded subset of the space $L^2(0,T;V)$. Hence, there exists a subsequence $\{y_\chi\}$ that weakly converges to the element $z \in L^2(0,T;V)$. Without loss of generality, it is stated that the whole sequence $\{y_m\}$ weakly converges to z.

Rewrite equality (1.16) as

$$\frac{d}{dt} a_0 (y_m, w_j) + a (y_m, w_j) = (f, w_j) + \int_\Gamma \beta w_j \, d\Gamma, \quad j = \overline{1,m},$$

multiply its both sides by the function

$$\varphi(t) \in C^1([0,T]), \quad \varphi(T) = 0,$$ (1.24)

and find the integral from 0 to T of the result:

$$\int_0^T \left\{ -a_0 (y_m(\cdot,t), \varphi_j'(\cdot,t)) + a (y_m, \varphi_j) \right\} dt =$$

$$= \int_0^T (f, \varphi_j) \, dt + \int_0^T \int_\Gamma \beta \varphi_j \, d\Gamma + a_0 (y_m, \varphi_j)(0);$$ (1.25)

in this case, $\varphi_j(x,t) = \varphi(t) w_j(x)$, $\varphi_j'(x,t) = \dfrac{d\varphi(t)}{dt} w_j(x)$.

By virtue of the aforesaid weak convergence, it is possible to pass in equality (1.25) to the limit under $m \to \infty$, and the following equality is obtained:

$$\int_0^T \left\{ -a_0\left(z, \varphi'_j\right) + a\left(z, \varphi_j\right) \right\} dt = \int_0^T \left(f, \varphi_j\right) dt +$$

$$+ \int_0^T \int_\Gamma \beta \varphi_j \, d\Gamma dt + a_0\left(z, \varphi_j\right)(0). \tag{1.26}$$

Consider the assumptions as for $\{w_j\}$, and it can be seen that the matrix M_m^0 from condition (1.19) is diagonal $\left(M_{mii}^0 = a_0\left(w_i, w_i\right), \ M_{mij}^0 = 0\right.$ under $i \neq j, \ i, j = \overline{1, m}\right)$. The equality $g_{im}(0) = a_0\left(y_0, w_i\right) / a_0\left(w_i, w_i\right)$ follows from condition (1.19), i.e. $g_{im}\left(i = \overline{1, m}\right)$ are the Fourier coefficients for the function y_0. By virtue of [49]:

$$y_m(x, 0) = \sum_{i=1}^m g_{im}(0) w_i(x) \to y_0(x) \text{ under } m \to \infty.$$

Hence: $z(x, 0) = y_0(x)$.

Equality (1.26) is true for the arbitrary function φ that meets conditions (1.24). Thus, the following can be assumed: $\varphi \in D(0, T)$ [58]. Therefore, the equality

$$\int_0^T \left\{ -a_0\left(z, w_j\right) \varphi' + a\left(z, \varphi_j\right) \right\} dt = \int_0^T \left(f, \varphi_j\right) dt + \int_0^T \int_\Gamma \beta \varphi_j \, d\Gamma dt$$

follows from equality (1.26). Hence:

$$\int_0^T \left\{ \frac{d}{dt} a_0\left(z, w_j\right) + a\left(z, w_j\right) - \left(f, w_j\right) - \int_\Gamma \beta w_j \, d\Gamma \right\} \varphi(t) \, dt = 0,$$

i.e.:

$$a_0\left(\frac{dz}{dt}, w_j\right) + a(z, w_j) = (f, w_j) + \int_\Gamma \beta w_j \, d\Gamma, \ t \in (0, T). \quad (1.26')$$

Take equality (1.26′), the space V_0 and assumptions as for the functions w_j into account, and it is stated that the equality

$$a_0\left(\frac{dz}{dt}, w\right) + a(z, w) = (f, w) + \int_\Gamma \beta w \, d\Gamma \quad (1.27)$$

is true $\forall w \in V_0$. The following equality is obtained from relation (1.17):

$$a_0\left(z(\cdot, 0), w(\cdot)\right) = a_0\left(y_0(\cdot), w(\cdot)\right), \ \forall w \in V_0. \quad (1.28)$$

Therefore, the function $z \in L^2(0, T; V)$ is the solution to problem (1.12), (1.13) $\forall f \in L^2(0, T; V')$ and under $Bu = 0$, i.e. to problem (1.27), (1.28).

Illustrate the uniqueness of the solution to problem (1.27), (1.28) by contradiction. Let there exist two solutions: $z_1(x, t)$ and $z_2(x, t) \in \in L^2(0, T; V)$. Then, on the basis of equality (1.27), the equality

$$\frac{d}{dt} a_0(\bar{z}, \bar{z}) + 2a(\bar{z}, \bar{z}) = 0 \quad (1.29)$$

is obtained, where $\bar{z} = z_1 - z_2 \neq 0$.

Consider equality (1.28), and the contradiction

$$0 < a_0(\bar{z}, \bar{z})(T) + \alpha_0 \int_0^T \|\bar{z}\|_V^2 \, dt \leq 0, \ \alpha_0 = \text{const} > 0,$$

follows from equality (1.29).

Therefore, the validity of the following statement is proved.

Theorem 1.1. *Initial boundary-value problem (1.1)–(1.5) has a unique generalized solution $y(x, t) \in L^2(0, T; V)$.*

Proceed from equality (1.29), and it is easy to see that $y(u_1) \neq y(u_2)$ under $u_1 \neq u_2$ $(Bu_1 \neq Bu_2)$. Let $\tilde{y}' = \tilde{y}(u')$ and $\tilde{y}'' = \tilde{y}(u'')$ be solutions from $L^2(0,T;V)$ to problem (1.12), (1.13) under $f = 0$ and $\beta = 0$ and under a function $u = u(x,t)$ that is equal, respectively, to u' and u''. Then:

$$\frac{1}{2}\frac{d}{dt}a_0\left(\tilde{y}' - \tilde{y}'', \tilde{y}' - \tilde{y}''\right) + \alpha_0\left\|\tilde{y}' - \tilde{y}''\right\|_V^2 \le \left\|u' - u''\right\|\left\|\tilde{y}' - \tilde{y}''\right\|_V. \quad (1.30)$$

Therefore, the inequality

$$\left\|\tilde{y}' - \tilde{y}''\right\|_{V \times L_2} \le \frac{1}{\alpha_0}\left\|u' - u''\right\|_{L_2 \times L_2} \quad (1.30')$$

is obtained, where

$$\left\|\varphi\right\|_{L_2 \times L_2}^2 = \int\limits_0^T \left\|\varphi\right\|_{L_2}^2 dt, \quad \left\|\varphi\right\|_{L_2}^2 = \left\|\varphi\right\|^2 = \int\limits_\Omega \varphi^2(x,t)\,dx, \quad \left\|\varphi\right\|_{V \times L_2}^2 = \int\limits_0^T \left\|\varphi\right\|_V^2 dt.$$

Rewrite functional (1.10) as

$$J(u) = \pi(u,u) - 2L(u) + \int\limits_0^T \left\|z_g - y(0)\right\|^2 dt, \quad (1.31)$$

where

$$\pi(u,v) = \left(y(u) - y(0),\ y(v) - y(0)\right)_{\mathcal{H}} + (\bar{a}\,u, v)_{\mathcal{U}}, \quad (1.32)$$

$$L(v) = \left(z_g - y(0),\ y(v) - y(0)\right)_{\mathcal{H}};$$

in this case,

$$(z,v)_{\mathcal{H}} = (z,v)_{\mathcal{U}} = \int\limits_0^T (z,v)\,dt, \quad (z,v) = \int\limits_\Omega zv\,dx.$$

Inequality (1.30′) provides the continuity of the linear functional $L(\cdot)$ and bilinear form $\pi(\cdot,\cdot)$ on $\mathcal{U}$.

On the basis of [58, Chapter 1, Theorem 1.1], the validity of the following statement is proved.

Theorem 1.2. *Let a system state be determined as a solution to problem (1.12), (1.13). Then, there exists a unique element* u *of a convex set* $\mathcal{U}_\partial$ *that is closed in* $\mathcal{U}$, *and*

$$J(u) = \inf\limits_{v \in \mathcal{U}_\partial} J(v) \quad (1.33)$$

takes place for u.

The control $u \in \mathcal{U}_\partial$ is optimal if and only if the inequality

$$\langle J'(u), v-u \rangle \geq 0, \ \forall v \in \mathcal{U}_\partial,$$ (1.33')

is true, i.e. under

$$\left(y(u) - z_g, y(v) - y(u) \right)_{\mathcal{H}} + (\mathcal{N}u, v-u)_{\mathcal{U}} \geq 0.$$ (1.34)

As for the control $v \in \mathcal{U}$, the conjugate state $p(v)$ is specified by the relations

$$\sum_{i,j=1}^{n} \frac{\partial}{\partial x_i} \left(a_{ij} \frac{\partial^2 p}{\partial x_j \partial t} \right) - a \frac{\partial}{\partial t} p - \sum_{i,j=1}^{n} \frac{\partial}{\partial x_i} \left(k_{ij} \frac{\partial p}{\partial x_j} \right) = y(v) - z_g, \ (x,t) \in \Omega_T,$$

$$\sum_{i,j=1}^{n} \left(-a_{ij} \frac{\partial^2 p}{\partial x_j \partial t} + k_{ij} \frac{\partial p}{\partial x_j} \right) \cos(v, x_i) = -\alpha p, \ (x,t) \in \Gamma_T,$$

$$\left[\sum_{i,j=1}^{n} \left(-a_{ij} \frac{\partial^2 p}{\partial x_j \partial t} + k_{ij} \frac{\partial p}{\partial x_j} \right) \cos(v, x_i) \right] = 0, \ (x,t) \in \gamma_T,$$ (1.35)

$$\left\{ \sum_{i,j=1}^{n} \left(-a_{ij} \frac{\partial^2 p}{\partial x_j \partial t} + k_{ij} \frac{\partial p}{\partial x_j} \right) \cos(v, x_i) \right\}^{\pm} = r[p], \ (x,t) \in \gamma_T,$$

$$p(x, T) = 0, \ x \in \bar{\Omega}_1 \cup \bar{\Omega}_2.$$

Substitute a time $T-t$ for the time t, proceed from Theorem 1.1, and it is concluded that initial boundary-value problem (1.35) has the unique generalized solution $p(v) \in L^2(0, T; V)$ as the unique one to the following equality system:

$$-a_0 \left(\frac{d}{dt} p(v), w \right) + a(p, w) = \left(y(v) - z_g, w \right), \ \forall w \in V_0,$$ (1.36)

$$a_0(p, w) = 0, \ t = T.$$ (1.37)

Choose the difference $y(v) - y(u)$ instead of w, consider equation (1.12), the equality

$$\int_0^T a_0 \left(\frac{dp(u)}{dt}, y(v) - y(u) \right) dt = -\int_0^T a_0 \left(p(u), \frac{d}{dt} (y(v) - y(u)) \right) dt, \quad (1.37')$$

and the equality

$$\int_0^T \left(y(u) - z_g, y(v) - y(u) \right) dt = \int_0^T \left(p(u), v - u \right) dt \qquad (1.38)$$

is obtained from equality (1.36).

Therefore, inequality (1.34) has the form

$$\int_0^T \left(p(u) + \bar{a}u, v - u \right) dt \geq 0, \quad \forall v \in \mathcal{U}_\partial. \qquad (1.39)$$

Thus, the optimal control $u \in \mathcal{U}_\partial$ is specified by relations (1.12), (1.13), (1.36), (1.37) and (1.39).

If the constraints are absent, i.e. when $\mathcal{U}_\partial = \mathcal{U}$, then the equality

$$u = -p/\bar{a}, \quad (x,t) \in \Omega_T, \qquad (1.40)$$

is obtained from inequality (1.39).

If the solution $(y, p)^T$ to problem (1.12), (1.13), (1.36), (1.37), (1.40) is smooth enough on $\bar{\Omega}_{IT}$, viz., $y, p|_{\bar{\Omega}_{IT}} \in C^{1,0}(\bar{\Omega}_{IT}) \cap C^{2,0}(\Omega_{IT}) \cap \cap C^{0,1}(\Omega_{IT})$, $l = 1,2$, then the differential problem of finding the vector-function $(y, p)^T$, that satisfies the equalities

$$-\sum_{i,j=1}^n \frac{\partial}{\partial x_i} \left(a_{ij} \frac{\partial^2 y}{\partial x_j \partial t} \right) + a \frac{\partial y}{\partial t} - \sum_{i,j=1}^n \frac{\partial}{\partial x_i} \left(k_{ij} \frac{\partial y}{\partial x_j} \right) + p/\bar{a} = f, \quad (x,t) \in \Omega_T,$$

$$\sum_{i,j=1}^{n} \frac{\partial}{\partial x_i}\left(a_{ij}\frac{\partial^2 p}{\partial x_j \partial t}\right) - a\frac{\partial p}{\partial t} - \sum_{i,j=1}^{n}\frac{\partial}{\partial x_i}\left(k_{ij}\frac{\partial p}{\partial x_j}\right) - y = -z_g, \quad (x,t) \in \Omega_T,$$

$$\sum_{i,j=1}^{n}\left(a_{ij}\frac{\partial^2 y}{\partial x_j \partial t} + k_{ij}\frac{\partial y}{\partial x_j}\right)\cos(\nu, x_i) = -\alpha y + \beta, \quad (x,t) \in \Gamma_T,$$

$$\sum_{i,j=1}^{n}\left(-a_{ij}\frac{\partial^2 p}{\partial x_j \partial t} + k_{ij}\frac{\partial p}{\partial x_j}\right)\cos(\nu, x_i) = -\alpha p, \quad (x,t) \in \Gamma_T,$$

$$\left[\sum_{i,j=1}^{n}\left(a_{ij}\frac{\partial^2 y}{\partial x_j \partial t} + k_{ij}\frac{\partial y}{\partial x_j}\right)\cos(\nu, x_i)\right] = 0, \quad (x,t) \in \gamma_T,$$

$$\left\{\sum_{i,j=1}^{n}\left(a_{ij}\frac{\partial^2 y}{\partial x_j \partial t} + k_{ij}\frac{\partial y}{\partial x_j}\right)\cos(\nu, x_i)\right\}^{\pm} = r[y], \quad (x,t) \in \gamma_T,$$

$$\left[\sum_{i,j=1}^{n}\left(-a_{ij}\frac{\partial^2 p}{\partial x_j \partial t} + k_{ij}\frac{\partial p}{\partial x_j}\right)\cos(\nu, x_i)\right] = 0, \quad (x,t) \in \gamma_T,$$

$$\left\{\sum_{i,j=1}^{n}\left(-a_{ij}\frac{\partial^2 p}{\partial x_j \partial t} + k_{ij}\frac{\partial p}{\partial x_j}\right)\cos(\nu, x_i)\right\}^{\pm} = r[p], \quad (x,t) \in \gamma_T,$$

$$y(x,0) = y_0(x), \quad x \in \bar{\Omega}_1 \cup \bar{\Omega}_2,$$

and

$$p(x,T) = 0, \quad x \in \bar{\Omega}_1 \cup \bar{\Omega}_2,$$

corresponds to problem (1.12), (1.13), (1.36), (1.37), (1.40).

7.2   CONTROL UNDER CONJUGATION CONDITION WITH OBSERVATION THROUGHOUT A WHOLE DOMAIN

Assume that equation (1.1) is specified in the domain Ω_T. On the boundary Γ_T, the boundary condition has the form of expression (1.2).

For every control $u \in \mathcal{U} = L_2(\gamma_T)$, determine a state $y = y(u)$ as a generalized solution to the initial boundary-value problem specified, in its turn, by equation (1.1), boundary condition (1.2), initial condition (1.5) and the conjugation conditions

$$[y] = 0, \ (x,t) \in \gamma_T,$$

$$\left[\sum_{i,j=1}^{n} \left(a_{ij} \frac{\partial^2 y}{\partial x_j \partial t} + k_{ij} \frac{\partial y}{\partial x_j} \right) \cos(\nu, x_i) \right] = \omega + u, \ (x,t) \in \gamma_T, \quad (2.1)$$

where $\omega = \omega(x,t) \in L_2(\gamma_T)$.

Since there exists the generalized solution $y(u) \in W(0,T)$ to initial boundary-value problem (1.1), (1.2), (1.5), (2.1), then such solution is reasonable on $\bar{\Omega}_{lT}$ $(l = 1,2)$.

The generalized problem corresponds to initial boundary-value problem (1.1), (1.2), (1.5), (2.1) and means to find a function $y(x,t;u) \in W(0,T)$ that satisfies the following equations $\forall w(x) \in V_0 = \left\{ v : v|_{\Omega_i} \in W_2^1(\Omega_i), \right.$ $i = 1,2; \ [v] = 0 \right\}$:

$$a_0 \left(\frac{dy}{dt}, w \right) + a(y,w) = (f,w) - \int_\gamma \omega w \, d\gamma - \int_\gamma u w \, d\gamma +$$

$$+ \int_\Gamma \beta w \, d\Gamma, \ \forall t \in (0,T), \quad (2.2)$$

and

$$a_0\left(y(\cdot,0;u),w(\cdot)\right)=a_0\left(y_0(\cdot),w(\cdot)\right),\ t=0; \tag{2.3}$$

in this case, $V=\left\{v(x,t):\ v|_{\Omega_i}\in W_2^1(\Omega_i),\ i=1,2;\ [v]=0,\ \forall t\in(0,T)\right\}$, the bilinear form $a_0(\cdot,\cdot)$ is specified by formula (1.13'), and

$$a(\varphi,\psi)=\int\limits_{\Omega}\sum_{i,j=1}^{n}k_{ij}\frac{\partial\varphi}{\partial x_j}\frac{\partial\psi}{\partial x_i}dx+\int\limits_{\Gamma}\alpha\varphi\psi\,d\Gamma. \tag{2.3'}$$

The following statement takes place.

Theorem 2.1. *Initial boundary-value problem (1.1), (1.2), (1.5), (2.1) has a unique generalized solution* $y(x,t;u)\in W(0,T)\ \forall u\in\mathcal{U}$.

The validity of Theorem 2.1 is stated by analogy with the proof of Theorem 1.1.

Proceed from equations (2.2) and (2.3), and it is easy to see that $y(u_1)\neq y(u_2)$ under $u_1\neq u_2$. If $\tilde{y}'=\tilde{y}(u')$ and $\tilde{y}''=\tilde{y}(u'')$ are solutions from $W(0,T)$ to problem (2.2), (2.3) under f, β and $\omega=0$ and under a function u that is equal, respectively, to u' and u'', then the inequality

$$\frac{1}{2}\frac{d}{dt}a_0\left(\tilde{y}'-\tilde{y}'',\tilde{y}'-\tilde{y}''\right)+\bar{\alpha}_0\left\|\tilde{y}'-\tilde{y}''\right\|_V^2\le$$

$$\le\left\|u'-u''\right\|_{L_2(\gamma)}\left\|\tilde{y}'-\tilde{y}''\right\|_{L_2(\gamma)}\le c_0\left\|u'-u''\right\|_{L_2(\gamma)}\left\|\tilde{y}'-\tilde{y}''\right\|_V$$

is derived, from which the inequality

$$\frac{1}{2}a_0\left(\tilde{y}'-\tilde{y}'',\tilde{y}'-\tilde{y}''\right)(T)+\bar{\alpha}_0\left\|\tilde{y}'-\tilde{y}''\right\|_{V\times L_2}^2\le$$

$$\le c_0\left\|u'-u''\right\|_{L_2(\gamma)\times L_2}\left\|\tilde{y}'-\tilde{y}''\right\|_{V\times L_2}$$

follows. Therefore, the inequality

$$\left\|\tilde{y}'-\tilde{y}''\right\|_{L_2\times L_2}\le\left\|\tilde{y}'-\tilde{y}''\right\|_{V\times L_2}\le\frac{c_0}{\bar{\alpha}_0}\left\|u'-u''\right\|_{L_2(\gamma)\times L_2}, \tag{2.4}$$

where $\|\varphi\|^2_{L_2(\gamma)\times L_2} = \int_0^T \|\varphi\|^2_{L_2(\gamma)} dt,$ $\|\varphi\|^2_{L_2(\gamma)} = \int_\gamma \varphi^2 d\gamma,$ is obtained that

provides the continuity of the linear functional $L(\cdot)$ and bilinear form $\pi(\cdot,\cdot)$ on $\mathcal{U}$. In this case, the linear functional $L(\cdot)$ and bilinear form $\pi(\cdot,\cdot)$

are specified by expressions (1.32), where $(\varphi,\psi)_{\mathcal{H}} = \int_0^T (\varphi,\psi)dt,$

$$(\varphi,\psi)_{\mathcal{U}} = \int_0^T \int_\gamma \varphi\psi \, d\gamma \, dt \,.$$

Specify the observation in the form of expression (1.8), where $Cy(u) \equiv y(u)$. Bring a value of cost functional (1.10), now in the form

$$J(u) = \int_0^T \int_\Omega \left(y(u) - z_g\right)^2 dx \, dt + \int_0^T \int_\gamma \bar{a}u^2 d\gamma dt \qquad (2.5)$$

in correspondence with every control $u \in \mathcal{U}$; in this case, z_g is a known

element from $L^2(0,T;V),$ $0 < a_0 \le \bar{a}(x) \le a_1 < \infty,$ $a_0, a_1 = \text{const},$ $\bar{a} \in L_2(\gamma)$.

On the basis of [58, Chapter 1, Theorem 1.1], the validity of the following statement is proved.

Theorem 2.2. *If a system state is determined as a solution to problem (2.2), (2.3), then there exists a unique element u of a convex set $\mathcal{U}_\partial$ that is closed in $\mathcal{U}$, and relation (1.11) takes place for u, where the cost functional has the form of expression (2.5).*

As for the control $v \in \mathcal{U}$, the conjugate state $p(v)$ is specified by the relations

$$\sum_{i,j=1}^n \frac{\partial}{\partial x_i}\left(a_{ij}\frac{\partial^2 p}{\partial x_j \partial t}\right) - a\frac{\partial p}{\partial t} - \sum_{i,j=1}^n \frac{\partial}{\partial x_i}\left(k_{ij}\frac{\partial p}{\partial x_j}\right) = y(v) - z_g, \quad (x,t) \in \Omega_T,$$

$$\sum_{i,j=1}^{n}\left(-a_{ij}\frac{\partial^2 p}{\partial x_j \partial t}+k_{ij}\frac{\partial p}{\partial x_j}\right)\cos(v,x_i)=-\alpha p, \quad (x,t)\in\Gamma_T,$$

$$[p]=0, \quad (x,t)\in\gamma_T, \tag{2.6}$$

$$\left[\sum_{i,j=1}^{n}\left(-a_{ij}\frac{\partial^2 p}{\partial x_j \partial t}+k_{ij}\frac{\partial p}{\partial x_j}\right)\cos(v,x_i)\right]=0, \quad (x,t)\in\gamma_T,$$

$$p(x,T)=0, \quad x\in\bar{\Omega}_1\cup\bar{\Omega}_2.$$

Problem (2.6) has the unique generalized solution $p(v)\in L^2(0,T;V)$ as the unique one to equality system like (1.36), (1.37), where the bilinear form $a_0(\cdot,\cdot)$ is written as expression (1.13′) and $a(\cdot,\cdot)$ is specified by expression (2.3′). Assume that $w=y(v)-y(u)$, consider equations (2.2) and (2.3), and the equality

$$\int_0^T\left(y(u)-z_g,y(v)-y(u)\right)dt=-\int_0^T\int_\gamma p(u)(v-u)\,d\gamma\,dt$$

is obtained from equality (1.36). Therefore, the control $u\in\mathcal{U}_\partial$ is optimal if and only if the following inequality is true:

$$\int_0^T\int_\gamma\left(-p(u)+\bar{a}\,u\right)(v-u)\,d\gamma dt\geq 0, \quad \forall v\in\mathcal{U}_\partial. \tag{2.7}$$

Thus, the optimal control $u\in\mathcal{U}_\partial$ is specified by relations (1.36), (1.37), (2.2), (2.3) and (2.7), where the bilinear forms $a_0(\cdot,\cdot)$ and $a(\cdot,\cdot)$ are specified, in their turn, respectively, by expressions (1.13′) and (2.3′). If the constraints are absent, i.e. when $\mathcal{U}_\partial=\mathcal{U}$, then the equality

$$-p(u)+\bar{a}u=0, \quad (x,t)\in\gamma_T,$$

is obtained from condition (2.7). The optimal control

$$u=p/\bar{a}, \quad (x,t)\in\gamma_T, \tag{2.8}$$

is found from the obtained equality.

If the solution $(y, p)^T$ to problem (1.36), (1.37), (2.2), (2.3), (2.8) is smooth enough on $\bar{\Omega}_{lT}$, viz., $y, p|_{\bar{\Omega}_{lT}} \in C^{1,0}(\bar{\Omega}_{lT}) \cap C^{2,0}(\Omega_{lT}) \cap C^{0,1}(\Omega_{lT})$, $l = 1, 2$, then the differential problem of finding the vector-function $(y, p)^T$, that satisfies the equalities

$$-\sum_{i,j=1}^{n} \frac{\partial}{\partial x_i}\left(a_{ij}\frac{\partial^2 y}{\partial x_j \partial t}\right) + a\frac{\partial y}{\partial t} - \sum_{i,j=1}^{n}\frac{\partial}{\partial x_i}\left(k_{ij}\frac{\partial y}{\partial x_j}\right) = f, \ (x,t) \in \Omega_T,$$

$$\sum_{i,j=1}^{n} \frac{\partial}{\partial x_i}\left(a_{ij}\frac{\partial^2 p}{\partial x_j \partial t}\right) - a\frac{\partial p}{\partial t} - \sum_{i,j=1}^{n}\frac{\partial}{\partial x_i}\left(k_{ij}\frac{\partial p}{\partial x_j}\right) - y = -z_g, \ (x,t) \in \Omega_T,$$

$$\sum_{i,j=1}^{n}\left(a_{ij}\frac{\partial^2 y}{\partial x_j \partial t} + k_{ij}\frac{\partial y}{\partial x_j}\right)\cos(v, x_i) = -\alpha y + \beta, \ (x,t) \in \Gamma_T,$$

$$\sum_{i,j=1}^{n}\left(-a_{ij}\frac{\partial^2 p}{\partial x_j \partial t} + k_{ij}\frac{\partial p}{\partial x_j}\right)\cos(v, x_i) = -\alpha p, \ (x,t) \in \Gamma_T,$$

$$[y] = 0, \ [p] = 0, \ (x,t) \in \gamma_T,$$

$$\left[\sum_{i,j=1}^{n}\left(a_{ij}\frac{\partial^2 y}{\partial x_j \partial t} + k_{ij}\frac{\partial y}{\partial x_j}\right)\cos(v, x_i)\right] = \omega + p/\bar{a}, \ (x,t) \in \gamma_T,$$

$$\left[\sum_{i,j=1}^{n}\left(-a_{ij}\frac{\partial^2 p}{\partial x_j \partial t} + k_{ij}\frac{\partial p}{\partial x_j}\right)\cos(v, x_i)\right] = 0, \ (x,t) \in \gamma_T,$$

$$y(x,0) = y_0(x), \ p(x,T) = 0, \ x \in \bar{\Omega}_1 \cup \bar{\Omega}_2,$$

and

$$u = p/\bar{a}, \ (x,t) \in \gamma_T,$$

corresponds to problem (1.36), (1.37), (2.2), (2.3), (2.8).

7.3 CONTROL UNDER CONJUGATION CONDITION WITH BOUNDARY OBSERVATION

Assume that equation (1.1) is specified in the domain Ω_T. On the boundary Γ_T, the boundary condition has the form of expression (1.2). For every control $u \in \mathscr{U} = L_2(\gamma_T)$, determine a system state $y = y(u)$ as a generalized solution to the initial boundary-value problem specified by equation (1.1), boundary condition (1.2), initial condition (1.5) and the conjugation conditions

$$[y] = 0, \ (x,t) \in \gamma_T, \tag{3.1}$$

and

$$\left[\sum_{i,j=1}^{n} \left(a_{ij} \frac{\partial^2 y}{\partial x_j \partial t} + k_{ij} \frac{\partial y}{\partial x_j} \right) \cos(\nu, x_i) \right] = \omega + u, \ (x,t) \in \gamma_T, \tag{3.2}$$

where $\omega = \omega(x,t) \in L_2(\gamma_T)$.

Specify the cost functional by the expression

$$J(u) = \int_0^T \int_\Gamma \left(y(u) - z_g \right)^2 d\Gamma dt + \int_0^T \int_\gamma \bar{a} u^2 d\gamma dt. \tag{3.3}$$

The generalized problem corresponds to initial boundary-value problem (1.1), (1.2), (1.5), (3.1), (3.2) and means to find a function $y(x,t;u) \in$ $\in W(0,T)$ that satisfies equations (2.2) and (2.3) $\forall w(x) \in V_0$; the bilinear forms $a_0(\cdot,\cdot)$ and $a(\cdot,\cdot)$ and spaces V (in $W(0,T)$) and V_0 are specified in point 7.2.

According to Theorem 2.1, initial boundary-value problem (1.1), (1.2), (1.5), (3.1), (3.2) has the unique generalized solution $y(x,t;u) \in W(0,T)$ $\forall u \in \mathscr{U}$.

Proceed from equations (2.2) and (2.3), and it is easy to see that $y(u_1) \neq y(u_2)$ under $u_1 \neq u_2$. Let $\tilde{y}' = \tilde{y}(u')$ and $\tilde{y}'' = \tilde{y}(u'')$ be solutions from $W(0,T)$ to problem (2.2), (2.3) under f, β and $\omega = 0$ and under a function u that is equal, respectively, to u' and u''. Then, consider the embedding theorems, and it is stated on the basis of inequality (2.4) that the inequality

$$\|\tilde{y}' - \tilde{y}''\|_{L_2(\Gamma)\times L_2} \leq c_2 \|u' - u''\|_{L_2(\gamma)\times L_2}$$

is true that provides the continuity of the linear functional $L(\cdot)$ and bilinear form $\pi(\cdot,\cdot)$ (1.32) on $\mathcal{U}$ for the representation of cost functional (3.3) like the representation of cost functional (1.31).

On the basis of [58, Chapter 1, Theorem 1.1], the validity of the following statement is proved.

Theorem 3.1. *Let a system state be determined as a solution to problem (2.2), (2.3). Then, there exists a unique element u of a convex set $\mathcal{U}_\partial$ that is closed in $\mathcal{U}$, and relation (1.11) is true for u, where the cost functional has the form of expression (3.3).*

As for the control $v \in \mathcal{U}$, the conjugate state $p(v)$ is specified by the equalities

$$\sum_{i,j=1}^{n} \frac{\partial}{\partial x_i}\left(a_{ij}\frac{\partial^2 p}{\partial x_j \partial t}\right) - a\frac{\partial p}{\partial t} - \sum_{i,j=1}^{n} \frac{\partial}{\partial x_i}\left(k_{ij}\frac{\partial p}{\partial x_j}\right) = 0, \ (x,t) \in \Omega_T,$$

$$\sum_{i,j=1}^{n}\left(-a_{ij}\frac{\partial^2 p}{\partial x_j \partial t} + k_{ij}\frac{\partial p}{\partial x_j}\right)\cos(v,x_i) = -\alpha p + y(v) - z_g, \ (x,t) \in \Gamma_T,$$

$$[p] = 0, \ (x,t) \in \gamma_T, \tag{3.4}$$

$$\left[\sum_{i,j=1}^{n}\left(-a_{ij}\frac{\partial^2 p}{\partial x_j \partial t} + k_{ij}\frac{\partial p}{\partial x_j}\right)\cos(v,x_i)\right] = 0, \ (x,t) \in \gamma_T,$$

$$p(x,T) = 0, \ x \in \bar{\Omega}_1 \cup \bar{\Omega}_2.$$

Problem (3.4) has the unique generalized solution $p(v) \in W(0,T)$ as the unique one to the following equality system:

$$-a_0\left(\frac{d}{dt}p(v), w\right) + a(p,w) = \int_{\Gamma}\left(y(v) - z_g\right)w\,d\Gamma, \ \forall w \in V_0, \tag{3.5}$$

$$a_0(p,w) = 0, \quad \forall w \in V_0, \ t = T. \tag{3.6}$$

Choose the difference $y(v) - y(u)$ instead of w, consider equality (1.37′) and equation (2.2), and the equality

$$-\int_0^T \int_\gamma p(u)(v-u)d\gamma dt = \int_0^T \int_\Gamma \big(y(u)-z_g\big)\big(y(v)-y(u)\big)d\Gamma dt$$

is obtained from equality (3.5). Therefore, when applied to the considered optimization problem, inequality (1.34) has the form

$$\int_0^T \int_\gamma (-p+\bar{a}u)(v-u)d\gamma dt \geq 0, \ \ \forall v \in \mathcal{U}_\partial. \tag{3.7}$$

Thus, the optimal control $u \in \mathcal{U}_\partial$ is specified by relations (2.2), (2.3), (3.5), (3.6) and (3.7), where the bilinear forms $a_0(\cdot,\cdot)$ and $a(\cdot,\cdot)$ are specified, in their turn, respectively, by expressions (1.13′) and (2.3′). If the constraints are absent, i.e. when $\mathcal{U}_\partial = \mathcal{U}$, then the equality

$$-p+\bar{a}u = 0, \ \ (x,t) \in \gamma_T,$$

i.e.

$$u = p/\bar{a}, \ \ (x,t) \in \gamma_T, \tag{3.8}$$

is obtained from condition (3.7).

If the solution $(y,p)^{\mathrm{T}}$ to problem (2.2), (2.3), (3.5), (3.6), (3.8) is smooth enough on $\bar{\Omega}_{lT}$, $l=1, 2$, then the differential problem of finding the vector-function $(y,p)^{\mathrm{T}}$, that satisfies the equalities

$$-\sum_{i,j=1}^n \frac{\partial}{\partial x_i}\left(a_{ij}\frac{\partial^2 y}{\partial x_j \partial t}\right) + a\frac{\partial y}{\partial t} - \sum_{i,j=1}^n \frac{\partial}{\partial x_i}\left(k_{ij}\frac{\partial y}{\partial x_j}\right) = f, \ (x,t) \in \Omega_T,$$

$$\sum_{i,j=1}^n \frac{\partial}{\partial x_i}\left(a_{ij}\frac{\partial^2 p}{\partial x_j \partial t}\right) - a\frac{\partial p}{\partial t} - \sum_{i,j=1}^n \frac{\partial}{\partial x_i}\left(k_{ij}\frac{\partial p}{\partial x_j}\right) = 0, \ (x,t) \in \Omega_T,$$

$$\sum_{i,j=1}^{n}\left(a_{ij}\frac{\partial^2 y}{\partial x_j\partial t}+k_{ij}\frac{\partial y}{\partial x_j}\right)\cos(\nu,x_i)=-\alpha y+\beta,\ (x,t)\in\Gamma_T,$$

$$\sum_{i,j=1}^{n}\left(-a_{ij}\frac{\partial^2 p}{\partial x_j\partial t}+k_{ij}\frac{\partial p}{\partial x_j}\right)\cos(\nu,x_i)=-\alpha\,p+y-z_g,\ (x,t)\in\Gamma_T,$$

$$[y]=0,\ [p]=0,\ (x,t)\in\gamma_T,$$

$$\left[\sum_{i,j=1}^{n}\left(a_{ij}\frac{\partial^2 y}{\partial x_j\partial t}+k_{ij}\frac{\partial y}{\partial x_j}\right)\cos(\nu,x_i)\right]=\omega+p/\bar{a},\ (x,t)\in\gamma_T,$$

$$\left[\sum_{i,j=1}^{n}\left(-a_{ij}\frac{\partial^2 p}{\partial x_j\partial t}+k_{ij}\frac{\partial p}{\partial x_j}\right)\cos(\nu,x_i)\right]=0,\ (x,t)\in\gamma_T,$$

and

$$y(x,0)=y_0(x),\ p(x,T)=0,\ x\in\bar{\Omega}_1\cup\bar{\Omega}_2,$$

corresponds to problem (2.2), (2.3), (3.5), (3.6), (3.8).

7.4   CONTROL UNDER CONJUGATION CONDITION WITH FINAL OBSERVATION

For every control $u\in\mathcal{U}=L_2(\gamma_T)$, determine a system state $y=y(u)$ as a generalized solution to the initial boundary-value problem specified by equation (1.1), boundary condition (1.2), initial condition (1.5) and conjugation conditions (3.1) and (3.2).

The cost functional is

$$J(u)=a_0\big(y(x,T;u)-z_g,y(x,T;u)-z_g\big)+\int_0^T\int_\gamma\bar{a}u^2d\gamma\,dt,\qquad(4.1)$$

where $0<a_0\le\bar{a}(x)\le a_1<\infty,\ a_0,a_1=\mathrm{const},\ z_g\in V_0,$ and it may be rewritten as

$$J(u) = \pi(u,u) - 2L(u) + a_0\left(z_g(\cdot) - y(\cdot,T;0), z_g(\cdot) - y(\cdot,T;0)\right), \quad (4.1')$$

where

$$\pi(u,v) = a_0\left(y(\cdot,T;u) - y(\cdot,T;0), y(\cdot,T;v) - y(\cdot,T;0)\right) + \int_0^T \int_\gamma \bar{a}\, u v\, d\gamma\, dt$$

and

$$L(v) = a_0\left(z_g(\cdot) - y(\cdot,T;0), y(\cdot,T;v) - y(\cdot,T;0)\right).$$

The generalized problem corresponds to initial boundary-value problem (1.1), (1.2), (1.5), (3.1), (3.2) and means to find a function $y(x,t;u) \in$ $\in W(0,T)$ that satisfies equations (2.2) and (2.3) $\forall w(x) \in V_0$; in this case, the bilinear forms $a_0(\cdot,\cdot)$ and $a(\cdot,\cdot)$ are specified, respectively, by expressions (1.13′) and (2.3′) and the space V_0 is specified, in its turn, in point 7.2.

According to Theorem 2.1, there exists the unique generalized solution to problem (1.1), (1.2), (1.5), (3.1), (3.2). It is stated in point 7.2 that $y(u_1) \neq y(u_2)$ under $u_1 \neq u_2$. If $\tilde{y}' = \tilde{y}(u')$ and $\tilde{y}'' = \tilde{y}(u'')$ are solutions from $W(0,T)$ to problem (2.2), (2.3), where the bilinear forms $a_0(\cdot,\cdot)$ and $a(\cdot,\cdot)$ are specified by formulas (1.13′) and (2.3′), under $f = 0$, $\beta = 0$ and $\omega = 0$ and under a function u that is equal, respectively, to u' and u'', then the inequalities

$$\frac{1}{2}a_0\left(\tilde{y}' - \tilde{y}'', \tilde{y}' - \tilde{y}''\right)(T) + \bar{\alpha}_0\left\|\tilde{y}' - \tilde{y}''\right\|_{V \times L_2}^2 \le$$

$$\le c_0\left\|u' - u''\right\|_{L_2(\gamma)\times L_2}\left\|\tilde{y}' - \tilde{y}''\right\|_{V \times L_2}, \qquad (4.2)$$

$$\bar{\alpha}_0\left\|\tilde{y}' - \tilde{y}''\right\|_{V \times L_2}^2 \le c_0\left\|u' - u''\right\|_{L_2(\gamma)\times L_2}\left\|\tilde{y}' - \tilde{y}''\right\|_{V \times L_2}$$

are true. Proceed from them, and the inequality

$$\left\|\tilde{y}' - \tilde{y}''\right\|_{V \times L_2} \le \frac{c_0}{\bar{\alpha}_0}\left\|u' - u''\right\|_{L_2(\gamma)\times L_2} \qquad (4.3)$$

is obtained. Consider it, and the inequality

$$a_0\left(\tilde{y}' - \tilde{y}'', \tilde{y}' - \tilde{y}''\right)(T) \le \frac{2c_0^2}{\bar{\alpha}_0}\left\|u' - u''\right\|_{L_2(\gamma)\times L_2}^2$$

follows from inequality (4.2).

Then, the inequality

$$\left\|\tilde{y}' - \tilde{y}''\right\|_V(T) \le c_0'\left\|u' - u''\right\|_{L_2(\gamma)\times L_2}, \quad c_0' = \text{const},$$

is derived that provides the continuity of the linear functional $L(\cdot)$ and bilinear form $\pi(\cdot,\cdot)$ on $\mathcal{U}$.

On the basis of [58, Chapter 1, Theorem 1.1], the validity of the following statement is proved.

Theorem 4.1. *If a system state is determined as a solution to problem (2.2), where the bilinear forms $a_0(\cdot,\cdot)$ and $a(\cdot,\cdot)$ are specified, respectively, by formulas (1.13′) and (2.3′), then there exists a unique element u of a convex set $\mathcal{U}_\partial$ that is closed in $\mathcal{U}$, and relation (1.11) takes place for u, where the cost functional has the form of expression (4.1).*

As for the control $v \in \mathcal{U}$, the conjugate state $p(v)$ is specified by the relations

$$\sum_{i,j=1}^{n}\frac{\partial}{\partial x_i}\left(a_{ij}\frac{\partial^2 p}{\partial x_j \partial t}\right) - a\frac{\partial p}{\partial t} - \sum_{i,j=1}^{n}\frac{\partial}{\partial x_i}\left(k_{ij}\frac{\partial p}{\partial x_j}\right) = 0, \ (x,t)\in\Omega_T,$$

$$\sum_{i,j=1}^{n}\left(-a_{ij}\frac{\partial^2 p}{\partial x_j \partial t} + k_{ij}\frac{\partial p}{\partial x_j}\right)\cos(v,x_i) = -\alpha p, \ (x,t)\in\Gamma_T,$$

$$[p] = 0, \ (x,t)\in\gamma_T, \tag{4.4}$$

$$\left[\sum_{i,j=1}^{n}\left(-a_{ij}\frac{\partial^2 p}{\partial x_j \partial t} + k_{ij}\frac{\partial p}{\partial x_j}\right)\cos(v,x_i)\right] = 0, \ (x,t)\in\gamma_T,$$

$$p(x,T;v) = y(x,T;v) - z_g, \quad x \in \bar{\Omega}_1 \cup \bar{\Omega}_2.$$

Problem (4.4) has the unique generalized solution $p(v) \in W(0,T)$ as the unique one to the following system:

$$-a_0\left(\frac{d}{dt}p(v), w\right) + a(p,w) = 0, \quad \forall w \in V_0, \tag{4.5}$$

$$a_0\left(p(\cdot,T;v), w\right) = a_0\left(y(v) - z_g, w\right), \quad \forall w \in V_0. \tag{4.6}$$

Choose the difference $y(v) - y(u)$ instead of w, consider equation (2.2), the equality

$$-\int_0^T a_0\left(\frac{dp(u)}{dt}, y(v) - y(u)\right)dt =$$

$$= \int_0^T a_0\left(p(u), \frac{d}{dt}(y(v) - y(u))\right)dt - a_0\left(p, y(v) - y(u)\right)\big|_{t=T},$$

and the equality

$$-\int_0^T \int_\gamma p(u)(v - u)d\gamma dt - a_0\left(p(u), y(v) - y(u)\right)\big|_{t=T} = 0$$

is obtained from equality (4.5). Consider also equality (4.6), and the equality

$$a_0\left(y(u) - z_g, y(v) - y(u)\right) = -\int_0^T \int_\gamma p(u)(v - u)d\gamma dt, \quad \forall v \in \mathcal{U}_\partial, \tag{4.7}$$

is derived. Take it into account when applied to the considered optimization problem, and inequality (1.33') has the form

$$\int_0^T \int_\gamma (-p(u) + \bar{a}u)(v - u)d\gamma dt \geq 0, \quad \forall v \in \mathcal{U}_\partial. \tag{4.8}$$

Thus, the optimal control $u \in \mathcal{U}_\partial$ is specified by relations (2.2), (2.3), (4.5), (4.6) and (4.8).

If the constraints are absent, i.e. when $\mathcal{U}_\partial = \mathcal{U}$, then the equality

$$-p + \bar{a}u = 0, \ (x,t) \in \gamma_T, \tag{4.8'}$$

is obtained from condition (4.8) along with the optimal control

$$u = p/\bar{a}, \ (x,t) \in \gamma_T. \tag{4.9}$$

If the solution $(y,p)^{\mathrm{T}}$ to problem (2.2), (2.3), (4.5), (4.6), (4.8′) is smooth enough on $\bar{\Omega}_{lT}$, $l = 1, 2$, then the differential problem, specified by equalities (1.1), (1.2), (1.5), (3.1), (3.2) and (4.4), corresponds to problem (2.2), (2.3), (4.5), (4.6), (4.8′), where the control u is found by formula (4.9).

7.5 CONTROL UNDER BOUNDARY CONDITION WITH FINAL OBSERVATION

For every control $u \in L_2(\Gamma_T)$, determine a system state $y = y(u)$ as a generalized solution to the initial boundary-value problem specified by equation (1.1), conjugation conditions (1.3) and (1.4), initial condition (1.5) and the boundary condition

$$\sum_{i,j=1}^{n} \left(a_{ij} \frac{\partial^2 y}{\partial x_j \partial t} + k_{ij} \frac{\partial y}{\partial x_j} \right) \cos(\nu, x_i) = -\alpha y + \beta + u, \ (x,t) \in \Gamma_T. \tag{5.1}$$

The cost functional is

$$J(u) = a_0 \left(y(x,T;u) - z_g, y(x,T;u) - z_g \right) + \int_0^T \int_\Gamma \bar{a}u^2 d\Gamma dt. \tag{5.1'}$$

The generalized problem corresponds to initial boundary-value problem (1.1), (1.3)–(1.5), (5.1) and means to find a function $y(x,t;u) \in W(0,T)$ that satisfies the following equations $\forall w(x) \in V_0$:

$$a_0\left(\frac{dy}{dt}, w\right) + a(y, w) = (f, w) + \int_\Gamma \beta w d\Gamma + \int_\Gamma u w d\Gamma \qquad (5.2)$$

and

$$a_0\left(y(\cdot, 0; u), w(\cdot)\right) = a_0\left(y_0(\cdot), w(\cdot)\right), \quad t = 0; \qquad (5.3)$$

in this case, the spaces $W(0, T)$ and V_0 are specified in point 7.1 and the bilinear forms $a_0(\cdot, \cdot)$ and $a(\cdot, \cdot)$ are specified, in their turn, respectively, by expressions (1.13$'$) and (1.13$''$).

Theorem 5.1. *Initial boundary-value problem (1.1), (1.3)–(1.5), (5.1) has a unique solution* $y(x, t; u)$ $\forall u \in \mathcal{U}$.

Proceed from equation (5.2), and it is easy to see that $y(u_1) \neq y(u_2)$ under $u_1 \neq u_2$. If $\tilde{y}' = \tilde{y}(u')$ and $\tilde{y}'' = \tilde{y}(u'')$ are solutions from $W(0, T)$ to problem (5.2), (5.3) under f, β and $\omega = 0$ and under a function that is equal, respectively, to u' and u'', then the inequality

$$\frac{1}{2}\frac{d}{dt} a_0\left(\tilde{y}' - \tilde{y}'', \tilde{y}' - \tilde{y}''\right) + \bar{\alpha}_0 \|\tilde{y}' - \tilde{y}''\|_V^2 \leq c_0 \|u' - u''\|_{L_2(\Gamma)} \|\tilde{y}' - \tilde{y}''\|_V$$

is derived, from which the inequality

$$\frac{1}{2} a_0\left(\tilde{y}' - \tilde{y}'', \tilde{y}' - \tilde{y}''\right)(T) + \bar{\alpha}_0 \|\tilde{y}' - \tilde{y}''\|_{V \times L_2}^2 \leq$$

$$\leq c_0 \|u' - u''\|_{L_2(\Gamma) \times L_2} \|\tilde{y}' - \tilde{y}''\|_{V \times L_2}$$

follows. Therefore, the inequality

$$\|\tilde{y}' - \tilde{y}''\|_V (T) \leq c_1 \|u' - u''\|_{L_2(\Gamma) \times L_2}$$

is obtained that provides the continuity of the linear functional

$$L(v) = a_0\left(z_g(\cdot) - y(\cdot, T; 0), y(\cdot, T; v) - y(\cdot, T; 0)\right)$$

and bilinear form

$$\pi(u, v) = a_0\left(y(\cdot, T; u) - y(\cdot, T; 0), y(\cdot, T; v) - y(\cdot, T; 0)\right) + \int_0^T \int_\Gamma \bar{a} u v d\Gamma dt$$

on $\mathcal{U}$ for representation (4.1′) of cost functional (5.1′).

On the basis of [58, Chapter 1, Theorem 1.1], the validity of the following statement is proved.

Theorem 5.2. *Let a system state be determined as a solution to problem (5.2), (5.3). Then, there exists a unique element u of a convex set $\mathcal{U}_\partial$ that is closed in $\mathcal{U}$, and relation (1.11) is true for u, where the cost functional has the form of expression (5.1′).*

As for the control $v \in \mathcal{U}$, the conjugate state $p(v)$ is specified by the equalities

$$\sum_{i,j=1}^{n} \frac{\partial}{\partial x_i}\left(a_{ij}\frac{\partial^2 p}{\partial x_j \partial t}\right) - a\frac{\partial p}{\partial t} - \sum_{i,j=1}^{n}\frac{\partial}{\partial x_i}\left(k_{ij}\frac{\partial p}{\partial x_j}\right) = 0, \ (x,t) \in \Omega_T,$$

$$\sum_{i,j=1}^{n}\left(-a_{ij}\frac{\partial^2 p}{\partial x_j \partial t} + k_{ij}\frac{\partial p}{\partial x_j}\right)\cos(v,x_i) = -\alpha\, p, \ (x,t) \in \Gamma_T,$$

$$\left[\sum_{i,j=1}^{n}\left(-a_{ij}\frac{\partial^2 p}{\partial x_j \partial t} + k_{ij}\frac{\partial p}{\partial x_j}\right)\cos(v,x_i)\right] = 0, \ (x,t) \in \gamma_T, \qquad (5.4)$$

$$\left\{\sum_{i,j=1}^{n}\left(-a_{ij}\frac{\partial^2 p}{\partial x_j \partial t} + k_{ij}\frac{\partial p}{\partial x_j}\right)\cos(v,x_i)\right\}^{\pm} = r[p], \ (x,t) \in \gamma_T,$$

$$p(x,T;v) = y(x,T;v) - z_g, \ x \in \bar{\Omega}_1 \cup \bar{\Omega}_2.$$

Problem (5.4) has the unique generalized solution as the unique one to the following system:

$$-a_0\left(\frac{d}{dt}p(v), w\right) + a(p,w) = 0, \quad \forall w \in V_0, \qquad (5.5)$$

$$a_0\left(p(\cdot,T;v), w\right) = a_0\left(y(v) - z_g, w\right), \quad \forall w \in V_0. \qquad (5.6)$$

Choose the difference $y(v) - y(u)$ instead of w, consider equations (5.2) and (5.3), and the equality

$$-a_0\left(p, y(v)-y(u)\right)(T)+\int\limits_{0}^{T}\int\limits_{\Gamma} p(v-u)\,d\Gamma\,dt = 0$$

is obtained. Consider also equality (5.6), and the equality

$$a_0\left(y(u)-z_g, y(v)-y(u)\right)\Big|_{t=T} = \int\limits_{0}^{T}\int\limits_{\Gamma} p(u)(v-u)\,d\Gamma\,dt$$

is derived. Therefore, when applied to the considered optimization problem, inequality (1.33′) has the form

$$\int\limits_{0}^{T}\int\limits_{\Gamma}\left(p(u)+\bar{a}u\right)(v-u)\,d\Gamma\,dt \ge 0, \quad \forall v \in \mathcal{U}_{\partial}. \tag{5.7}$$

Thus, the optimal control $u \in \mathcal{U}_{\partial}$ is specified by relations (5.2), (5.3), (5.5), (5.6) and (5.7). If the constraints are absent, i.e. when $\mathcal{U}_{\partial} = \mathcal{U}$, then the equality

$$p(u)+\bar{a}u = 0, \quad (x,t) \in \Gamma_T,$$

is obtained from inequality (5.7) along with the control

$$u = -p/\bar{a}, \quad (x,t) \in \Gamma_T. \tag{5.8}$$

If the solution $(y,p)^{\mathrm{T}}$ to problem (5.2), (5.3), (5.5), (5.6), (5.8) is smooth enough on $\bar{\Omega}_{lT}$, $l=1,2$, then the differential problem, specified by equalities (1.1), (1.3), (1.4), (1.5), (5.1) and (5.4), corresponds to problem (5.2), (5.3), (5.5), (5.6), (5.8), where the optimal control u is found by formula (5.8).

7.6 CONTROL UNDER BOUNDARY CONDITION WITH OBSERVATION UNDER CONJUGATION CONDITION

For every control $u \in L_2(\Gamma_T)$, determine a system state $y=y(u)$ as a generalized solution to the boundary-value problem specified by equation (1.1), initial condition (1.5), boundary condition (5.1) and the conjugation conditions

$$[y] = 0, \quad (x,t) \in \gamma_T, \tag{6.1}$$

and

$$\left[\sum_{i,j=1}^{n} \left(a_{ij} \frac{\partial^2 y}{\partial x_j \partial t} + k_{ij} \frac{\partial y}{\partial x_j} \right) \cos(\nu, x_i) \right] = \omega, \quad (x,t) \in \gamma_T. \tag{6.2}$$

The cost functional is

$$J(u) = \int_0^T \int_\gamma \left(y(u) - z_g \right)^2 d\gamma dt + \int_0^T \int_\Gamma \bar{a} u^2 d\Gamma dt. \tag{6.3}$$

The generalized problem corresponds to initial boundary-value problem (1.1), (1.5), (5.1), (6.1), (6.2) and means to find a function $y(x,t;u) \in W(0,T)$ that satisfies the following equations $\forall w(x) \in V_0$:

$$a_0 \left(\frac{dy}{dt}, w \right) + a(y, w) = (f, w) + \int_\Gamma \beta w d\Gamma + \int_\Gamma u w d\Gamma \tag{6.4}$$

and

$$a_0 \left(y(\cdot, 0; u), w(\cdot) \right) = a_0 \left(y_0(\cdot), w(\cdot) \right), \quad t = 0; \tag{6.5}$$

in this case, the spaces $W(0,T)$ and V_0 are specified in point 7.2 and the bilinear forms $a_0(\cdot, \cdot)$ and $a(\cdot, \cdot)$ are specified, in their turn, respectively, by expressions (1.13') and (2.3').

Theorem 6.1. *Initial boundary-value problem (1.1), (1.5), (5.1), (6.1), (6.2) has a unique generalized solution* $y(x,t;u) \in W(0,T)$ $\forall u \in \mathcal{U}$.

Proceed from equation (6.4), and it is easy to see that $y(u_1) \ne y(u_2)$ under $u_1 \ne u_2$. If $\tilde{y}' = \tilde{y}(u')$ and $\tilde{y}'' = \tilde{y}(u'')$ are solutions from $W(0,T)$ to problem (6.4), (6.5) under f, β and $\omega = 0$ and under a function u that is equal, respectively, to u' and u'', then:

$$\frac{1}{2} \frac{d}{dt} a_0 \left(\tilde{y}' - \tilde{y}'', \tilde{y}' - \tilde{y}'' \right) + \bar{\alpha}_0 \left\| \tilde{y}' - \tilde{y}'' \right\|_V \le c_0 \left\| u' - u'' \right\|_{L_2(\Gamma)} \left\| \tilde{y}' - \tilde{y}'' \right\|_V.$$

Therefore, the inequality

$$\left\| \tilde{y}' - \tilde{y}'' \right\|_{L_2(\gamma) \times L_2} \le c_1 \left\| u' - u'' \right\|_{L_2(\Gamma) \times L_2}$$

is obtained that provides the continuity of the linear functional

$$L(v) = \left(z_g - y(0), y(v) - y(0)\right)_{L_2(\gamma) \times L_2}$$

and bilinear form

$$\pi(u,v) = \left(y(u) - y(0), y(v) - y(0)\right)_{L_2(\gamma) \times L_2} + \int_0^T \int_\Gamma \bar{a}uv\, d\Gamma dt$$

for the representation

$$J(u) = \pi(u,u) - 2L(u) + \left\| z_g - y(0) \right\|_{L_2(\gamma) \times L_2}^2$$

of cost functional (6.3) on $\mathcal{U}$.

On the basis of [58, Chapter 1, Theorem 1.1], the validity of the following statement is proved.

Theorem 6.2. *Let a system state be determined as a solution to problem (6.4), (6.5). Then, there exists a unique element u of a convex set $\mathcal{U}_\partial$ that is closed in $\mathcal{U}$, and relation (1.11) takes place for u, where the cost functional has the form of expression (6.3).*

As for the control $v \in \mathcal{U}$, the conjugate state $p(v)$ is specified by the equalities

$$\sum_{i,j=1}^n \frac{\partial}{\partial x_i}\left(a_{ij}\frac{\partial^2 p}{\partial x_j \partial t}\right) - a\frac{\partial p}{\partial t} - \sum_{i,j=1}^n \frac{\partial}{\partial x_i}\left(k_{ij}\frac{\partial p}{\partial x_j}\right) = 0, \ (x,t) \in \Omega_T,$$

$$\sum_{i,j=1}^n \left(-a_{ij}\frac{\partial^2 p}{\partial x_j \partial t} + k_{ij}\frac{\partial p}{\partial x_j}\right)\cos(v,x_i) = -\alpha p, \ (x,t) \in \Gamma_T,$$

$$[p] = 0, \ (x,t) \in \gamma_T, \tag{6.6}$$

$$\left[\sum_{i,j=1}^n \left(-a_{ij}\frac{\partial^2 p}{\partial x_j \partial t} + k_{ij}\frac{\partial p}{\partial x_j}\right)\cos(v,x_i)\right] = y(v) - z_g, \ (x,t) \in \gamma_T,$$

$$p(x,T;v) = 0, \ x \in \bar{\Omega}_1 \cup \bar{\Omega}_2.$$

Problem (6.6) has the unique generalized solution as the unique one to the following system:

$$-a_0\left(\frac{d}{dt}p(v),w\right)+a(p,w)=$$

$$=-\left(y(v)-z_g,w\right)_{L_2(\gamma)}, \quad \forall w\in V_0, \quad t\in(0,T),$$ (6.7)

$$a_0\left(p(\cdot,T;v),w(\cdot)\right)=0, \quad \forall w\in V_0.$$ (6.8)

Choose the difference $y(v)-y(u)$ instead of w, consider equations (6.4) and (6.5), and the equality

$$\int_0^T\int_\Gamma p(u)(v-u)d\Gamma dt = -\int_0^T\int_\gamma\left(y(u)-z_g\right)\left(y(v)-y(u)\right)d\gamma dt$$

is obtained. Therefore, when applied to the considered optimization problem, inequality (1.33′) has the form

$$\int_0^T\int_\Gamma\left(-p+\bar a u\right)(v-u)\,d\Gamma dt \ge 0, \quad \forall v\in\mathcal{U}_\partial.$$ (6.9)

Thus, the optimal control $u\in\mathcal{U}_\partial$ is specified by relations (6.4), (6.5) and (6.7)–(6.9). If the constraints are absent, i.e. when $\mathcal{U}_\partial=\mathcal{U}$, then the equality

$$-p+\bar a u = 0, \quad (x,t)\in\Gamma_T,$$

is obtained from inequality (6.9) along with the control

$$u = p/\bar a, \quad (x,t)\in\Gamma_T.$$ (6.10)

If the solution $(y,p)^{\mathrm{T}}$ to problem (6.4), (6.5), (6.7), (6.8), (6.10) is smooth enough on $\bar\Omega_{lT}$, $l=1,2$, then the differential problem, specified by equalities (1.1), (1.5), (5.1), (6.1), (6.2) and (6.6), corresponds to problem (6.4), (6.5), (6.7), (6.8) and (6.10), where the optimal control u is found by formula (6.10).

CONTROL OF A SYSTEM DESCRIBED BY A HYPERBOLIC EQUATION UNDER CONJUGATION CONDITIONS

Let there be the following denotations: Ω is a domain that consists of two open, non-intersecting and strictly Lipschitz domains Ω_1 and Ω_2 from an n-dimensional real linear space R^n; $\Gamma = (\partial\Omega_1 \cup \partial\Omega_2)\backslash\gamma$ $(\gamma = \partial\Omega_1 \cap \cap\partial\Omega_2 \neq \varnothing)$ is a boundary of a domain $\bar{\Omega}$, $\partial\Omega_i$ is a boundary of a domain Ω_i, $i = 1,2$; $\Omega_T = \Omega\times(0,T)$ is a complicated cylinder; $\Gamma_T = \Gamma\times(0,T)$ is the lateral surface of a cylinder $\Omega_T \cup \gamma_T$, $\gamma_T = \gamma\times(0,T)$.

Consider such spaces V and H that $V \subset H$, and V is separable and dense in H.

Identify H with a space dual with respect to it, denote by V' a space that is dual with respect to V, and the following can be written: $V \subset H \subset V'$. By analogy [58], introduce a space $L^2(0,T;V)$ of functions $t \to f(t)$ that map an interval $(0,T)$ into the space V of measurable functions, namely, of such ones that

$$\left(\int_0^T \|f(t)\|_V^2\, dt\right)^{1/2} < \infty.$$

8.1   DISTRIBUTED CONTROL

Assume that the hyperbolic equation

$$\frac{\partial^2 y}{\partial t^2} = \sum_{i,j=1}^{n} \frac{\partial}{\partial x_i}\left(k_{ij}(x)\frac{\partial y}{\partial x_j} \right) + f(x,t) \tag{1.1}$$

is specified in the domain Ω_T, where

$$k_{ij}\big|_{\overline{\Omega}_l} = k_{ji}\big|_{\overline{\Omega}_l} \in C(\overline{\Omega}_l) \cap C^1(\Omega_l), \quad i,j = \overline{1,n};$$

$$\sum_{i,j=1}^{n} k_{ij}\,\xi_i\,\xi_j \geq \alpha_0 \sum_{i=1}^{n} \xi_i^2, \quad \forall \xi_i,\xi_j \in R^1,\ \forall x \in \Omega,\ \alpha_0 = \text{const} > 0;$$

$$f\big|_{\Omega_{lT}} \in C(\Omega_{lT}),\ \Omega_{lT} = \Omega_l \times (0,T),\ l = 1,2;\ |f| < \infty \ .$$

The third boundary condition

$$\sum_{i,j=1}^{n} k_{ij}\frac{\partial y}{\partial x_j}\cos(\nu,x_i) = -\alpha\, y + \beta \tag{1.2}$$

is specified, in its turn, on the boundary Γ_T, where $\alpha = \alpha(x) \geq \alpha^0 > 0$; $\alpha, \beta \in L_2(\Gamma)$, $\alpha^0 = \text{const}$, and ν is a unit vector of an outer normal or simply an outer normal to Γ.

On γ_T, the conjugation conditions are

$$\left[\sum_{i,j=1}^{n} k_{ij}\frac{\partial y}{\partial x_j}\cos(\nu,x_i) \right] = 0 \tag{1.3}$$

and

$$\left\{ \sum_{i,j=1}^{n} k_{ij}\frac{\partial y}{\partial x_j}\cos(\nu,x_i) \right\}^{\pm} = r[y], \tag{1.4}$$

where $0 \leq r = r(x) \leq r_1 < \infty,\ r_1 = \text{const}$, $[\varphi] = \varphi^+ - \varphi^-;\ \varphi^+ = \{\varphi\}^+ = $ $= \varphi(x,t)$ under $(x,t) \in \gamma_T^+ = (\partial\Omega_2 \cap \gamma) \times (0,T);\ \varphi^- = \{\varphi\}^- = \varphi(x,t)$ under

$(x,t) \in \gamma_T^- = (\partial\Omega_1 \cap \gamma) \times (0,T)$; v is an ort of a normal to γ called simply a normal to γ and it is directed into the domain Ω_2.

The initial conditions

$$y(x,0) = y_0(x), \quad x \in \bar{\Omega}_1 \cup \bar{\Omega}_2, \tag{1.5}$$

and

$$\left.\frac{\partial y}{\partial t}\right|_{t=0} = y_1(x), \quad x \in \bar{\Omega}_1 \cup \bar{\Omega}_2, \tag{1.6}$$

where $y_0 \in V_0$ and $y_1 \in H$ are specified under $t = 0$.

Let there be a control Hilbert space $\mathcal{U}$ and operator $B \in \mathcal{L}\left(\mathcal{U}; L^2(0,T;H)\right)$. For every control $u \in \mathcal{U}$, determine a system state $y = y(u) = y(x,t;u)$ as a generalized solution to the problem specified by the equation

$$\frac{\partial^2 y}{\partial t^2} = \sum_{i,j=1}^{n} \frac{\partial}{\partial x_i}\left(k_{ij}\frac{\partial y}{\partial x_j}\right) + f + Bu \tag{1.7}$$

and by conditions (1.2)–(1.6). Further on, without loss of generality, assume the following: $Bu \equiv u$.

Specify the observation by the expression

$$Z(u) = C\,y(u), \quad C \in \mathcal{L}\left(W(0,T); \mathcal{H}\right), \tag{1.8}$$

where $\quad W(0,T) = \left\{v \in L^2(0,T;V): \dfrac{dv}{dt}, \dfrac{d^2v}{dt^2} \in L^2\left(0,T; L_2(\Omega)\right)\right\}$, $V =$

$= \left\{v(x,t): v|_{\Omega_l} \in W_2^1(\Omega_l), \; l=1,2, \; \forall t \in (0,T)\right\}$ and $W_2^1(\Omega_l)$ is the space of the Sobolev functions specified, in their turn, on the domain Ω_l. Specify the operator

$$\mathcal{N} \in \mathcal{L}(\mathcal{U}; \mathcal{U}); \quad (\mathcal{N}u,u)_{\mathcal{U}} \geq v_0 \|u\|_{\mathcal{U}}^2, \quad v_0 = \text{const} > 0. \tag{1.9}$$

Assume the following: $\mathcal{N}u = \bar{a}u$; in this case, $\bar{a}\big|_{\Omega_l} \in C(\Omega_l)$,
$l = 1,2$; $0 < a_0 \leq \bar{a} \leq a_1 < \infty$; a_0, $a_1 = \text{const}$. The cost functional is

$$J(u) = \|C\,y(u) - z_g\|_{\mathcal{H}}^2 + (\mathcal{N}u, u)_{\mathcal{U}}, \qquad (1.10)$$

where z_g is a known element of the space $\mathcal{H}$.

The optimal control problem is to find such an element $u \in \mathcal{U}$ that the condition

$$J(u) = \inf_{v \in \mathcal{U}_\partial} J(v) \qquad (1.11)$$

is met, where $\mathcal{U}_\partial$ is some convex closed subset in $\mathcal{U}$.

Definition 1.1. If an element $u \in \mathcal{U}_\partial$ meets condition (1.11), it is called an optimal control.

The generalized problem corresponds to initial boundary-value problem (1.7), (1.2)–(1.6) and means to find a function $y(x,t;u) \in W(0,T)$ that satisfies the following equations $\forall w(x) \in V_0 = \left\{ v : v\big|_{\Omega_i} \in W_2^1(\Omega_i),\ i = 1,2 \right\}$:

$$\int_\Omega \frac{d^2 y}{dt^2} w\,dx + \int_\Omega \sum_{i,j=1}^n k_{ij} \frac{\partial y}{\partial x_j} \frac{\partial w}{\partial x_i}\,dx + \int_\gamma r[y][w]\,d\gamma +$$

$$+ \int_\Gamma \alpha y w\,d\Gamma = (f, w) + (Bu, w) + \int_\Gamma \beta w\,d\Gamma, \quad \forall t \in (0,T), \qquad (1.12)$$

$$\int_\Omega y(x,0;\cdot)\, w(x)dx = \int_\Omega y_0(x)\, w(x)dx,\ t = 0, \qquad (1.13)$$

and

$$\int_\Omega \frac{dy}{dt}(x,0;\cdot)\, w(x)dx = \int_\Omega y_1(x)\, w(x)dx,\ t = 0. \qquad (1.14)$$

Use the previous results [58, 41, 49, 55, 64, 21] and consider the existence and uniqueness of the solution to problem (1.12)–(1.14). Since Bu and $f \in L^2(0,T; H)$ $\left(H = \{v(x,t) : v \in L_2(\Omega),\ \forall t \in (0,T)\} \right)$, then, without loss of generality, assume the following: $Bu \equiv 0$.

The space V_0 is complete, separable and reflexive [41, 49, 55], and $V_0 \subset L_2(\Omega)$. Choose an arbitrary fundamental system of linearly independent functions $w_k(x)$, $k = 1, 2, \ldots,$ in V_0. For the sake of simplicity, let this system be orthonormal in $L_2(\Omega)$ so that $(w_k, w_l) = \delta_k^l$, where $\delta_k^k = 1$ and $\delta_k^l = 0$ under $k \neq l$.

Assume the following:

$$y_{0m} = \sum_{i=1}^{m} \xi_{im}^0 w_i, \quad y_{0m} \to y_0 \text{ under } m \to \infty,$$

and

$$y_{1m} = \sum_{i=1}^{m} \xi_{im}^1 w_i, \quad y_{1m} \to y_1 \text{ under } m \to \infty.$$

Specify the approximate solution to problem (1.1)–(1.6) by the relations

$$y_m(x, t) = \sum_{i=1}^{m} g_{im}(t) w_i(x), \tag{1.15}$$

$$\left(\frac{\partial^2 y_m}{\partial t^2}, w_j \right) + a(y_m, w_j) = (f, w_j) + \int_{\Gamma} \beta w_j \, d\Gamma, \quad j = \overline{1, m}, \tag{1.16}$$

$$\left(y_m(\cdot, 0), w_j(\cdot) \right) = (y_0, w_j), \quad j = \overline{1, m}, \tag{1.17}$$

and

$$\left(\frac{\partial y_m}{\partial t}(\cdot, 0), w_j(\cdot) \right) = (y_1, w_j), \quad j = \overline{1, m}. \tag{1.18}$$

It is easy to see that Cauchy problem (1.16)–(1.18) has the unique solution for the system of m linear ordinary differential equations as for $g_{im}(t)$; in this case,

$$a(v, z) = \int_{\Omega} \sum_{i,j=1}^{n} k_{ij} \frac{\partial v}{\partial x_j} \frac{\partial z}{\partial x_i} \, dx + \int_{\gamma} r[v][z] \, d\gamma + \int_{\Gamma} \alpha vz \, d\Gamma. \tag{1.19}$$

To obtain an a priori estimate of the function $y_m(x,t)$, multiply, by analogy [64], equality (1.16) by $g'_{jm}(t)$ and find the sum over j for the result. Then:

$$\left(y''_m(\cdot,t), y'_m(\cdot,t)\right) + a\left(y_m, y'_m\right) = \left(f, y'_m\right) + \left(\beta, y'_m\right)_{L_2(\Gamma)},$$

i.e.:

$$\frac{d}{dt}\left\{\left\|y'_m(\cdot,t)\right\|^2 + a\left(y_m, y_m\right)\right\} = 2\left(f, y'_m\right) + 2\left(\beta, y'_m\right)_{L_2(\Gamma)}.$$

Therefore,

$$\left\|y'_m(\cdot,t)\right\|^2 + a\left(y_m(\cdot,t),\ y_m(\cdot,t)\right) = \left\|y'_m\right\|^2(0) + 2\int_0^t \left(f(\cdot,\tau), y'_m(\cdot,\tau)\right)d\tau +$$

$$+2\int_0^t (\beta, y'_m)_{L_2(\Gamma)}\, d\tau + a\left(y_m(\cdot,0),\ y_m(\cdot,0)\right). \tag{1.20}$$

The inequality

$$\left\|\frac{\partial y_m}{\partial t}(\cdot,0)\right\|^2 = \left(y_1(\cdot), \frac{\partial y_m}{\partial t}(\cdot,0)\right) \le \|y_1\|\cdot\left\|\frac{\partial y_m}{\partial t}(\cdot,0)\right\|$$

follows from equality (1.18), i.e.:

$$\left\|\frac{\partial y_m}{\partial t}(\cdot,0)\right\| \le \|y_1\|. \tag{1.21}$$

Let $w_j(x)$, $j=1,2,\ldots$, be eigenfunctions of the spectral problem that means to find

$$\{\lambda, w\} \in R^1 \times V_0:\ a(w,z) = \lambda(w,z),\ \forall z \in V_0,$$

and they are such eigenfunctions that meet the condition $(w_i, w_j) = \delta_i^j$. Then, it is easy to see the following:

$$a\left(y_m(\cdot,0),\ y_m(\cdot,0)\right) \le a\left(y_0, y_0\right). \tag{1.22}$$

Introduce the denotation

$$z_m(t) = \left\|y'_m\right\|^2(t) + \left\|y_m\right\|_V^2(t), \tag{1.23}$$

where $\|\varphi\|_V^2 = \sum\limits_{i=1}^{2} \|\varphi\|_{W_2^1(\Omega_i)}^2$. Consider relations (1.21)–(1.23), the Cauchy-Bunyakovsky and generalized Friedrichs inequalities [21], and the inequality

$$z_m(t) \le c_1 \left\{ \|y_1\|^2 + a(y_0, y_0) + \int_0^t \|f(\cdot, \tau)\|^2 d\tau \right\} + \|\beta\|_{L_2(\Gamma)}^2 + c_1 \int_0^t z_m(\tau) d\tau$$

is obtained from equality (1.20). The inequalities

$$\| y_m' \|^2 (t) + \| y_m \|_V^2 (t) \le c,$$

$$\| y_m' \|^2 (t) + \| y_m \|^2 (t) \le c, \tag{1.24}$$

where $c = \text{const} > 0$, follow from the latter inequality by virtue of the Gronwall lemma [41, 64].

Hence, y_m and y_m' are still in the bounded sets, respectively, in $L^2(0, T; V)$ and $L^2(0, T; H)$. That is why such subsequence y_ν of the sequence y_m can be chosen, that the convergences

$$y_\nu \to y \text{ in } L^2(0, T; V),$$

$$y_\nu' \to z \text{ in } L^2(0, T; H) \tag{1.25}$$

are weak. By virtue of [64], $z = y'$. It follows from expressions (1.25) that $y_\nu(0) \to y(0)$ is weak and, since $y_\nu(0) = y_{0\nu} \to y_0$ in V_0, then $y(0) = y_0$.

It remains to show that, when constructed in such a way, the function y is the solution to problem (1.12)–(1.14).

Assume the following: $\varphi \in C^1([0, T])$ and $\varphi(T) = 0$. Introduce the denotation $\varphi_j(t) = \varphi(t)\omega_j(x)$. Multiply equality (1.16) by the function $\varphi(t)$ and find the integral from 0 to T of the result:

$$\int\limits_0^T \left\{ -\left(y'_m, \varphi'_j \right) + a\left(y_m, \varphi_j \right) \right\} dt =$$

$$= \int\limits_0^T \left(f, \varphi_j \right) dt + \int\limits_0^T \left(\beta, \varphi_j \right)_{L_2(\Gamma)} dt + \left(y_{1m}, \varphi_j \right)(0).$$

Pass to the limit under $m \to \infty$, and the following equality is obtained:

$$\int\limits_0^T \left\{ -\left(y', w_j \right) \varphi' + a\left(y, w_j \right) \varphi \right\} dt = \int\limits_0^T \left(f, w_j \right) \varphi \, dt +$$

$$+ \int\limits_0^T \left(\beta, w_j \right)_{L_2(\Gamma)} \varphi \, dt + \left(y_1, w_j \right) \varphi(0). \tag{1.26}$$

If $\varphi \in D\left((0,T) \right)$ is used [58], then the equality

$$\left(\frac{d^2 y}{dt^2}, w_j \right) + a\left(y, w_j \right) = \left(f, w_j \right) + \left(\beta, w_j \right)_{L_2(\Gamma)}, \quad j = 1, 2, \ldots,$$

is derived from equality (1.26).

Proceed from the obtained relations, and the equality

$$\left(\frac{d^2 y}{dt^2}, \tilde{w}_n \right) + a\left(y, \tilde{w}_n \right) = \left(f, \tilde{w}_n \right) + \left(\beta, \tilde{w}_n \right)_{L_2(\Gamma)} \tag{1.27}$$

follows that is true $\forall \tilde{w}_n \in \bigcup\limits_{n=1}^\infty V_{0n}$ and, therefore, $\forall w \in V_0$, where V_{0n} is

the n-dimensional subspace of the space V_0, and the functions of the set

$\{ w_i(x) \}_{i=1}^n$ make up the basis of V_{0n}. Hence, the equalities

$$\left(\frac{d^2 y}{dt^2}, w \right) + a(y, w) = (f, w) + (\beta, w)_{L_2(\Gamma)}, \quad \forall w \in V_0, \, t \in (0,T), \tag{1.28}$$

$$\int\limits_\Omega y' w \, dx = \int\limits_\Omega y_1 w \, dx, \quad \forall w \in V_0, \, t = 0, \tag{1.29}$$

and

$$\int_\Omega yw\,dx = \int_\Omega y_0 w\,dx, \quad \forall w \in V_0, \; t=0, \tag{1.30}$$

take place.

Thus, $y(x,t)$ is the generalized solution to initial boundary-value problem (1.1)–(1.6). Illustrate its uniqueness. Let $y(x,t)$ be the generalized solution to the problem. Assume the following in equality (1.28): $w = \dfrac{dy}{dt}$. Find the integral of equality (1.28) over $\tau \in (0,t)$ now, and the equality

$$z^2(t) = 2\int_0^t \left(f, \frac{dy}{dt}\right)d\tau + 2\int_0^t \left(\beta, \frac{dy}{dt}\right)_{L_2(\Gamma)}(\tau)\,d\tau + z^2(0) \tag{1.31}$$

is obtained, where $z^2(t) = \left\|\dfrac{dy}{dt}\right\|^2 + a(y,y)$, $z^2(0) = \|y_1\|^2 + a(y_0,y_0)$.

Transform the second addend in the right-hand side of equality (1.31). Consider the embedding theorems and ε-inequality, and the inequality

$$2\int_0^t \left(\beta, \frac{dy}{dt}\right)_{L_2(\Gamma)}(\tau)\,d\tau \le 2\left|(\beta,y)_{L_2(\Gamma)}\right|_0^t \le$$

$$\le 2\varepsilon\, c_0\|y\|_V^2(t) + \frac{1}{2\varepsilon}\|\beta\|_{L_2(\Gamma)}^2(t) + \|\beta\|_{L_2(\Gamma)}^2(0) + \|y_0\|_{L_2(\Gamma)}^2$$

is derived. Use it, and the inequality

$$\tilde{z}^2(t) \le c_0'\left(\int_0^T \|f\|^2 d\tau + \|\beta\|_{L_2(\Gamma)}^2 + \|y_1\|^2 + \right.$$

$$\left. + a(y_0,y_0) + \|y_0\|_{L_2(\Gamma)}^2 + \int_0^t \tilde{z}^2(\tau)\,d\tau \right), \tag{1.32}$$

where $c_0' = \text{const} > 0$, $\tilde{z}^2(t) = \left\| \dfrac{dy}{dt} \right\|^2 (t) + \| y \|_V^2$, is obtained from equality (1.31).

The inequality

$$\| y' \|^2 (t) + \| y \|_V^2 (t) \leq$$

$$\leq c_1 \left\{ \int_0^T \| f \|^2 d\tau + \| \beta \|_{L_2(\Gamma)}^2 + \| y_1 \|^2 + a(y_0, y_0) + \| y_0 \|_{L_2(\Gamma)}^2 \right\} \quad (1.33)$$

follows from inequality (1.32) by virtue of the Gronwall lemma.

Hence, the unique generalized solution to problem (1.1)–(1.6) is zero under $f = 0$, $\beta = 0$, $y_0 = 0$ and $y_1 = 0$. Therefore, the validity of the following statement is proved.

Theorem 1.1. *Problem (1.1)–(1.6) has a unique generalized solution* $y(x,t) \in W(0,T)$.

It is easy to see that $y(u_1) \neq y(u_2)$ under $u_1 \neq u_2$ $(Bu_1 \neq Bu_2)$. Let $\tilde{y}' = \tilde{y}(u')$ and $\tilde{y}'' = \tilde{y}(u'')$ be solutions from $W(0,T)$ to problem (1.12)–(1.14) under $f = 0$ and $\beta = 0$ and under a function $u = u(x,t)$ that is equal, respectively, to u' and u''. Then, the inequality

$$\| \tilde{y}' - \tilde{y}'' \|_{L_2 \times L_2} \leq c \| u' - u'' \|_{L_2 \times L_2} \quad (1.34)$$

is obtained from inequality (1.33).

Rewrite cost functional (1.10) as

$$J(v) = \pi(v,v) - 2L(v) + \int_0^T \| z_g - y(0) \|^2 dt, \quad (1.35)$$

where

$$\pi(u,v) = \big(y(u) - y(0), \ y(v) - y(0) \big)_{\mathcal{H}} + (\overline{a}\, u, v)_{\mathcal{U}},$$

$$L(v) = \big(z_g - y(0), \ y(v) - y(0) \big)_{\mathcal{H}}; \quad (1.36)$$

in this case, $(z,v)_{\mathcal{H}} = \int\limits_0^T (z,v)dt, \quad (z,v)_{\mathcal{U}} = \int\limits_0^T (z,v)dt, \quad (z,v) = \int\limits_\Omega zvdx.$

Inequality (1.34) provides the continuity of the linear functional $L(\cdot)$ and bilinear form $\pi(\cdot,\cdot)$ on $\mathcal{U}$.

On the basis of [58, Chapter 1, Theorem 1.1], the validity of the following statement is proved.

Theorem 1.2. *Let a system state be determined as a solution to problem (1.12)–(1.14). Then, there exists a unique element u of a convex set $\mathcal{U}_\partial$ that is closed in $\mathcal{U}$, and*

$$J(u) = \inf_{v \in \mathcal{U}_\partial} J(v) \tag{1.37}$$

takes place for u.

A control $u \in \mathcal{U}_\partial$ is optimal if and only if the inequality

$$\langle J'(u), v-u \rangle \geq 0, \quad \forall v \in \mathcal{U}_\partial,$$

is true, i.e.:

$$\left(y(u) - z_g, \; y(v) - y(u) \right)_{\mathcal{H}} + (\mathcal{N} u, v - u)_{\mathcal{U}} \geq 0. \tag{1.38}$$

As for the control $v \in \mathcal{U}$, the conjugate state $p(v)$ is specified by the relations

$$\frac{\partial^2 p}{\partial t^2} - \sum_{i,j=1}^n \frac{\partial}{\partial x_i}\left(k_{ij} \frac{\partial p}{\partial x_j} \right) = y(v) - z_g, \; (x,t) \in \Omega_T,$$

$$\sum_{i,j=1}^n k_{ij} \frac{\partial p}{\partial x_j} \cos(v, x_i) = -\alpha p, \; (x,t) \in \Gamma_T,$$

$$\left[\sum_{i,j=1}^n k_{ij} \frac{\partial p}{\partial x_j} \cos(v, x_i) \right] = 0, \; (x,t) \in \gamma_T, \tag{1.39}$$

$$\left\{ \sum_{i,j=1}^n k_{ij} \frac{\partial p}{\partial x_j} \cos(v, x_i) \right\}^{\pm} = r[p], \; (x,t) \in \gamma_T,$$

$$p(x,T;v) = 0, \quad \frac{\partial p}{\partial t}(x,T;v) = 0, \quad x \in \bar{\Omega}_1 \cup \bar{\Omega}_2.$$

By virtue of Theorem 1.1, problem (1.39) has the unique generalized solution $p(v) \in L^2(0,T;V)$, $\dfrac{dp(v)}{dt}, \dfrac{d^2 p(v)}{dt^2} \in L^2(0,T;H)$ as the unique one to the following equality system:

$$\left(\frac{d^2 p}{dt^2}, w\right) + a(p,w) = \left(y - z_g, w\right), \quad \forall w \in V_0, \ t \in (0,T),$$

$$\int_{\Omega} \frac{dp}{dt} w\, dx = 0, \quad \forall w \in V_0, \ t = T, \qquad (1.40)$$

$$\int_{\Omega} p w\, dx = 0, \quad \forall w \in V_0, \ t = T.$$

Use the difference $y(v) - y(u)$ instead of w in the first equality of system (1.40) under $v = u$, find the integral from 0 to T of the result, and the equality

$$\int_0^T \left(\frac{d^2 p(u)}{dt^2}, y(v) - y(u)\right) dt + \int_0^T a\left(p(u), y(v) - y(u)\right) dt =$$

$$= \int_0^T \left(y(u) - z_g, y(v) - y(u)\right) dt \qquad (1.41)$$

is obtained.

Under $\varphi, \psi \in L^2(0,T;V)$; $\varphi', \psi' \in L^2(0,T;H)$, $\varphi'', \psi'' \in L^2(0,T;V')$, the equality [58]

$$\int_0^T (\varphi'', \psi)\, dt = (\varphi'(T), \psi(T)) - (\varphi'(0), \psi(0)) -$$

$$- (\varphi(T), \psi'(T)) + (\varphi(0), \psi'(0)) + \int_0^T (\varphi, \psi'')\, dt \qquad (1.42)$$

is derived. Take it and equation (1.12) into account, and the equality

$$\int_0^T \left(y(u) - z_g, y(v) - y(u) \right) dt = \int_0^T \left(p(u), (v-u) \right) dt$$

follows from equality (1.41).

Therefore, optimality condition (1.38) for the control $u \in \mathcal{U}_\partial$ is equivalent to the inequality

$$\int_0^T \left(p(u) + \bar{a}u, v - u \right) dt \geq 0, \quad \forall v \in \mathcal{U}_\partial. \tag{1.43}$$

Thus, the optimal control $u \in \mathcal{U}_\partial$ is specified by relations (1.12)–(1.14), (1.40) and (1.43). If the constraints are absent, i.e. when $\mathcal{U}_\partial = \mathcal{U}$, then the equality

$$p(u) + \bar{a}\, u = 0, \quad (x,t) \in \Omega_T, \tag{1.44}$$

follows from inequality (1.43). The control

$$u = -p/\bar{a}, \quad (x,t) \in \Omega_T, \tag{1.45}$$

is found from equality (1.44).

If the solution $(y,p)^{\mathrm{T}}$ to problem (1.12)–(1.14), (1.40), (1.44) is smooth enough on $\bar{\Omega}_{lT}$, $l = 1,2$, viz., $y|_{\bar{\Omega}_{lT}}$, $p|_{\bar{\Omega}_{lT}} \in C^{1,0}(\bar{\Omega}_{lT}) \cap$ $\cap C^{2,0}(\Omega_{lT}) \cap C^{0,2}(\Omega_{lT})$, $l = 1,2$, then the differential problem of finding the vector-function $(y,p)^{\mathrm{T}}$, that satisfies the equalities

$$\frac{\partial^2 y}{\partial t^2} - \sum_{i,j=1}^n \frac{\partial}{\partial x_i}\left(k_{ij} \frac{\partial y}{\partial x_j} \right) + p/\bar{a} = f, \quad (x,t) \in \Omega_T,$$

$$\frac{\partial^2 p}{\partial t^2} - \sum_{i,j=1}^n \frac{\partial}{\partial x_i}\left(k_{ij} \frac{\partial p}{\partial x_i} \right) - y = -z_g, \quad (x,t) \in \Omega_T,$$

$$\sum_{i,j=1}^{n} k_{ij} \frac{\partial y}{\partial x_j} \cos(v, x_i) = -\alpha y + \beta, \quad (x,t) \in \Gamma_T,$$

$$\sum_{i,j=1}^{n} k_{ij} \frac{\partial p}{\partial x_j} \cos(v, x_i) = -\alpha p, \quad (x,t) \in \Gamma_T,$$

$$\left[\sum_{i,j=1}^{n} k_{ij} \frac{\partial y}{\partial x_j} \cos(v, x_i)\right] = 0, \quad \left[\sum_{i,j=1}^{n} k_{ij} \frac{\partial p}{\partial x_j} \cos(v, x_i)\right] = 0, \quad (x,t) \in \gamma_T,$$

$$\left\{\sum_{i,j=1}^{n} k_{ij} \frac{\partial y}{\partial x_j} \cos(v, x_i)\right\}^{\pm} = r[y], \quad (x,t) \in \gamma_T,$$

$$\left\{\sum_{i,j=1}^{n} k_{ij} \frac{\partial p}{\partial x_j} \cos(v, x_i)\right\}^{\pm} = r[p], \quad (x,t) \in \gamma_T,$$

$$y(x,0) = y_0(x), \quad p(x,T) = 0, \quad x \in \bar{\Omega}_1 \cup \bar{\Omega}_2,$$

$$\frac{\partial y}{\partial t}(x,0) = y_1(x), \quad \frac{\partial p}{\partial t}(x,T) = 0, \quad x \in \bar{\Omega}_1 \cup \bar{\Omega}_2,$$

and

$$u = -p/\bar{a}, \quad (x,t) \in \Omega_T,$$

corresponds to problem (1.12)–(1.14), (1.40), (1.44).

8.2   CONTROL UNDER CONJUGATION CONDITION

Assume that equation (1.1) is specified in the domain Ω_T. On the boundary Γ_T, the boundary condition has the form of expression (1.2).

For every control $u \in \mathcal{U} = L_2(\gamma)$, determine a state $y = y(u)$ as a generalized solution to the initial boundary-value problem specified, in its turn, by equation (1.1), boundary condition (1.2), initial conditions (1.5) and (1.6) and the conjugation conditions

$$[y] = 0, \quad (x,t) \in \gamma_T, \tag{2.1}$$

and

$$\left[\sum_{i,j=1}^{n} k_{ij} \frac{\partial y}{\partial x_j} \cos(\nu, x_i) \right] = \omega + u, \quad (x,t) \in \gamma_T, \tag{2.2}$$

where $\omega = \omega(x) \in L_2(\gamma)$.

Since there exists the generalized solution $y(x,t;u)$ to initial boundary-value problem (1.1), (1.2), (1.5), (1.6), (2.1), (2.2), then such solution is reasonable on $\bar{\Omega}_{lT}, l = 1,2$. The generalized problem corresponds to initial boundary-value problem (1.1), (1.2), (1.5), (1.6), (2.1), (2.2) and means to find a function $y(x, t; u) \in W(0,T)$ that satisfies the following equalities

$$\forall w(x) \in V_0 = \left\{ v(x) : v|_{\Omega_i} \in W_2^1(\Omega_i), \ i = 1,2; \ [v]|_\gamma = 0 \right\} :$$

$$\int_\Omega \frac{d^2 y}{dt^2} w \, dx + \int_\Omega \sum_{i,j=1}^{n} k_{ij} \frac{\partial y}{\partial x_j} \frac{\partial w}{\partial x_i} dx + \int_\Gamma \alpha y w d\Gamma =$$

$$= (f, w) - \int_\gamma \omega w d\gamma - \int_\gamma u w d\gamma + \int_\Gamma \beta w d\Gamma,$$

$$\int_\Omega y(x,0;u) w(x) dx = \int_\Omega y_0(x) w(x) dx, \tag{2.3}$$

$$\int_\Omega \frac{dy}{dt}(x,0;u) w(x) dx = \int_\Omega y_1(x) w(x) dx;$$

in this case,

$$V = \left\{ v(x,t) : v|_{\Omega_i} \in W_2^1(\Omega_i), \frac{dv}{dt}\Big|_{\Omega_i}, \frac{d^2 v}{dt^2}\Big|_{\Omega_i} \in L_2(\Omega_i), \ \forall t \in [0,T], \ i = 1,2; \right.$$

$$\left. [v]|_\gamma = 0, \ \forall t \in [0,T] \right\}.$$

The forthcoming statement takes place.

Theorem 2.1. *Initial boundary-value problem (1.1), (1.2), (1.5), (1.6), (2.1), (2.2) has a unique generalized solution* $y(x,t;u) \in W(0,T)$ $\forall u \in \mathcal{U}$.

The validity of Theorem 2.1 is stated by analogy with the proof of Theorem 1.1.

Proceed from equalities (2.3), and it is easy to see on the basis of the first one that $y(u_1) \neq y(u_2)$ under $u_1 \neq u_2$. If $\tilde{y}' = \tilde{y}(u')$ and $\tilde{y}'' = \tilde{y}(u'')$ are solutions in $W(0,T)$ to problem (2.3) under f, ω and $\beta = 0$ and under a function u that is equal, respectively, to u' and u'', then, assume $w = \dfrac{d}{dt}(\tilde{y}' - \tilde{y}'')$ in the first equality of system (2.3), find the integral of its first equality over $\tau \in (0,t)$ now, and the equality

$$z^2(t) = -2\int_0^t \left((u'-u''), \frac{d}{dt}(\tilde{y}'-\tilde{y}'')\right)_{L_2(\gamma)} dt + z^2(0) \qquad (2.4)$$

is obtained, where

$$z^2(t) = \left\|\frac{d}{dt}(\tilde{y}'-\tilde{y}'')\right\|^2 + a(\tilde{y}'-\tilde{y}'', \ \tilde{y}'-\tilde{y}'')(t)$$

and

$$a(y,w) = \int_\Omega \sum_{i,j=1}^n k_{ij}\frac{\partial y}{\partial x_j}\frac{\partial w}{\partial x_i}dx + \int_\Gamma \alpha y w d\Gamma.$$

Since the equality

$$\int_0^t \left(\varphi, \frac{d}{dt}\psi\right)_{L_2(\gamma)} d\tau = (\varphi,\psi)_{L_2(\gamma)}\Big|_0^t - \int_0^t \left(\frac{d}{dt}\varphi,\psi\right)_{L_2(\gamma)}(\tau)d\tau \qquad (2.4')$$

takes place and the ε- and Cauchy-Bunyakovsky inequalities and embedding theorems are taken into account, then the inequality

$$\tilde{z}^2(t) \leq c_1\|u'-u''\|^2_{L_2(\gamma)}, \qquad (2.5)$$

where $\tilde{z}^2(t) = \left\|\dfrac{d}{dt}(\tilde{y}'-\tilde{y}'')\right\|^2 + \|\tilde{y}'-\tilde{y}''\|^2_V$ is derived from equality (2.4).

Therefore,

$$\left\|\frac{d}{dt}(\tilde{y}'-\tilde{y}'')\right\|^2 + \|\tilde{y}'-\tilde{y}''\|_V^2(t) \le c_1 \|u'-u''\|_{L_2(\gamma)}^2, \tag{2.6}$$

i.e.

$$\|\tilde{y}'-\tilde{y}''\|_V^2(t) \le c_1 \|u'-u''\|_{L_2(\gamma)}^2. \tag{2.7}$$

The continuity of the linear functional $L(\cdot)$ and bilinear form $\pi(\cdot,\cdot)$ for cost functional (1.35) follows from inequality (2.7); in this case, $(z,v)_{\mathcal{U}} =$

$$= \int_0^T (z,v)_{L_2(\gamma)} dt.$$ Specify the observation in the form of expression (1.8),

where $C y(u) \equiv y(u)$. Bring a value of cost functional (1.10) in correspondence with every control $u \in \mathcal{U}$; in this case, z_g is a known element from $L^2(0,T;V)$,

$$J(u) = \int_0^T\int_\Omega \big(y(u)-z_g\big)^2 dx\,dt + \int_0^T\int_\gamma \bar{a}u^2 d\gamma\,dt$$

and

$$0 < a_0 \le \bar{a} \le a_1 < \infty, \; a_0, a_1 = \text{const}, \; \bar{a} \in L_2(\gamma).$$

On the basis of [58, Chapter 1, Theorem 1.1], the validity of the following statement is proved.

Theorem 2.2. *If a system state is determined as a solution to problem (2.3), then there exists a unique element u of a convex set $\mathcal{U}_\partial$ that is closed in $\mathcal{U}$, and relation (1.37) takes place for u.*

As for the control $v \in \mathcal{U}$, the conjugate state $p(v)$ is specified by the relations

$$\frac{\partial^2 p}{\partial t^2} - \sum_{i,j=1}^n \frac{\partial}{\partial x_i}\left(k_{ij}\frac{\partial p}{\partial x_j}\right) = y(v)-z_g, \; (x,t)\in\Omega_T,$$

$$\sum_{i,j=1}^n k_{ij}\frac{\partial p}{\partial x_j}\cos(v,x_i) = -\alpha\,p, \; (x,t)\in\Gamma_T,$$

$$[p] = 0, \quad (x,t) \in \gamma_T,$$

$$\left[\sum_{i,j=1}^{n} k_{ij} \frac{\partial p}{\partial x_j} \cos(\nu, x_i) \right] = 0, \quad (x,t) \in \gamma_T, \tag{2.8}$$

$$p(x,T) = 0, \quad x \in \bar{\Omega}_1 \cup \bar{\Omega}_2,$$

$$\frac{\partial p}{\partial t}(x,T) = 0, \quad x \in \bar{\Omega}_1 \cup \bar{\Omega}_2.$$

Problem (2.8) has the unique generalized solution $p(v) \in L^2(0,T;V)$ as the unique one to equality system like (1.40), where

$$a(v,z) = \int_{\Omega} \sum_{i,j=1}^{n} k_{ij} \frac{\partial v}{\partial x_j} \frac{\partial z}{\partial x_i} dx + \int_{\Gamma} \alpha v z \, d\Gamma. \tag{2.9}$$

If the difference $y(v) - y(u)$ is used instead of w in the first equality of system (1.40), where $v = u$ and the bilinear form $a(\cdot,\cdot)$ is specified by expression (2.9), then equality (1.41) is present after taking the integral from 0 to T of the result. Consider equality (1.42), the first equality of system (2.3) and expression (2.9), and the equality

$$\int_0^T \big(y(u) - z_g, y(v) - y(u) \big) dt = - \int_0^T \int_{\gamma} p(u) \, (v - u) d\gamma \, dt \tag{2.10}$$

is obtained from equality (1.41).

Therefore, the control $u \in \mathcal{U}_\partial$ is optimal if and only if the following inequality is true:

$$\int_0^T \int_{\gamma} \big(-p(u) + \bar{a}u \big)(v - u) d\gamma \, dt \geq 0, \quad \forall v \in \mathcal{U}_\partial. \tag{2.11}$$

Thus, the optimal control $u \in \mathcal{U}_\partial$ is specified by inequality (2.11) and relations (2.3) and (1.40), where the bilinear form $a(\cdot,\cdot)$ is specified, in its turn, by expression (2.9).

8.3 CONTROL UNDER CONJUGATION CONDITION WITH BOUNDARY OBSERVATION

Assume that equation (1.1) is specified in the domain Ω_T. On the boundary Γ_T, the boundary condition has the form of expression (1.2). For every control $u \in \mathcal{U} = L_2(\gamma)$, determine a state $y(x,t;u)$ as a generalized solution to initial boundary-value problem specified by equation (1.1), boundary condition (1.2), initial conditions (1.5) and (1.6) and the conjugation conditions

$$[y] = 0, \quad (x,t) \in \gamma_T, \tag{3.1}$$

and

$$\left[\sum_{i,j=1}^{n} k_{ij} \frac{\partial y}{\partial x_j} \cos(v, x_i) \right] = \omega + u, \quad (x,t) \in \gamma_T, \tag{3.2}$$

where $\omega = \omega(x) \in L_2(\gamma)$.

The cost functional is

$$J(u) = \int_0^T \int_\Gamma \left(y(u) - z_g \right)^2 d\Gamma\, dt + \int_0^T \int_\gamma \bar{a} u^2 d\gamma\, dt. \tag{3.3}$$

The generalized problem corresponds to initial boundary-value problem (1.1), (1.2), (1.5), (1.6), (3.1), (3.2) and means to find a function $y(x,t;u) \in W(0,T)$ that satisfies equation system (2.3) $\forall w(x) \in V_0$; the spaces $W(0,T)$ and V_0 are specified in point 8.2.

According to Theorem 2.1, initial boundary-value problem (1.1), (1.2), (1.5), (1.6), (3.1), (3.2) has the unique generalized solution $y(x,t;u) \in W(0,T) \ \forall u \in \mathcal{U}$.

Proceed from the first equality of system (2.3), and it is easy to see that $y(u_1) \neq y(u_2)$ under $u_1 \neq u_2$. Let $\tilde{y}' = \tilde{y}(u')$ and $\tilde{y}'' = \tilde{y}(u'')$ be solutions from $W(0,T)$ to problem (2.3) under f, ω and $\beta = 0$ and under a function u that is equal, respectively, to u' and u''. Inequality (2.7) is true, from which, by virtue of the embedding theorems, the inequality

$$\int_0^T \int_\Gamma (\tilde{y}' - \tilde{y}'')^2 \, d\Gamma \, dt \le c_1' \int_0^T \|u' - u''\|_{L_2(\gamma)}^2 \, dt$$

is obtained that is the evidence of the fact that the linear functional $L(\cdot)$ and bilinear form $\pi(\cdot,\cdot)$ of the cost functional

$$J(u) = \int_0^T \int_\Gamma \left(y(u) - z_g \right)^2 d\Gamma \, dt + \int_0^T \int_\gamma \bar{a} u^2 \, d\gamma \, dt =$$

$$= \pi(u,u) - 2L(u) + \int_0^T \|z_g - y(0)\|_{L_2(\Gamma)}^2 \, dt \qquad (3.4)$$

are continuous on $\mathcal{U}$; in this case, $0 < a_0 \le \bar{a}(x) \le a_1$; $a_0, a_1 =$ $= \mathrm{const}$, $\bar{a}(x) \in L_2(\gamma)$,

$$\pi(u,v) = \left(y(u) - y(0), \ y(v) - y(0) \right)_{L_2(\Gamma) \times L_2} + (\bar{a} u, v)_{L_2(\gamma) \times L_2}$$

and

$$L(v) = \left(z_g - y(0), \ y(v) - y(0) \right)_{L_2(\Gamma) \times L_2}.$$

On the basis of [58, Chapter 1, Theorem 1.1], the validity of the following statement is proved.

Theorem 3.1. *Let a system state be determined as a solution to problem (2.3). Then, there exists a unique element u of a convex set $\mathcal{U}_\partial$ that is closed in $\mathcal{U}$, and relation (1.37) takes place for u, where cost functional has the form of expression (3.3).*

As for the control $v \in \mathcal{U}$, the conjugate state $p(v)$ is specified by the relations

$$\frac{\partial^2 p}{\partial t^2} - \sum_{i,j=1}^n \frac{\partial}{\partial x_i} \left(k_{ij} \frac{\partial p}{\partial x_j} \right) = 0, \ (x,t) \in \Omega_T,$$

$$\sum_{i,j=1}^n k_{ij} \frac{\partial p}{\partial x_j} \cos(\nu, x_i) = -\alpha \, p + y(v) - z_g, \ (x,t) \in \Gamma_T,$$

$$[p] = 0, \ (x,t) \in \gamma_T, \qquad (3.5)$$

$$\left[\sum_{i,j=1}^{n} k_{ij} \frac{\partial p}{\partial x_j} \cos(v, x_i) \right] = 0, \ (x,t) \in \gamma_T,$$

$$p(x,T) = 0, \ x \in \bar{\Omega}_1 \cup \bar{\Omega}_2,$$

$$\frac{\partial p}{\partial t}(x,T) = 0, \ x \in \bar{\Omega}_1 \cup \bar{\Omega}_2.$$

Problem (3.5) has the unique generalized solution $p(v) \in L^2(0,T;V)$ as the unique one to the equality system

$$\int_{\Omega} \frac{d^2 p}{dt^2} w\, dx + \int_{\Omega} \sum_{i,j=1}^{n} k_{ij} \frac{\partial p}{\partial x_j} \frac{\partial w}{\partial x_i} dx + \int_{\Gamma} \alpha p\, w\, d\Gamma =$$

$$= \int_{\Gamma} \left(y(v) - z_g \right) w\, d\Gamma, \ \forall w \in V_0, \ t \in (0,T), \tag{3.6}$$

$$\int_{\Omega} \frac{dp}{dt} w\, dx = 0, \ \forall w \in V_0, \ t = T,$$

$$\int_{\Omega} p\, w\, dx = 0, \ \forall w \in V_0, \ t = T;$$

in this case, the spaces V_0 and V are specified in point 8.2.

The following statement takes place.

Theorem 3.2. *Initial boundary-value problem (3.5) has a unique generalized solution.*

Use the difference $y(v) - y(u)$ instead of w in the first equality of system (3.6), where $v = u$, find the integral from 0 to T of the result, and the equality

$$\int_0^T \left(\frac{d^2 p(u)}{dt^2}, y(v) - y(u) \right) dt + \int_0^T a\left(p(u), y(v) - y(u) \right) dt =$$

$$= \int_0^T \int_\Gamma \left(y(u) - z_g \right) \left(y(v) - y(u) \right) d\Gamma dt \tag{3.7}$$

is obtained, where the bilinear form $a(\cdot,\cdot)$ is specified by expression (2.9).

Consider equality (1.42), the first equality of system (2.3) and expression (2.9), and the equality

$$\int_0^T \int_\Gamma \left(y(u) - z_g \right) \left(y(v) - y(u) \right) d\Gamma dt = - \int_0^T \int_\gamma p(u) \, (v - u) d\gamma \, dt \tag{3.8}$$

is derived from equality (3.7). Therefore, the control $u \in \mathcal{U}_\partial$ is optimal if and only if the following inequality is true:

$$\left(-p(u) + \bar{a} \, u, v - u \right)_{L_2(\gamma) \times L_2} \geq 0, \ \forall v \in \mathcal{U}_\partial. \tag{3.9}$$

Thus, the optimal control $u \in \mathcal{U}_\partial$ is specified by equalities (2.3) and (3.6) and inequality (3.9). If the solution $(y,p)^{\mathrm{T}}$ to problem (2.3), (3.6), (3.9) is smooth enough on $\bar{\Omega}_{lT}$, then the problem of finding the vector-function $(y,p)^{\mathrm{T}}$, that satisfies inequality (3.9) and the equalities

$$\frac{\partial^2 y}{\partial t^2} - \sum_{i,j=1}^n \frac{\partial}{\partial x_i} \left(k_{ij} \frac{\partial y}{\partial x_j} \right) = f(x,t), \ (x,t) \in \Omega_T,$$

$$\frac{\partial^2 p}{\partial t^2} - \sum_{i,j=1}^n \frac{\partial}{\partial x_i} \left(k_{ij} \frac{\partial p}{\partial x_j} \right) = 0, \ (x,t) \in \Omega_T,$$

$$\sum_{i,j=1}^n k_{ij} \frac{\partial y}{\partial x_j} \cos(v, x_i) = -\alpha y + \beta, \ (x,t) \in \Gamma_T,$$

$$\sum_{i,j=1}^n k_{ij} \frac{\partial p}{\partial x_j} \cos(v, x_i) = -\alpha p + y - z_g, \ (x,t) \in \Gamma_T,$$

$$[y] = 0, \ [p] = 0, \ (x,t) \in \gamma_T,$$

$$\left[\sum_{i,j=1}^{n} k_{ij} \frac{\partial y}{\partial x_j} \cos(\nu, x_i)\right] = \omega + u, \quad (x,t) \in \gamma_T,$$

$$\left[\sum_{i,j=1}^{n} k_{ij} \frac{\partial p}{\partial x_j} \cos(\nu, x_i)\right] = 0, \quad (x,t) \in \gamma_T,$$

$$y(x,0) = y_0(x), \quad p(x,T) = 0, \quad x \in \bar{\Omega}_1 \cup \bar{\Omega}_2,$$

and

$$\frac{\partial y}{\partial t}(x,0) = y_1(x), \quad \frac{\partial p}{\partial t}(x,T) = 0, \quad x \in \bar{\Omega}_1 \cup \bar{\Omega}_2,$$

corresponds to problem (2.3), (3.6), (3.9).

8.4 CONTROL UNDER CONJUGATION CONDITION WITH FINAL OBSERVATION

Assume that equation (1.1) is specified in the domain Ω_T. On the boundary Γ_T, the boundary condition has the form of expression (1.2). For every control $u \in \mathcal{U} = L_2(\gamma)$, determine a state $y(x,t;u)$ as a generalized solution to the initial boundary-value problem specified by equation (1.1), boundary condition (1.2), initial conditions (1.5) and (1.6) and conjugation conditions (2.1) and (2.2).

8.4.1 Final Observation with Taking Sight on a State

The cost functional is

$$J(u) = \int_{\Omega} \left(y(x,T;u) - z_g(x)\right)^2 dx + \int_0^T \int_{\gamma} \bar{a} u^2 \, d\gamma \, dt, \tag{4.1}$$

where $0 < a_0 \le \bar{a} \le a_1 < \infty$, $a_0, a_1 = \text{const}$, and it may be rewritten as

$$J(u) = \pi(u,u) - 2L(u) + \int_\Omega \left(z_g(x) - y(x,T;0) \right)^2 dx ; \qquad (4.1')$$

in this case,

$$\pi(u,v) = \left(y(\cdot,T;u) - y(\cdot,T;0), \; y(\cdot,T;v) - y(\cdot,T;0) \right) + \int_0^T \int_\gamma \bar{a} u v \, d\gamma \, dt$$

and

$$L(v) = \left(z_g(\cdot) - y(\cdot,T;0), \; y(\cdot,T;v) - y(\cdot,T;0) \right).$$

The generalized problem corresponds to initial boundary-value problem (1.1), (1.2), (1.5), (1.6), (2.1), (2.2) and means to find a function $y(x,t;u) \in W(0,T)$ that satisfies equality system (2.3) $\forall w \in V_0$; the spaces $W(0,T)$ and V_0 are specified in point 8.2.

Theorem 2.1 takes place. It is stated in point 8.3 that $y(u_1) \ne y(u_2)$ under $u_1 \ne u_2$. If $\tilde{y}' = \tilde{y}(u')$ and $\tilde{y}'' = \tilde{y}(u'')$ are solutions from $W(0,T)$ to problem (2.3) under f, β and $\omega = 0$ and under a function u that is equal, respectively, to u' and u'', then inequality (2.6) is true. Consider the generalized Friedrichs inequality [21], and the inequality

$$\mu \| \tilde{y}' - \tilde{y}'' \|^2 (T) \le a \left(\tilde{y}' - \tilde{y}'', \tilde{y}' - \tilde{y}'' \right) \le c_0 \| u' - u'' \|^2_{L_2(\gamma) \times L_2}$$

follows from inequality (2.6). Therefore, the linear functional $L(\cdot)$ and bilinear form $\pi(\cdot, \cdot)$ in representation (4.1') of cost functional (4.1) are continuous on $\mathcal{U}$.

On the basis of [58, Chapter 1, Theorem 1.1], the validity of the following statement is proved.

Theorem 4.1. *If a system state is determined as a solution to problem (1.1), (1.2), (1.5), (1.6), (2.1), (2.2), then there exists a unique element u of a convex set $\mathcal{U}_\partial$ that is closed in $\mathcal{U}$, and relation (1.37) takes place for u, where the cost functional has the form of expression (4.1).*

As for the control $v \in \mathcal{U}$, the conjugate state $p(v)$ is specified by the equalities

$$\frac{\partial^2 p}{\partial t^2} - \sum_{i,j=1}^{n} \frac{\partial}{\partial x_i}\left(k_{ij}\frac{\partial p}{\partial x_j}\right) = 0, \quad (x,t) \in \Omega_T,$$

$$\sum_{i,j=1}^{n} k_{ij}\frac{\partial p}{\partial x_j}\cos(v,x_i) = -\alpha p, \quad (x,t) \in \Gamma_T,$$

$$[p] = 0, \quad (x,t) \in \gamma_T,$$

$$\left[\sum_{i,j=1}^{n} k_{ij}\frac{\partial p}{\partial x_j}\cos(v,x_i)\right] = 0, \quad (x,t) \in \gamma_T, \tag{4.2}$$

$$p(x,T;v) = 0, \quad x \in \bar{\Omega}_1 \cup \bar{\Omega}_2,$$

$$\frac{\partial p}{\partial t}(x,T;v) = y(x,T;v) - z_g, \quad x \in \bar{\Omega}_1 \cup \bar{\Omega}_2.$$

The generalized solution to problem (4.2) is the solution to the equality system

$$\left(\frac{d^2 p}{dt^2}, w\right) + a(p,w) = 0, \quad \forall w \in V_0, \quad t \in (0,T),$$

$$\left(\frac{dp}{dt}, w\right) = \left(y(\cdot,T;u) - z_g, w(\cdot)\right), \quad \forall w \in V_0, \quad t = T, \tag{4.3}$$

$$\left(p(\cdot,T;v), w(\cdot)\right) = 0, \quad \forall w \in V_0, \quad t = T,$$

where $a(z,v) = \int_\Omega \sum_{i,j=1}^{n} k_{ij}\frac{\partial z}{\partial x_j}\frac{\partial v}{\partial x_i}dx + \int_\Gamma \alpha z v \, d\Gamma$.

Use the difference $y(v) - y(u)$ instead of w in the first equality of system (4.3) under $v = u$, consider equality (1.42), and the equality

$$\left(y(\cdot,t;u)-z_g,y(v)-y(u)\right)(T)-\int_0^T\int_\gamma p(u)(v-u)d\gamma\,dt=0$$

is obtained, i.e. $\left(y(\cdot,t;u)-z_g,y(v)-y(u)\right)(T)=\left(p(u),v-u\right)_{L_2(\gamma)\times L_2}.$

Therefore, the control $u\in\mathcal{U}_\partial$ is optimal if and only if the following inequality is true:

$$\int_0^T\int_\gamma\left(p(u)+\bar{a}u\right)(v-u)d\gamma\,dt\geq 0,\ \forall v\in\mathcal{U}_\partial. \tag{4.4}$$

Thus, the optimal control $u\in\mathcal{U}_\partial$ is specified by relations (2.3), (4.3) and (4.4).

8.4.2 Final Observation with Taking Sight on a System State Changing Rate

The cost functional is

$$J(u)=\int_\Omega\left(\frac{d}{dt}y(x,T;u)-z_g(x)\right)^2dx+\int_0^T\int_\gamma\bar{a}u^2d\gamma\,dt \tag{4.5}$$

and it may be rewritten as

$$J(u)=\pi(u,u)-2L(u)+\int_\Omega\left(z_g(x)-y'(x,T;0)\right)^2dx,$$

where

$$\pi(u,v)=\left(y'(\cdot,T;u)-y'(\cdot,T;0),\ y'(\cdot,T;v)-y'(\cdot,T;0)\right)+\int_0^T\int_\gamma\bar{a}uv\,d\gamma\,dt,$$

$$L(v)=\left(z_g(\cdot)-y'(\cdot,T;0),\ y'(\cdot,T;v)-y'(\cdot,T;0)\right). \tag{4.5'}$$

The generalized problem is specified in point 8.4.1 and corresponds to initial boundary-value problem (1.1), (1.2), (1.5), (1.6), (2.1), (2.2). Theorem 2.1 takes place. It is stated in point 8.3 that $y(u_1) \neq y(u_2)$ under $u_1 \neq u_2$. If $\tilde{y}' = \tilde{y}(u')$ and $\tilde{y}'' = \tilde{y}(u'')$ are solutions from $W(0,T)$ to problem (2.3) under f, β and $\omega = 0$ and under a function u that is equal, respectively, to u' and u'', then inequality (2.6) is true under $t = T$.

The obtained inequality shows that the function $\dfrac{d\tilde{y}}{dt}$ is continuously dependent on the control $u \in \mathcal{U}$. Therefore, the linear functional $L(\cdot)$ and bilinear form $\pi(\cdot, \cdot)$ of cost functional (4.5) are continuous on $\mathcal{U}$.

On the basis of [58, Chapter 1, Theorem 1.1], the validity of the following statement is proved.

Theorem 4.2. *If a system state is determined as a generalized solution to problem (1.1), (1.2), (1.5), (1.6), (2.1), (2.2), then there exists a unique element u of a convex set $\mathcal{U}_{\partial}$ that is closed in $\mathcal{U}$, and relation (1.37) takes place for u, where cost functional has the form of expression (4.5).*

As for the control $v \in \mathcal{U}$, the conjugate state $p(v)$ is specified by the first four equalities of system (4.2) and by the conditions

$$p(x, T; v) = \frac{d}{dt} y(x, T; v) - z_g, \quad x \in \bar{\Omega}_1 \cup \bar{\Omega}_2, \qquad (4.6)$$

and

$$\frac{dp}{dt}(x, T; v) = 0, \quad x \in \bar{\Omega}_1 \cup \bar{\Omega}_2. \qquad (4.7)$$

The present differential problem has the unique solution as the unique one to the following equality system:

$$\left(\frac{d^2 p}{dt^2}, w \right) + a(p, w) = 0, \quad \forall w \in V_0, \quad t \in (0, T),$$

$$(p(\cdot, T; v), w(\cdot)) = \left(\frac{d}{dt} y(\cdot, T; v) - z_g, w(\cdot) \right), \quad \forall w \in V_0, \quad t = T, \qquad (4.8)$$

$$\left(\frac{dp}{dt}(\cdot, T; v), w(\cdot) \right) = 0, \quad \forall w \in V_0, \quad t = T.$$

Use the difference $y(v) - y(u)$ instead of w in the first equality of system (4.8) under $v = u$, and the equality

$$-\left(\frac{d}{dt} y(\cdot,t;u) - z_g, \frac{d\left(y(v) - y(u)\right)}{dt}\right)(T) - \int_0^T \int_\gamma p(u)(v - u)\,d\gamma\,dt = 0$$

is obtained.

Therefore, the optimality condition for the control $u \in \mathcal{U}_\partial$ is

$$\int_0^T \int_\gamma (-p + \bar{a}u)\,(v - u)d\gamma\,dt \geq 0, \quad \forall v \in \mathcal{U}_\partial. \tag{4.9}$$

Thus, the optimal control $u \in \mathcal{U}_\partial$ is specified by the relations (2.3), (4.8) and (4.9).

8.5 CONTROL UNDER BOUNDARY CONDITION WITH OBSERVATION ON A THIN INCLUSION

For every control $u \in \mathcal{U} = L_2(\Gamma)$, determine a state $y(x,t;u)$ as a generalized solution to the initial boundary-value problem specified by equation (1.1), initial conditions (1.5) and (1.6), the conjugation conditions

$$[y] = 0, \quad (x,t) \in \gamma_T,$$

and

$$\left[\sum_{i,j=1}^n k_{ij} \frac{\partial y}{\partial x_j} \cos(\nu, x_i)\right] = \omega, \quad (x,t) \in \gamma_T, \tag{5.1}$$

and the boundary condition

$$\sum_{i,j=1}^n k_{ij} \frac{\partial y}{\partial x_j} \cos(\nu, x_i) = -\alpha y + \beta + u, \quad (x,t) \in \Gamma_T. \tag{5.1'}$$

The cost functional is

$$J(u) = \int_0^T \int_\gamma \left(y(u) - z_g \right)^2 d\gamma\, dt + \int_0^T \int_\Gamma \bar{a} u^2 d\Gamma dt, \qquad (5.2)$$

where $0 < a_0 \le \bar{a} \le a_1 < \infty$; $a_0, a_1 = \text{const}$, and it may be rewritten as

$$J(u) = \pi(u,u) - 2L(u) + \int_0^T \int_\gamma \left(z_g - y(\cdot,t;0) \right)^2 d\gamma\, dt; \qquad (5.2')$$

in this case,

$$\pi(u,v) = \int_0^T \left(y(\cdot,t;u) - y(\cdot,t;0), y(\cdot,t;v) - y(\cdot,t;0) \right)_{L_2(\gamma)} dt + \int_0^T \int_\Gamma \bar{a} u v d\Gamma dt$$

and

$$L(v) = \left(z_g - y(\cdot,t;0), y(\cdot,t;v) - y(\cdot,t;0) \right)_{L_2(\gamma) \times L_2}.$$

The generalized problem corresponds to initial boundary-value problem (1.1), (1.5), (1.6), (5.1), (5.1') and means to find a function $y(x,t;u) \in W(0,T)$ that satisfies the following equality system $\forall w \in V_0$:

$$\left(\frac{d^2 y}{dt^2}, w \right) + a(y,w) = (f,w) - \int_\gamma \omega w d\gamma +$$

$$+ \int_\Gamma \beta w d\Gamma + \int_\Gamma u w d\Gamma, \quad \forall w \in V_0, \ t \in (0,T),$$

$$\left(y(\cdot,0;u), w(\cdot) \right) = \left(y_0(\cdot), w(\cdot) \right), \quad \forall w \in V_0, \qquad (5.3)$$

$$\left(\frac{dy}{dt}(\cdot,0;u), w(\cdot) \right) = \left(y_1(\cdot), w(\cdot) \right), \quad \forall w \in V_0.$$

The forthcoming statement takes place.

Theorem 5.1. *Initial boundary-value problem* (1.1), (1.5), (1.6), (5.1), *(5.1') has a unique generalized solution* $y(x,t;u) \in W(0,T)$ $\forall u \in \mathcal{U}$.

Proceed from the first equality of system (5.3), and it is easy to see that $y(u_1) \ne y(u_2)$ under $u_1 \ne u_2$. Let $\tilde{y}' = \tilde{y}(u')$ and $\tilde{y}'' = \tilde{y}(u'')$ be solutions from $W(0,T)$ to problem (5.3) under f, ω and $\beta = 0$ and under a function u that is equal, respectively, to u' and u''. The inequality

$$z^2(t) \leq c_1 \|u' - u''\|^2_{L_2(\Gamma)},\qquad(5.4)$$

where $z^2(t) = \left\|\dfrac{d}{dt}(\tilde{y}' - \tilde{y}'')\right\|^2 + \|\tilde{y}' - \tilde{y}''\|^2_V$, is true and it shows that the

function $\tilde{y}$ linearly depends upon the control $u \in \mathcal{U}$ on γ.

On the basis of [58, Chapter 1, Theorem 1.1], the validity of the following statement is proved.

Theorem 5.2. *If a system state is determined as a generalized solution to problem (1.1), (1.5), (1.6), (5.1), (5.1'), then there exists a unique element u of a convex set $\mathcal{U}_\partial$ that is closed in $\mathcal{U}$, and relation (1.37) takes place for u, where the cost functional has the form of expression (5.2).*

As for the control $v \in \mathcal{U}$, the conjugate state $p(v)$ is specified by the first three equalities of system (4.2) and by the conditions

$$\left[\sum_{i,j=1}^n k_{ij} \frac{\partial p}{\partial x_j} \cos(v, x_i)\right] = y(v) - z_g, \ (x,t) \in \gamma_T,$$

$$p(x,T) = 0, \ x \in \bar{\Omega}_1 \cup \bar{\Omega}_2,\qquad(5.5)$$

$$\frac{\partial p}{\partial t}(x,T) = 0, \ x \in \bar{\Omega}_1 \cup \bar{\Omega}_2.$$

The present initial boundary-value problem has the unique generalized solution $p(v) \in W(0,T)$ as the unique one to the following equality system:

$$\left(\frac{d^2 p}{dt^2}, w\right) + a(p, w) = -\left(y(v) - z_g, w\right)_{L_2(\gamma)}, \ \forall w \in V_0, \ t \in (0,T),$$

$$(p, w) = 0, \ \forall w \in V_0, \ t = T,\qquad(5.6)$$

$$\left(\frac{dp}{dt}, w\right) = 0, \ \forall w \in V_0, \ t = T;$$

in this case, the spaces V_0 and $W(0,T)$ are specified in point 8.2.

Use the difference $y(v) - y(u)$ instead of w in the first equality of system (5.6) under $v = u$, find the integral from 0 to T of the result, consider equality (1.42) and system (5.3), and the equality

$$\left(p(u), v-u\right)_{L_2(\Gamma)\times L_2} = -\left(y(u)-z_g, y(v)-y(u)\right)_{L_2(\gamma)\times L_2}$$

is obtained.

Therefore, the optimality condition for the control $u \in \mathcal{U}$ is

$$\left(-p+\bar{a}u, v-u\right)_{L_2(\Gamma)\times L_2} \geq 0, \quad \forall v \in \mathcal{U}_\partial. \qquad (5.7)$$

Thus, the optimal control $u \in \mathcal{U}_\partial$ is specified by relations (5.3), (5.6) and (5.7).

8.6 CONTROL UNDER BOUNDARY CONDITION WITH FINAL OBSERVATION

For every control $u \in \mathcal{U} = L_2(\Gamma)$, determine a system state $y(x,t;u)$ as a generalized solution to initial boundary-value problem specified by equalities (1.1), (1.5), (1.6), (5.1) and (5.1').

8.6.1 Taking Sight on a System State

The cost functional is

$$J(u) = \int_\Omega \left(y(x,T;u) - z_g(x)\right)^2 dx + \int_0^T \int_\Gamma \bar{a}u^2 d\Gamma dt \qquad (6.1)$$

and it may be represented by expression (4.1'); in this case,

$$\pi(u,v) = \left(y(\cdot,T;u) - y(\cdot,T;0), \; y(\cdot,T;v) - y(\cdot,T;0)\right) + \int_0^T \int_\Gamma \bar{a}uv \, d\Gamma \, dt$$

and
$$L(v) = \left(z_g(\cdot) - y(\cdot, T; 0), \ y(\cdot, T; v) - y(\cdot, T; 0)\right).$$

The generalized problem corresponds to initial boundary-value problem (1.1), (1.5), (1.6), (5.1), (5.1') and means to find a function $y(x, t; u) \in W(0, T)$ that satisfies equality system (5.3) $\forall w \in V_0$.

Proceed from the first equality of system (5.3), and the inequality

$$\left\|\frac{d}{dt}(\tilde{y}' - \tilde{y}'')\right\|^2 + \left\|\tilde{y}' - \tilde{y}''\right\|_V^2 (t) \leq c_1 \left\|u' - u''\right\|_{L_2(\Gamma)}^2 \qquad (6.2)$$

is obtained, i.e.

$$\left\|\tilde{y}' - \tilde{y}''\right\|^2 (T) \leq c_1 \left\|u' - u''\right\|_{L_2(\Gamma)}^2. \qquad (6.3)$$

Therefore, the linear functional $L(\cdot)$ and bilinear form $\pi(\cdot, \cdot)$ of cost functional (6.1) are continuous on $\mathcal{U}$.

On the basis of [58, Chapter 1, Theorem 1.1], the validity of the following statement is proved.

Theorem 6.1. *If a system state is determined as a generalized solution to problem (1.1), (1.5), (1.6), (5.1), (5.1'), then there exists a unique element u of a convex set $\mathcal{U}_\partial$ that is closed in $\mathcal{U}$, and relation (1.37) is true for u, where the cost functional has the form of expression (6.1).*

As for the control $v \in \mathcal{U}$, the conjugate state $p(v)$ is specified by system (4.2). The generalized solution to problem (4.2) is the solution to equality system (4.3).

Use the difference $y(v) - y(u)$ instead of w in the first equality of system (4.3) under $v = u$, consider equality (1.42), and the equality

$$\left(y(\cdot, t; u) - z_g(\cdot), y(v) - y(u)\right)(T) + \left(p(u), v - u\right)_{L_2(\Gamma) \times L_2} = 0$$

is obtained, i.e.

$$\left(y(\cdot, t; u) - z_g(\cdot), y(v) - y(u)\right)(T) = -\left(p(u), v - u\right)_{L_2(\Gamma) \times L_2}, \forall v \in \mathcal{U}_\partial.$$

Therefore, the control $u \in \mathcal{U}_\partial$ is optimal if and only if the following inequality is true:

$$\int_0^T \int_\Gamma (-p(u) + \bar{a}\,u)(v - u)d\Gamma\,dt \geq 0, \ \forall v \in \mathcal{U}_\partial. \tag{6.4}$$

Thus, the optimal control $u \in \mathcal{U}_\partial$ is specified by relations (4.3), (5.3) and (6.4).

8.6.2 Taking Sight on a System State Changing Rate

The cost functional is specified by the expression

$$J(u) = \int_\Omega \left(\frac{d}{dt} y(x, T; u) - z_g(x) \right)^2 dx + \int_0^T \int_\Gamma \bar{a}u^2 d\Gamma\,dt. \tag{6.4'}$$

Inequality like (5.4) is true. Therefore, the inequality

$$\left\| \frac{d}{dt}(\tilde{y}' - \tilde{y}'') \right\|^2 (T) \leq c_1 \|u' - u''\|^2_{L_2(\Gamma) \times L_2}$$

is obtained that shows the continuous dependence of the function $\dfrac{d\tilde{y}}{dt}$ on the control $u \in \mathcal{U}$, and the continuity of the linear functional $L(\cdot)$ and bilinear form $\pi(\cdot, \cdot)$ of cost functional (6.4') is thus provided on $\mathcal{U}$.

On the basis of [58, Chapter 1, Theorem 1.1], the validity of the following statement is proved.

Theorem 6.2. *If a system state is determined as a generalized solution to problem (1.1), (1.5), (1.6), (5.1), (5.1'), then there exists a unique element u of a convex set $\mathcal{U}_\partial$ that is closed in $\mathcal{U}$, and relation (1.37) takes place for u, where the cost functional has the form of expression (6.4').*

As for the control $v \in \mathcal{U}$, the conjugate state $p(v)$ is specified as a solution to the initial boundary-value problem specified, in its turn, by the first four equalities of system (4.2) and conditions (4.6) and (4.7). The generalized problem is written by equalities (4.8) and corresponds to such initial boundary-value problem.

Use the difference $y(v) - y(u)$ instead of w in the first equality of system (4.8) under $v = u$, and

$$-\left(\frac{d}{dt} y(\cdot, t; u) - z_g, \frac{d(y(v) - y(u))}{dt} \right)(T) +$$

$$+ \int_0^T \int_\Gamma p(u)(v - u) d\Gamma dt, \quad \forall v \in \mathcal{U}_\partial.$$

Therefore, the optimality condition for the control $u \in \mathcal{U}_\partial$ is

$$\int_0^T \int_\Gamma (p(u) + \bar{a}u)(v - u) d\Gamma dt \geq 0, \quad \forall v \in \mathcal{U}_\partial. \tag{6.5}$$

Thus, the optimal control $u \in \mathcal{U}_\partial$ is specified by relations (4.8), (5.3) and (6.5).

CONTROL OF A SYSTEM DESCRIBED BY A PSEUDOHYPERBOLIC EQUATION UNDER CONJUGATION CONDITIONS

Let there be the following denotations: Ω is a domain that consists of two open, non-intersecting and strictly Lipschitz domains Ω_1 and Ω_2 from an n-dimensional real linear space R^n; $\Gamma = (\partial\Omega_1 \cup \partial\Omega_2)/\gamma$ $(\gamma = \partial\Omega_1 \cap \cap\partial\Omega_2 \neq \varnothing)$ is a boundary of a domain $\bar\Omega$, $\partial\Omega_i$ is a boundary of a domain Ω_i, $i = 1,2$; $\Omega_T = \Omega \times (0,T)$ is a complicated cylinder; $\Gamma_T = \Gamma \times (0,T)$ is the lateral surface of a cylinder $\Omega_T \cup \gamma_T$, $\gamma_T = \gamma \times (0,T)$.

Consider such spaces V and H that $V \subset H$, and V is separable and dense in H. By analogy [58], introduce a space $L^2(0,T;V)$ of functions $t \to f(t)$ that map an interval $(0,T)$ into the space V of measurable functions, namely, of such ones that

$$\left(\int_0^T \|f\|_V^2 \, dt \right)^{1/2} < \infty.$$

9.1 DISTRIBUTED CONTROL

Assume that the pseudohyperbolic equation [67]

$$\frac{\partial^2 y}{\partial t^2} - \sum_{i,j=1}^n \frac{\partial}{\partial x_i}\left(a_{ij}(x)\frac{\partial^2 y}{\partial x_j \partial t} \right) + a(x)\frac{\partial y}{\partial t} -$$

$$-\sum_{i,j=1}^{n}\frac{\partial}{\partial x_i}\left(k_{ij}(x)\frac{\partial y}{\partial x_j}\right)+b(x)y=f(x,t) \qquad (1.1)$$

is specified in the domain Ω_T, where

$$a_{ij}=a_{ji}, \quad k_{ij}=k_{ji}; \quad a_{ij}\big|_{\bar{\Omega}_l}, \quad k_{ij}\big|_{\bar{\Omega}_l}\in C(\bar{\Omega}_l)\cap C^1(\Omega_l),$$

$$a\big|_{\Omega_l}\in C(\Omega_l),\; 0<a_0'\le a\le a_1<\infty,\; 0\le b\le b_1<\infty,\; a_0', a_1, b_1=\text{const},$$

$$\sum_{i,j=1}^{n}a_{ij}\xi_i\xi_j\ge\alpha_0\sum_{i=1}^{n}\xi_i^2,\; \sum_{i,j=1}^{n}k_{ij}\xi_i\xi_j\ge\alpha_1\sum_{i=1}^{n}\xi_i^2,\; \alpha_0,\alpha_1=\text{const}>0, \quad (1.1')$$

$$f\big|_{\Omega_{lT}}\in C(\Omega_{lT}),\; l=1,2,\; |f|<\infty.$$

The third boundary condition

$$\frac{\partial_L y}{\partial\nu}=-\alpha y+\beta \qquad (1.2)$$

is specified, in its turn, on the boundary Γ_T, where $0<\alpha^0\le\alpha=$ $=\alpha(x)\le\alpha_1^0$, $\alpha^0,\alpha_1^0=\text{const}$; the functions β and $\dfrac{\partial\beta}{\partial t}$ are continuous and bounded on $(\partial\Omega_i\setminus\gamma)\times(0,T)$, $i=1,2$. By analogy [67], the denotation

$$\frac{\partial_L y}{\partial\nu}=\sum_{i,j=1}^{n}\left(a_{ij}\frac{\partial^2 y}{\partial x_j\partial t}+k_{ij}\frac{\partial y}{\partial x_j}\right)\cos(\nu,x_i)$$

is used, and ν is an outer ort of a normal (or simply an outer normal) to Γ.

On a section γ_T, the conjugation conditions are

$$\left[\frac{\partial_L y}{\partial\nu}\right]=0 \qquad (1.3)$$

and

$$\left\{\frac{\partial_L y}{\partial\nu}\right\}^{\pm}=r[y], \qquad (1.4)$$

where $0 \le r = r(x) \le r_1 < \infty$, $r_1 = \text{const}$, $[\varphi] = \varphi^+ - \varphi^-$; $\varphi^+ = \{\varphi\}^+ = \varphi(x,t)$

under $(x,t) \in \gamma_T^+ = (\partial\Omega_2 \cap \gamma) \times (0,T)$; $\varphi^- = \{\varphi\}^- = \varphi(x,t)$ under $(x,t) \in$

$\in \gamma_T^- = (\partial\Omega_1 \cap \gamma) \times (0,T)$; ν is an ort of a normal to γ called simply a normal to γ and such normal is directed into the domain Ω_2.

The initial conditions

$$y(x,0) = y_0(x), \quad x \in \bar{\Omega}_1 \cup \bar{\Omega}_2, \tag{1.5}$$

and

$$\left.\frac{\partial y}{\partial t}\right|_{t=0} = y_1(x), \quad x \in \bar{\Omega}_1 \cup \bar{\Omega}_2, \tag{1.6}$$

where $y_0 \in V$ and $y_1 \in H$, are specified under $t = 0$.

Let there be a control Hilbert space $\mathcal{U}$ and operator $B \in$

$\in \mathcal{L}\left(\mathcal{U}; L^2(0,T;H)\right)$. For every control $u \in \mathcal{U}$, determine a system state $y = y(u) = y(x,t;u)$ as a generalized solution to the problem specified by the equation

$$\frac{\partial^2 y}{\partial t^2} + A\left(\frac{\partial y}{\partial t}\right) + K(y) + a\frac{\partial y}{\partial t} + by = f + Bu \tag{1.7}$$

and by conditions (1.2)–(1.6), and

$$A(y) = -\sum_{i,j=1}^{n} \frac{\partial}{\partial x_i}\left(a_{ij}\frac{\partial y}{\partial x_j}\right), \quad K(y) = -\sum_{i,j=1}^{n} \frac{\partial}{\partial x_i}\left(k_{ij}\frac{\partial y}{\partial x_j}\right).$$

Further on, without loss of generality, assume the following: $Bu \equiv u$. Specify the observation by the expression

$$Z(u) = Cy(u), \quad C \in \mathcal{L}\left(W(0,T);\mathcal{H}\right). \tag{1.8}$$

Specify the operator

$$\mathcal{N} \in \mathcal{L}(\mathcal{U};\mathcal{U}); \quad (\mathcal{N}u, u)_{\mathcal{U}} \geq v_0 \|u\|^2_{\mathcal{U}}, \quad v_0 = \text{const} > 0. \tag{1.9}$$

Assume the following: $\mathcal{N}u = \bar{a}u$; in this case, $\bar{a}|_{\Omega_l} \in C(\Omega_l)$, $0 < \bar{a}_0 \leq$ $\leq \bar{a} \leq \bar{a}_1$, $\bar{a}_0, \bar{a}_1 = \text{const}$. The cost functional is

$$J(u) = \|Cy(u) - z_g\|^2_{\mathcal{H}} + (\mathcal{N}u, u)_{\mathcal{U}}, \tag{1.10}$$

where z_g is a known element of the space $\mathcal{H}$.

The optimal control problem is to find such an element $u \in \mathcal{U}$ that the condition

$$J(u) = \inf_{v \in \mathcal{U}_\partial} J(v) \tag{1.11}$$

is met, where $\mathcal{U}_\partial$ is some convex closed subset in $\mathcal{U}$.

Definition 1.1. If an element $u \in \mathcal{U}_\partial$ meets condition (1.11), it is called an optimal control.

The generalized problem corresponds to initial boundary-value problem (1.7), (1.2)–(1.6) and means to find such a function $y(x,t;u) \in W(0,T)$, where

$$W(0,T) = \left\{ v \in L^2(0,T;V): \frac{dv}{dt} \in L^2(0,T;V), \frac{d^2v}{dt^2} \in L^2(0,T;L_2(\Omega)) \right\},$$

$V = \left\{ v(x,t): v|_{\Omega_l} \in W^1_2(\Omega_l), l = 1, 2; \forall t \in (0,T) \right\}$ and $W^1_2(\Omega_l)$ is the space of the Sobolev functions specified on the domain Ω_l, that satisfies the following equations $\forall w(x) \in V_0 = \left\{ v: v|_{\Omega_i} \in W^1_2(\Omega_i), i = 1,2 \right\}$:

$$\left(\frac{d^2y}{dt^2}, w \right) + a_0 \left(\frac{dy}{dt}, w \right) + a_1(y,w) = (f,w) +$$

$$+ (Bu,w) + \int_\Gamma \beta w \, d\Gamma, \quad \forall t \in (0,T), \tag{1.12}$$

$$a_0(y, w) = a_0(y_0, w), \quad t = 0, \tag{1.13}$$

and

$$\left(\frac{dy}{dt}, w \right) = (y_1, w), \quad t = 0; \tag{1.14}$$

in this case,

$$(\varphi, \psi) = \int_\Omega \varphi(x,t) \psi(x,t) \, dx, \quad a_0(\varphi, \psi) = \int_\Omega \left(\sum_{i,j=1}^n a_{ij} \frac{\partial \varphi}{\partial x_j} \frac{\partial \psi}{\partial x_i} + a\varphi\psi \right) dx$$

and

$$a_1(\varphi, \psi) = \int_\Omega \left(\sum_{i,j=1}^n k_{ij} \frac{\partial \varphi}{\partial x_j} \frac{\partial \psi}{\partial x_i} + b\varphi\psi \right) dx + \int_\gamma r[\varphi][\psi] \, d\gamma + \int_\Gamma \alpha\varphi\psi \, d\Gamma.$$

Use the previous results [41, 55, 32, 64, 49, 21] and consider the existence and uniqueness of the solution to problem (1.12)–(1.14). Since Bu and $f \in L^2(0,T;H)$ $\left(H = \{v(x,t) : v \in L_2(\Omega) \; \forall t \in (0,T)\} \right)$, then, without loss of generality, assume the following: $Bu \equiv 0$.

The space V_0 is complete, separable and reflexive [41, 55, 32], and $V_0 \subset L_2(\Omega)$. Choose an arbitrary fundamental system of linearly independent functions $w_k(x)$, $k = 1, 2, \ldots$, in V_0. For the sake of simplicity, let this system be orthogonal in $L_2(\Omega)$, i.e. $(w_k, w_l) = 0$ under $k \neq l$, $k, l = 1, 2, \ldots$.

Assume the following:

$$y_{0m} = \sum_{i=1}^m \xi_{im}^0 w_i, \quad y_{0m} \to y_0 \text{ under } m \to \infty,$$

and

$$y_{1m} = \sum_{i=1}^m \xi_{im}^1 w_i, \quad y_{1m} \to y_1 \text{ under } m \to \infty.$$

Specify the approximate generalized solution to problem (1.1)–(1.6) by the relations

$$y_m(x,t) = \sum_{i=1}^{m} g_{im}(t) w_i(x),$$ (1.15)

$$\left(\frac{\partial^2 y_m}{\partial t^2}, w_j \right) + a_0 \left(\frac{\partial y_m}{\partial t}, w_j \right) + a_1 (y_m, w_j) =$$

$$= (f, w_j) + \int_{\Gamma} \beta w_j \, d\Gamma, \quad j = \overline{1,m},$$ (1.16)

$$a_0 (y_m, w_j) = a_0 (y_0, w_j), \quad j = \overline{1,m}, \ t = 0,$$ (1.17)

and

$$\left(\frac{\partial y_m}{\partial t}, w_j \right) = (y_1, w_j), \quad j = \overline{1,m}, \ t = 0.$$ (1.18)

It is easy to see that Cauchy problem (1.16)–(1.18) has the unique solution for the system of m linear ordinary differential equations as for $g_{im}(t)$.

To obtain an a priori estimate of the functions $y_m(x,t)$ and $\dfrac{\partial y_m}{\partial t}$, multiply equality (1.16) by $g'_{im}(t)$ and find the sum over j for the result. Then:

$$(y''_m(\cdot,t), y'_m(\cdot,t)) + a_0 (y'_m, y'_m) + a_1 (y_m, y'_m) = (f, y'_m) + (\beta, y'_m)_{L_2(\Gamma)},$$

i.e.:

$$\frac{d}{dt} \left(\| y'_m(\cdot,t) \|^2 + a_1 (y_m, y_m) \right) + 2 a_0 (y'_m, y'_m) =$$

$$= 2(f, y'_m) + 2(\beta, y'_m)_{L_2(\Gamma)}.$$

Therefore,

$$\| y'_m(\cdot,t) \|^2 + a_1 (y_m(\cdot,t), y_m(\cdot,t)) + 2 \int_0^t a_0 (y'_m(\cdot,\tau), y'_m(\cdot,\tau)) d\tau =$$

$$= \|y'_m\|^2 (0) + a_1 (y_m, y_m)(0) + 2 \int_0^t (f(\cdot, \tau), y'_m(\cdot, \tau)) d\tau +$$

$$+ 2 \int_0^t (\beta, y'_m)_{L_2(\Gamma)} d\tau . \tag{1.19}$$

The inequality

$$\left\| \frac{\partial y_m}{\partial t}(\cdot, 0) \right\|^2 = \left(y_1(\cdot), \frac{\partial y_m}{\partial t}(\cdot, 0) \right) \le \|y_1\| \cdot \left\| \frac{\partial y_m}{\partial t}(\cdot, 0) \right\|$$

follows from equality (1.18), i.e.:

$$\left\| \frac{\partial y_m}{\partial t}(\cdot, 0) \right\| \le \|y_1\| . \tag{1.20}$$

Let $w_j(x)$, $j = 1, 2, \dots$, be eigenfunctions of the spectral problem that means to find $\{\lambda, w\} \in R^1 \times V_0$, $w \ne 0 : a_1(w, z) = \lambda(w, z)$, $\forall z \in V_0$, and they are such eigenfunctions that meet the condition $(w_i, w_j) = 0$ under $i \ne j$, $i, j = 1, 2, \dots$. Then, it is easy to see the following:

$$a_1(y_m(\cdot, 0), y_m(\cdot, 0)) \le a_1(y_0, y_0). \tag{1.21}$$

Consider the expression

$$\int_0^t (\beta, y'_m)_{L_2(\Gamma)} d\tau = (\beta, y_m)_{L_2(\Gamma)}(t) -$$

$$- (\beta, y_m)_{L_2(\Gamma)}(0) - \int_0^t (\beta', y_m)_{L_2(\Gamma)} d\tau . \tag{1.22}$$

The inequalities

$$\mu \|y_m\|_V^2 (0) \le a_1(y_0, y_0),$$

$$\|y_m\|_{L_2(\Gamma)}^2 (0) \le c_0 \|y_m\|_V^2 (0) \le c_1 a_1(y_m, y_m)(0) \le c_1 a_1(y_0, y_0), \tag{1.23}$$

where μ, c_0, $c_1 = \text{const} > 0$, $\|\varphi\|_V = \left\{ \sum_{i=1}^{2} \|\varphi\|_{W_2^1(\Omega_i)}^2 \right\}^{1/2}$, follow from inequality (1.21). Consider inequalities (1.20), (1.21) and (1.23), equality (1.22), the ε- and Cauchy-Bunyakovsky inequalities and embedding theorems, and the inequality

$$z_m(t) \le c_1 \left\{ \|y_1\|^2 + a_1(y_0, y_0) + \int_0^t \|f(\cdot, \tau)\|^2 d\tau + \right.$$

$$\left. + \sup_{t \in (0,T)} \|\beta\|_{L_2(\Gamma)}^2 + \int_0^t \|\beta'\|_{L_2(\Gamma)}^2 d\tau + \int_0^t z_m(\tau) d\tau \right\},$$

where $z_m(t) = \|y_m'\|^2(t) + \|y_m\|_V^2(t) + \mu \int_0^t \|y_m'\|_V^2 d\tau$, $\mu = \text{const} > 0$, is

obtained from equality (1.19). The inequalities

$$\|y_m'\|^2(t) + \|y_m\|_V^2(t) \le c,$$

$$\|y_m'\|^2(t) + \int_0^t \|y_m'\|_V^2 d\tau \le c, \tag{1.24}$$

$$\|y_m'\|_{V \times L_2}^2 \le c, \quad c = \text{const},$$

follow from the latter inequality by virtue of the Gronwall lemma [41, 64]. Hence, y_m and y_m' remain in the bounded set $L^2(0,T;V)$. That is why such subsequence y_ν of the sequence y_m can be chosen, that the convergences

$$y_\nu \to y \text{ in } L^2(0,T;V),$$

$$y_m' \to z \text{ in } L^2(0,T;V) \tag{1.25}$$

are weak. Without loss of generality, suppose that all the sequences y_m and y_m' converge in the sense of expressions (1.25). By virtue of [64],

$z = y'$. It follows from expressions (1.25) that $y_\nu(0) \to y(0)$ is weak and, since $y_\nu(0) = y_{0\nu} \to y_0$ in V_0, then $y(0) = y_0$. Consider inequalities (1.24), and it is seen that y'_m are still in the bounded set $L^2(0,T;V)$. Therefore: $z \in L^2(0,T;V)$.

It remains to show that, when constructed in such a way, the function y is the solution to problem (1.12)–(1.14).

Assume the following: $\varphi(t) \in C^1([0,T])$, $\varphi(T) = 0$. Introduce the denotation $\varphi_j(t) = \varphi(t)w_j(x)$. Multiply equality (1.16) by the function $\varphi(t)$ and find the integral from 0 to T of the result:

$$\int_0^T \left\{ -\left(y'_m, \varphi'_j\right) + a_0\left(y'_m, \varphi_j\right) + a_1\left(y_m, \varphi_j\right) \right\} dt =$$

$$= \int_0^T \left(f, \varphi_j\right) dt + \int_0^T \left(\beta, \varphi_j\right)_{L_2(\Gamma)} dt + \left(y_1, \varphi_j\right)(0).$$

Pass to the limit under $m \to \infty$, and the following equality is obtained:

$$\int_0^T \left\{ -\left(y', w_j\right)\varphi' + a_0\left(y', w_j\right)\varphi + a_1\left(y, w_j\right)\varphi \right\} dt =$$

$$= \int_0^T \left(f, w_j\right)\varphi \, dt + \int_0^T \left(\beta, w_j\right)_{L_2(\Gamma)} \varphi \, dt + \left(y_1, w_j\right)\varphi(0). \tag{1.26}$$

If $\varphi \in D((0,T))$ is used [58], then the equality

$$\left(\frac{d^2y}{dt^2}, w_j\right) + a_0\left(\frac{dy}{dt}, w_j\right) + a_1\left(y, w_j\right) = \left(f, w_j\right) + \left(\beta, w_j\right)_{L_2(\Gamma)} \tag{1.27}$$

is derived from equality (1.26). Equality (1.27) is true $\forall \tilde{w}_n \in \bigcup\limits_{n=1}^{\infty} V_{0n}$ and, therefore, $\forall w \in V_0$, where V_{0n} is the n-dimensional subspace of the space

V_0, and the functions of the set $\{w_i(x)\}_{i=1}^n$ make up the basis of V_{0n}. Hence, the equalities

$$\left(\frac{d^2y}{dt^2}, w\right) + a_0\left(\frac{dy}{dt}, w\right) + a_1(y, w) =$$

$$= (f, w) + (\beta, w)_{L_2(\Gamma)}, \quad \forall t \in (0, T), \tag{1.28}$$

$$a_0(y, w) = a_0(y_0, w), \quad t = 0, \quad \forall w \in V_0, \tag{1.29}$$

and

$$\left(\frac{dy}{dt}, w\right) = (y_1, w), \quad t = 0, \tag{1.30}$$

take place.

Thus, $y(x, t)$ is the generalized solution to initial boundary-value problem (1.1)–(1.6). Illustrate its uniqueness. Assume the following in equality (1.28): $w = \frac{dy}{dt}$. Find the integral of equality (1.28) over $\tau \in (0, t)$ now, and the equality

$$z^2(t) + 2\int_0^t a_0\left(\frac{dy}{dt}, \frac{dy}{dt}\right) d\tau = 2\int_0^t \left(f, \frac{dy}{dt}\right) d\tau +$$

$$+ 2\int_0^t \left(\beta, \frac{dy}{dt}\right)_{L_2(\Gamma)} d\tau + z^2(0) \tag{1.31}$$

is derived, where $z(t) = \left\|\frac{dy}{dt}\right\|^2 + a_1(y, y)$, $z^2(0) = \|y_1\|^2 + a_1(y_0, y_0)$.

Consider the ε-, Cauchy-Bunyakovsky and Friedrichs inequalities and embedding theorems, and the inequality

$$z^2(t) + (\mu - 2\varepsilon c_0) \int_0^t \left\|\frac{dy}{dt}\right\|_V^2 (\tau) \ d\tau \le$$

$$\leq \int_0^t \|f\|^2 (\tau) d\tau + \int_0^t \left\|\frac{dy}{dt}\right\|^2 d\tau + \frac{1}{2\varepsilon} \int_0^t \|\beta\|_{L_2(\Gamma)}^2 (\tau) d\tau + z^2(0),$$

where $\mu - 2\varepsilon c_0 = \text{const} > 0$, follows from equality (1.31).

Therefore,

$$\tilde{z}^2(t) \leq c_0' \left\{ \|f\|_{L_2 \times L_2}^2 + \|\beta\|_{L_2(\Gamma) \times L_2}^2 + \right.$$

$$\left. + \|y_1\|^2 + a_1(y_0, y_0) + \int_0^t \tilde{z}^2(\tau) dt \right\}, \tag{1.32}$$

where $\tilde{z}^2(t) = \left\|\frac{dy}{dt}\right\|^2 (t) + \|y\|_V^2 (t) + \int_0^t \left\|\frac{dy}{dt}\right\|_V^2 (\tau) \ d\tau.$

The inequality

$$\|y'\|^2 (t) + \sup_{\tau \in (0,T)} \|y\|_V^2 (\tau) + \|y'\|_{V \times L_2}^2 \leq$$

$$\leq c_1 \left\{ \|f\|_{L_2 \times L_2}^2 + \|\beta\|_{L_2(\Gamma) \times L_2}^2 + \|y_1\|^2 + a_1(y_0, y_0) \right\} \tag{1.33}$$

follows from inequality (1.32) by virtue of the Gronwall lemma.

Hence, the unique generalized solution to problem (1.1)–(1.6) is zero under $f = 0$, $\beta = 0$, $y_1 = 0$ and $y_0 = 0$. Therefore, the validity of the following statement is proved.

Theorem 1.1. *Problem (1.1)–(1.6) has a unique generalized solution* $y(x, t) \in W(0, T)$.

It is easy to see that $y(u_1) \neq y(u_2)$ under $u_1 \neq u_2$ $(Bu_1 \neq Bu_2)$. Let $\tilde{y}' = \tilde{y}(u')$ and $\tilde{y}'' = \tilde{y}(u'')$ be solutions from $W(0, T)$ to problem (1.12)–(1.14) under $f = 0$ and $\beta = 0$ and under a function $u = u(x, t)$ that is equal, respectively, to u' and u''. Then, the inequality

$$\|\tilde{y}' - \tilde{y}''\|_{L_2 \times L_2} \leq c \ \|u' - u''\|_{L_2 \times L_2} \tag{1.34}$$

is obtained from inequality (1.33). Rewrite cost functional (1.10) as

$$J(v) = \pi(v,v) - 2L(v) + \int_0^T \|z_g - y(0)\|^2 dt, \tag{1.35}$$

where

$$\pi(u,v) = \left(y(u) - y(0), \ y(v) - y(0)\right)_{\mathscr{H}} + (\bar{a}\, u, v)_{\mathscr{U}},$$

$$L(v) = \left(z_g - y(0), \ y(v) - y(0)\right)_{\mathscr{H}}; \tag{1.36}$$

in this case, $(z,v)_{\mathscr{H}} = \int_0^T (z,v) dt, \ (z,v)_{\mathscr{U}} = \int_0^T (z,v) dt, \ (z,v) = \int_\Omega zv dx$.

Inequality (1.34) provides the continuity of the linear functional $L(\cdot)$ and bilinear form $\pi(\cdot,\cdot)$ on $\mathscr{U}$.

On the basis of [58, Chapter 1, Theorem 1.1], the validity of the following statement is proved.

Theorem 1.2. *Let a system state be determined as a solution to problem (1.12)–(1.14). Then, there exists a unique element u of a convex set $\mathscr{U}_\partial$ that is closed in $\mathscr{U}$, and*

$$J(u) = \inf_{v \in \mathscr{U}_\partial} J(v) \tag{1.37}$$

takes place for u.

The control $u \in \mathscr{U}_\partial$ is optimal if and only if the inequality

$$\langle J'(u), \ v - u \rangle \geq 0, \ \forall v \in \mathscr{U}_\partial,$$

is true, i.e.:

$$\left(y(u) - z_g, \ y(v) - y(u)\right)_{\mathscr{H}} + (\mathscr{N} u, v - u)_{\mathscr{U}} \geq 0. \tag{1.38}$$

As for the control $v \in \mathscr{U}$, the conjugate state $p(v)$ is specified by the relations

$$\frac{\partial^2 p}{\partial t^2} - A\left(\frac{\partial p}{\partial t}\right) - a\frac{\partial p}{\partial t} + K(p) + bp = y(v) - z_g, \ (x,t) \in \Omega_T,$$

$$\sum_{i,j=1}^{n}\left(-a_{ij}\frac{\partial^2 p}{\partial x_j\partial t}+k_{ij}\frac{\partial p}{\partial x_j}\right)\cos(\nu,x_i)=-\alpha p,\ (x,t)\in\Gamma_T,$$

$$\left[\sum_{i,j=1}^{n}\left(-a_{ij}\frac{\partial^2 p}{\partial x_j\partial t}+k_{ij}\frac{\partial p}{\partial x_j}\right)\cos(\nu,x_i)\right]=0,\ (x,t)\in\gamma_T,\qquad(1.39)$$

$$\left\{\sum_{i,j=1}^{n}\left(-a_{ij}\frac{\partial^2 p}{\partial x_j\partial t}+k_{ij}\frac{\partial p}{\partial x_j}\right)\cos(\nu,x_i)\right\}^{\pm}=r[p],\ (x,t)\in\gamma_T,$$

$$p(x,T;\nu)=0,\ \frac{\partial p}{\partial t}(x,T;\nu)=0,\ x\in\bar{\Omega}_1\cup\bar{\Omega}_2.$$

By virtue of Theorem 1.1, problem (1.39) has the unique generalized solution $p(\nu)\in W(0,T)$ as the unique one to the equality system:

$$\left(\frac{d^2p}{dt^2},w\right)-a_0\left(\frac{dp}{dt},w\right)+a_1(p,w)=\left(y-z_g,w\right),\ \forall w\in V_0,\ t\in(0,T),$$

$$a_0(p,w)=0,\ t=T,\ \forall w\in V_0,\qquad(1.40)$$

$$\left(\frac{dp}{dt},w\right)=0,\ t=T,\ \forall w\in V_0.$$

Use the difference $y(\nu)-y(u)$ instead of w in the first equality of system (1.40) under $v=u$, find the integral from 0 to T of the result, and the equality

$$\int_0^T\left(\frac{d^2p(u)}{dt^2},\ y(\nu)-y(u)\right)dt-\int_0^T a_0\left(\frac{dp}{dt},\ y(\nu)-y(u)\right)dt+$$

$$+\int_0^T a_1(p,y(\nu)-y(u))dt=\int_0^T\left(y(u)-z_g,y(\nu)-y(u)\right)dt\qquad(1.41)$$

is obtained.

Consider the second addend in the left-hand side of equality (1.41):

$$\int_0^T a_0\left(\frac{dp}{dt}, y(v) - y(u)\right) dt = a_0\left(p, y(v) - y(u)\right)\Big|_0^T -$$

$$- \int_0^T a_0\left(p, \frac{d}{dt}(y(v) - y(u))\right) dt. \tag{1.42}$$

Take equality (1.42) and equations (1.12)–(1.14) into account, and the equality

$$\int_0^T \left(y(u) - z_g, y(v) - y(u)\right) dt = \int_0^T \left(p(u), v - u\right) dt$$

is derived from equality (1.41).

Therefore, optimality condition (1.38) for the control $u \in \mathcal{U}_\partial$ is equivalent to the inequality

$$\int_0^T \left(p(u) + \bar{a}u, v - u\right) dt \geq 0, \quad \forall v \in \mathcal{U}_\partial. \tag{1.43}$$

Thus, the optimal control $u \in \mathcal{U}_\partial$ is specified by relations (1.12)–(1.14), (1.40) and (1.43). If the constraints are absent, i.e. when $\mathcal{U}_\partial = \mathcal{U}$, then the equality

$$p(u) + \bar{a}\, u = 0, \quad (x,t) \in \Omega_T, \tag{1.44}$$

follows from inequality (1.43).

If the solution $(y, p)^T$ to problem (1.12)–(1.14), (1.40), (1.44) is smooth enough on $\overline{\Omega}_{lT}, l = 1, 2$, then the differential problem of finding the vector-function $(y, p)^T$, that satisfies conditions (1.2)–(1.6) and (1.39) and the equality

$$\frac{\partial^2 y}{\partial t^2} + A\left(\frac{\partial y}{\partial t}\right) + K(y) + a\frac{\partial y}{\partial t} + by + p/\bar{a} = f, \quad (x,t) \in \Omega_T, \tag{1.45}$$

corresponds to problem (1.12)–(1.14), (1.40), (1.44), where the optimal control is specified as

$$u = -p/\bar{a}, \ (x,t) \in \Omega_T. \tag{1.46}$$

9.2   CONTROL UNDER CONJUGATION CONDITION WITH OBSERVATION THROUGHOUT A WHOLE DOMAIN

Assume that equation (1.1) is specified in the domain Ω_T. On the boundary Γ_T, the boundary condition has the form of expression (1.2).

For every control $u \in \mathcal{U} \subset L_2(\gamma_T)$, determine a system state $y = y(u)$ as a generalized solution to the initial boundary-value problem specified, in its turn, by equation (1.1), boundary condition (1.2), initial conditions (1.5) and (1.6) and the conjugation conditions

$$[y] = 0, \ (x,t) \in \gamma_T, \tag{2.1}$$

and

$$\left[\frac{\partial_L y}{\partial \nu} \right] = \omega + u, \ (x,t) \in \gamma_T, \tag{2.2}$$

where $\omega = \omega(x,t) \in L_2(\gamma_T)$.

Since there exists the generalized solution $y(x,t;u)$ to initial boundary-value problem (1.1), (1.2), (1.5), (1.6), (2.1), (2.2), then such solution is reasonable on $\bar{\Omega}_{lT}$, $l = 1,2$. The generalized problem corresponds to initial boundary-value problem and means to find a function $y(x,t;u) \in W(0,T)$ that satisfies the following equalities $\forall w(x) \in V_0 = \{ v(x) : v|_{\Omega_i} \in$

$\in W_2^1(\Omega_i), \ i = 1,2; \ [v]|_\gamma = 0 \}$:

$$\left(\frac{d^2 y}{dt^2}, w \right) + a_0 \left(\frac{dy}{dt}, w \right) + a_1(y, w) =$$

$$= (f,w) - \int_\gamma \omega w \, d\gamma - \int_\gamma u \, w \, d\gamma + \int_\Gamma \beta w \, d\Gamma, \quad t \in (0,T),$$

$$a_0(y,w) = a_0(y_0,w), \quad t = 0, \qquad (2.3)$$

$$\left(\frac{dy}{dt}, w \right) = (y_1, w), \quad t = 0;$$

in this case,

$$V = \left\{ v(x,t): \; v|_{\Omega_l}, \; \frac{dv}{dt}\bigg|_{\Omega_l}, \; \in W_2^1(\Omega_l), \right.$$

$$\left. \frac{d^2 v}{dt^2} \in L_2(\Omega), \; \forall t \in [0,T], \; l = 1,2; \; [v]|_\gamma = 0 \; \forall t \in (0,T) \right\}$$

and

$$a_1(\varphi,\psi) = \int_\Omega \left(\sum_{i,j=1}^n a_{ij} \frac{\partial \varphi}{\partial x_j} \frac{\partial \psi}{\partial x_i} + b\varphi\psi \right) dx + \int_\Gamma \alpha\varphi\psi \, d\Gamma.$$

The following statement takes place.

Theorem 2.1. *Initial boundary-value problem (1.1), (1.2), (1.5), (1.6), (2.1), (2.2) has a unique generalized solution $y(x,t;u) \in W(0,T)$ $\forall u \in \mathcal{U}$.*

The validity of Theorem 2.1 is stated by analogy with the proof of Theorem 1.1.

Proceed from equalities (2.3), and it is easy to see on the basis of the first one that $y(u_1) \neq y(u_2)$ under $u_1 \neq u_2$. If $\tilde{y}' = \tilde{y}(u')$ and $\tilde{y}'' = \tilde{y}(u'')$ are solutions from $W(0,T)$ to problem (2.3) under f, ω and $\beta = 0$ and under a function u that is equal, respectively, to u' and u'', then, assume $w = \dfrac{d}{dt}(\tilde{y}' - \tilde{y}'')$ in the first equality of system (2.3), find the integral of its first equality over $\tau \in (0,t)$ now, and the equality

$$z^2(t) + 2 \int_0^t a_0 \left(\frac{d(\tilde{y}' - \tilde{y}'')}{dt}, \frac{d(\tilde{y}' - \tilde{y}'')}{dt} \right) dt =$$

$$= -2 \int\limits_0^t \left((u' - u''), \frac{d}{d\tau}(\tilde{y}' - \tilde{y}'') \right)_{L_2(\gamma)} d\tau + z^2(0) \qquad (2.4)$$

is obtained, where $z^2(t) = \left\| \frac{d}{dt}(\tilde{y}' - \tilde{y}'') \right\|^2 + a_1 (\tilde{y}' - \tilde{y}'', \tilde{y}' - \tilde{y}'')(t)$.

Consider the ε- and Cauchy-Bunyakovsky inequalities and embedding theorems, and the inequality

$$z^2(t) + c_1 \int\limits_0^t \left\| \frac{d}{dt}(\tilde{y}' - \tilde{y}'') \right\|_V^2 d\tau \le \frac{1}{2\varepsilon} \int\limits_0^T \|u' - u''\|_{L_2(\gamma)}^2 dt, \qquad (2.4')$$

where $c_1 = \text{const} > 0$, follows from equality (2.4).

Take the obtained inequality into account, and the inequality

$$\|\tilde{y}' - \tilde{y}''\|_{L_2 \times L_2} \le c_1' \|u' - u''\|_{L_2(\gamma) \times L_2}$$

is derived from equality (2.4).

Therefore, the linear functional $L(\cdot)$ and bilinear form $\pi(\cdot, \cdot)$ of cost functional (1.35), where $(z, v)_{\mathcal{U}} = \int\limits_0^T (z, v)_{L_2(\gamma)} dt$, are continuous on $\mathcal{U}$. The observation is specified here in the form of expression (1.8), where $Cy(u) \equiv y(u)$, and the value of cost functional (1.10) corresponds to every control $u \in \mathcal{U}$, where z_g is a known element from $L^2(0, T; V)$,

$$J(u) = \int\limits_0^T \int\limits_\Omega \left(y(u) - z_g \right)^2 dx\, dt + \int\limits_0^T \int\limits_\gamma \bar{a} u^2 d\gamma\, dt,$$

$$0 < a_0 \le \bar{a} \le a_1 < \infty, \quad a_0, a_1 = \text{const}, \quad \bar{a} \in L_2(\gamma).$$

On the basis of [58, Chapter 1, Theorem 1.1], the validity of the following statement is proved.

Theorem 2.2. *If a system state is determined as a generalized solution to problem (1.1), (1.2), (1.5), (1.6), (2.1), (2.2), then there exists a unique* *element u of a convex set $\mathcal{U}_\partial$ that is closed in $\mathcal{U}$, and relation (1.37) takes place for u.*

As for the control $v \in \mathcal{U}$, the conjugate state $p(v)$ is specified by the relations

$$\frac{\partial^2 p}{\partial t^2} - A\left(\frac{\partial p}{\partial t}\right) - a\frac{\partial p}{\partial t} + K(p) + bp = y(v) - z_g, \quad (x,t) \in \Omega_T,$$

$$\sum_{i,j=1}^{n}\left(-a_{ij}\frac{\partial^2 p}{\partial x_j \partial t} + k_{ij}\frac{\partial p}{\partial x_j}\right)\cos(v, x_i) = -\alpha\, p, \quad (x,t) \in \Gamma_T,$$

$$[p] = 0, \quad (x,t) \in \gamma_T, \tag{2.5}$$

$$\left[\sum_{i,j=1}^{n}\left(-a_{ij}\frac{\partial^2 p}{\partial x_j \partial t} + k_{ij}\frac{\partial p}{\partial x_j}\right)\cos(v, x_i)\right] = 0, \quad (x,t) \in \gamma_T,$$

$$p(x,T) = 0, \quad x \in \bar{\Omega}_1 \cup \bar{\Omega}_2,$$

$$\frac{\partial p}{\partial t}(x,T) = 0, \quad x \in \bar{\Omega}_1 \cup \bar{\Omega}_2.$$

Problem (2.5) has the unique generalized solution $p(v) \in L^2(0,T;V)$ as the unique one to equality system like (1.40), where

$$a_1(v,z) = \int_{\Omega} \sum_{i,j=1}^{n} k_{ij}\frac{\partial v}{\partial x_j}\frac{\partial z}{\partial x_i}\,dx + \int_{\Gamma}\alpha z v\,d\Gamma. \tag{2.6}$$

If the difference $y(v) - y(u)$ is used instead of w in the first equality of system (1.40), where $v = u$ and the bilinear form $a_1(\cdot,\cdot)$ is specified by expression (2.6), then equality (1.41) is present after taking the integral over $t \in (0,T)$ of the result. Consider equality (1.42), the first equality of system (2.3) and expression (2.6), and the equality

$$\int_0^T \left(y(u) - z_g, y(v) - y(u)\right)dt = -\int_0^T\int_\gamma p(u)\,(v - u)\,d\gamma\,dt \tag{2.7}$$

is obtained from equality (1.41). Therefore, the control $u \in \mathcal{U}_\partial$ is optimal if and only if the following inequality is true:

$$\int\limits_0^T \int\limits_\gamma (-p(u) + \bar{a}u)(v - u)\, d\gamma dt \geq 0, \quad \forall v \in \mathcal{U}_\partial . \qquad (2.8)$$

Thus, the optimal control $u \in \mathcal{U}_\partial$ is specified by inequality (2.8) and relations (2.3) and (1.40), where the bilinear form $a_1(\cdot,\cdot)$ is specified, in its turn, by expression (2.6). If the constraints are absent, i.e. when $\mathcal{U}_\partial = \mathcal{U}$, then the equality

$$u = p/\bar{a}, \quad (x,t) \in \gamma_T , \qquad (2.9)$$

follows from inequality (2.8).

If the solution $(y,p)^T$ to problem (2.3), (1.40), (2.9) is smooth enough on $\bar{\Omega}_{lT}$, $l = 1,2$, then the differential problem, specified by equalities (1.1), (1.2), (1.5), (1.6), (2.1) and (2.5) and by the condition

$$\left[\sum_{i,j=1}^n \left(a_{ij} \frac{\partial^2 y}{\partial x_j \partial t} + k_{ij} \frac{\partial y}{\partial x_j} \right) \cos(v, x_i) \right] = \omega + p/\bar{a}, \quad (x,t) \in \gamma_T , \quad (2.9')$$

corresponds to problem (2.3), (1.40), (2.9), where the optimal control u is found by formula (2.9).

9.3   CONTROL UNDER CONJUGATION CONDITION WITH BOUNDARY OBSERVATION

Assume that equation (1.1) is specified in the domain Ω_T. On the boundary Γ_T, the boundary condition has the form of expression (1.2). For every control $u \in \mathcal{U} = L_2(\gamma_T)$, determine a system state $y(x,t;u)$ as a generalized solution to initial-boundary-value problem specified by equation (1.1), boundary condition (1.2), initial conditions (1.5) and (1.6) and conjugation conditions (2.1) and (2.2). The cost functional is

$$J(u) = \int\limits_0^T \int\limits_\Gamma \big(y(u) - z_g \big)^2 d\Gamma dt + \int\limits_0^T \int\limits_\gamma \bar{a}u^2 d\gamma\, dt .$$

The generalized problem corresponds to initial boundary-value problem
(1.1), (1.2), (1.5), (1.6), (2.1), (2.2) and means to find a function
$y(x,t;u) \in W(0,T)$ that satisfies equation system (2.3) $\forall w(x) \in V_0$; the
space $W(0,T)$ is specified in point 9.1 and the spaces V and V_0 are
specified, in their turn, in point 9.2.

The following statement takes place.

Theorem 3.1. *Initial boundary-value problem (1.1), (1.2), (1.5), (1.6),
(2.1), (2.2) has a unique generalized solution* $y(x,t;u) \in W(0,T)$ $\forall u \in \mathcal{U}$.

The validity of Theorem 3.1 is stated by analogy with the proof of
Theorem 1.1.

Proceed from the first equality of system (2.3), and it is easy to see that
$y(u_1) \neq y(u_2)$ under $u_1 \neq u_2$. Let $\tilde{y}' = \tilde{y}(u')$ and $\tilde{y}'' = \tilde{y}(u'')$ be solutions
from $W(0,T)$ to problem (2.3) under f, ω and $\beta = 0$ and under a function
u that is equal, respectively, to u' and u'' . Inequality (2.4') is true, from
which, by virtue of the embedding theorems, the inequality

$$\int_0^T \int_\Gamma (\tilde{y}' - \tilde{y}'')^2 \, d\Gamma dt \leq c_1' \int_0^T \|u' - u''\|_{L_2(\gamma)}^2 dt$$

is obtained that is the evidence of the fact that the linear functional $L(\cdot)$
and bilinear form $\pi(\cdot,\cdot)$ of the cost functional

$$J(u) = \int_0^T \int_\Gamma \left(y(u) - z_g\right)^2 d\Gamma dt + \int_0^T \int_\gamma \bar{a} u^2 d\gamma dt =$$

$$= \pi(u,u) - 2L(u) + \int_0^T \left\|z_g - y(0)\right\|_{L_2(\Gamma)}^2 dt \tag{3.1}$$

are continuous; in this case, $a_0, a_1 = \text{const}$, $\bar{a}(x) \in L_2(\gamma)$, $0 < a_0 \leq$
$\leq \bar{a}(x) \leq a_1$;

$$\pi(u,v) = \left(y(u) - y(0), y(v) - y(0)\right)_{L_2(\Gamma) \times L_2} + (\bar{a} u, v)_{L_2(\gamma) \times L_2}$$

and

$$L(v) = \left(z_g - y(0), \ y(v) - y(0) \right)_{L_2(\Gamma) \times L_2}.$$

On the basis of [58, Chapter 1, Theorem 1.1], the validity of the following statement is proved.

Theorem 3.2. *Let a system state be determined as a solution to problem (2.3). Then, there exists a unique element u of a convex set $\mathcal{U}_\partial$ that is closed in $\mathcal{U}$, and relation (1.37) takes place for u, where the cost functional has the form of expression (3.1).*

As for the control $v \in \mathcal{U}$, the conjugate state $p(v)$ is specified by the relations

$$\frac{\partial^2 p}{\partial t^2} - A\left(\frac{\partial p}{\partial t} \right) - a \frac{\partial p}{\partial t} + K(p) + bp = 0, \ (x,t) \in \Omega_T,$$

$$\sum_{i,j=1}^{n} \left(-a_{ij} \frac{\partial^2 p}{\partial x_j \partial t} + k_{ij} \frac{\partial p}{\partial x_j} \right) \cos(v, x_i) = -\alpha p + y(v) - z_g, \ (x,t) \in \Gamma_T,$$

$$[p] = 0, \ (x,t) \in \gamma_T, \tag{3.2}$$

$$\left[\sum_{i,j=1}^{n} \left(-a_{ij} \frac{\partial^2 p}{\partial x_j \partial t} + k_{ij} \frac{\partial p}{\partial x_j} \right) \cos(v, x_i) \right] = 0, \ (x,t) \in \gamma_T,$$

$$p(x,T) = 0, \ x \in \bar{\Omega}_1 \cup \bar{\Omega}_2,$$

$$\frac{\partial p}{\partial t}(x,T) = 0, \ x \in \bar{\Omega}_1 \cup \bar{\Omega}_2.$$

Problem (3.2) has the unique generalized solution $p(v) \in W(0,T)$ as the unique one to the equality system

$$\int_\Omega \frac{d^2 p}{dt^2} w \, dt - a_0 \left(\frac{dp}{dt}, w \right) + a_1(p, w) =$$

$$= \int_\Gamma (y(v) - z_g) \, w \, d\Gamma, \ \forall w \in V_0, \ t \in (0,T),$$

$$\left(\frac{dp}{dt}, w \right) = 0, \ \forall w \in V_0, \ t = T, \tag{3.3}$$

$$a_0(p, w) = 0, \quad \forall w \in V_0, \quad t = T ;$$

the space V is included into the specification of $W(0, T)$ and specified, in its turn, in point 9.2 along with the space V_0.

The following statement takes place.

Theorem 3.3. *Initial boundary-value problem (3.2) has a unique generalized solution.*

Use the difference $y(v) - y(u)$ instead of w in the first inequality of system (3.3), where $v = u$, find the integral from 0 to T of the result, and the equality

$$\int_0^T \left(\frac{d^2 p(u)}{dt^2}, y(v) - y(u) \right) dt - \int_0^T a_0 \left(\frac{dp}{dt}, y(v) - y(u) \right) dt +$$

$$+ \int_0^T a_1 \left(p, y(v) - y(u) \right) dt = \int_0^T \int_\Gamma \left(y(u) - z_g \right) \left(y(v) - y(u) \right) d\Gamma dt \quad (3.4)$$

is obtained, where the bilinear form $a_1(\cdot, \cdot)$ is specified by expression (2.6).

Consider equality (1.42) and system (2.3), and the equality

$$\int_0^T \int_\Gamma \left(y(u) - z_g \right) \left(y(v) - y(u) \right) d\Gamma dt = - \int_0^T \int_\gamma p(u)(v - u) d\gamma dt$$

is derived from equality (3.4). Therefore, the control $u \in \mathcal{U}_\partial$ is optimal if and only if the following inequality is true:

$$\int_0^T \int_\gamma \left(-p(u) + \bar{a} u \right)(v - u) d\gamma dt \geq 0, \quad \forall v \in \mathcal{U}_\partial. \quad (3.5)$$

Thus, the optimal control $u \in \mathcal{U}_\partial$ is specified by equalities (2.3) and (3.3) and inequality (3.5). If the constraints are absent, i.e. when $\mathcal{U}_\partial = \mathcal{U}$, then the equality

$$u = p / \bar{a}, \quad (x, t) \in \gamma_T, \quad (3.6)$$

follows from inequality (3.5). If the solution $(y, p)^T$ to problem (2.3), (3.3), (3.6) is smooth enough on $\overline{\Omega}_{lT}$, $l = 1,2$, then the differential problem of finding the vector-function $(y, p)^T$, that satisfies equalities (1.1), (1.2), (1.5), (1.6), (2.1), (2.9'), (3.2) and (3.6), corresponds to problem (2.3), (3.3), (3.6).

9.4   CONTROL UNDER CONJUGATION CONDITION WITH FINAL OBSERVATION

Assume that equation (1.1) is specified in the domain Ω_T. On the boundary Γ_T, the boundary condition has the form of expression (1.2). For every control $u \in \mathcal{U} = L_2(\gamma_T)$, determine a system state $y(x, t; u)$ as a generalized solution to the initial boundary-value problem specified by equation (1.1), boundary condition (1.2), initial conditions (1.5) and (1.6) and conjugation conditions (2.1) and (2.2).

Let the cost functional be

$$J(u) = \int_{\Omega} \left(y(x, T; u) - z_g \right)^2 dx + \int_0^T \int_\gamma \bar{a} \, u^2 d\gamma \, dt, \qquad (4.1)$$

where $0 < a_0 \le \bar{a} \le a_1 < \infty$; $a_0, a_1 = \text{const}$, and it may be rewritten as

$$J(u) = \pi(u, u) - 2L(u) + \int_{\Omega} \left(z_g(x) - y(x, T; 0) \right)^2 dx; \qquad (4.1')$$

in this case,

$$\pi(u, v) = \left(y(\cdot, T; u) - y(\cdot, T; 0), \; y(\cdot, T; v) - y(\cdot, T; 0) \right) + \int_0^T \int_\gamma \bar{a} \, uv \, d\gamma \, dt$$

and

$$L(v) = \left(z_g(\cdot) - y(\cdot, T; 0), \; y(\cdot, T; v) - y(\cdot, T; 0) \right).$$

The generalized problem corresponds to initial boundary-value problem (1.1), (1.2), (1.5), (1.6), (2.1), (2.2) and means to find a function $y(x,t;u) \in W(0,T)$ that satisfies equality system (3.2) $\forall w(x) \in V_0$; the space $W(0,T)$ is specified in point 9.1 and the spaces V and V_0 are specified, in their turn, in point 9.2.

Theorem 3.1 takes place. It is stated in point 9.3 that $y(u_1) \neq y(u_2)$ under $u_1 \neq u_2$. If $\tilde{y}' = \tilde{y}(u')$ and $\tilde{y}'' = \tilde{y}(u'')$ are solutions from $W(0,T)$ to problem (2.3) under $f = 0$, $\beta = 0$ and $\omega = 0$ and under a function u that is equal, respectively, to u' and u'', then inequality (2.4′) is true. The inequality

$$a_1 \left(\tilde{y}' - \tilde{y}'', \, \tilde{y}' - \tilde{y}'' \right)(T) \leq \frac{1}{2\varepsilon} \|u' - u''\|_{L_2(\gamma) \times L_2}^2 \tag{4.2}$$

follows from inequality (2.4′). Consider the Friedrichs inequality, and the inequality

$$\|\tilde{y}' - \tilde{y}''\|(T) \leq c_0 \|u' - u''\|_{L_2(\gamma) \times L_2}$$

is obtained. Therefore, the linear functional $L(\cdot)$ and bilinear form $\pi(\cdot,\cdot)$ of cost functional (4.1) are continuous on $\mathcal{U}$.

On the basis of [58, Chapter 1, Theorem 1.1], the validity of the following statement is proved.

Theorem 4.1. *If s system state is determined as a generalized solution to problem (1.1), (1.2), (1.5), (1.6), (2.1), (2.2), then there exists a unique element u of a convex set $\mathcal{U}_\partial$ that is closed in $\mathcal{U}$, and relation (1.37) takes place for u.*

As for the control $v \in \mathcal{U}$, the conjugate state $p(v)$ is specified by the relations

$$\frac{\partial^2 p}{\partial t^2} - A\left(\frac{\partial p}{\partial t} \right) - a\frac{\partial p}{\partial t} + K(p) + bp = 0, \quad (x,t) \in \Omega_T,$$

$$\sum_{i,j=1}^{n} \left(-a_{ij} \frac{\partial^2 p}{\partial x_j \partial t} + k_{ij} \frac{\partial p}{\partial x_j} \right) \cos(v, x_i) = -\alpha\, p, \quad (x,t) \in \Gamma_T,$$

$$[p] = 0, \ (x,t) \in \gamma_T, \tag{4.3}$$

$$\left[\sum_{i,j=1}^{n} \left(-a_{ij} \frac{\partial^2 p}{\partial x_j \partial t} + k_{ij} \frac{\partial p}{\partial x_j} \right) \cos(v, x_i) \right] = 0, \ (x,t) \in \gamma_T,$$

$$p(x,T;v) = 0, \ x \in \bar{\Omega}_1 \cup \bar{\Omega}_2,$$

$$\frac{\partial p}{\partial t}(x,T;v) = y(x,T;v) - z_g, \ x \in \bar{\Omega}_1 \cup \bar{\Omega}_2.$$

The generalized solution to problem (4.3) is the solution to the equality system

$$\left(\frac{d^2 p}{dt^2}, w \right) - a_0 \left(\frac{dp}{dt}, w \right) + a_1(p,w) = 0, \ \forall w \in V_0, \ t \in (0,T),$$

$$a_0(p,w) = 0, \ t = T, \ \forall w \in V_0, \tag{4.4}$$

$$\left(\frac{dp}{dt}, w \right) = \left(y(\cdot,T;v) - z_g, w \right), \ t = T, \ \forall w \in V_0.$$

Use the difference $y(v) - y(u)$, instead of w in the first equality of system (4.4) under $v = u$, consider the relation

$$\int_0^T a_0 \left(\frac{dp}{dt}, y(v) - y(u) \right) dt = -\int_0^T a_0 \left(p, \frac{d}{dt}(y(v) - y(u)) \right) dt,$$

and the equality

$$\left(y(\cdot,T;u) - z_g, y(v) - y(u) \right) - \int_0^T \int_\gamma p(u)(v-u) d\gamma \, dt = 0$$

is obtained, i.e.

$$\left(y(\cdot,T;u) - z_g, y(v) - y(u) \right) = \left(p(u), v - u \right)_{L_2(\gamma) \times L_2}.$$

Therefore, the control $u \in \mathcal{U}_\partial$ is optimal if and only if the following inequality is true:

$$\int\limits_{0}^{T}\int\limits_{\gamma}(p(u)+\bar{a}\,u)(v-u)\,d\gamma dt \geq 0, \ \forall v \in \mathcal{U}_{\partial}.\tag{4.5}$$

Thus, the optimal control $u \in \mathcal{U}_{\partial}$ is specified by relations (2.3), (4.4) and (4.5). If the constraints are absent, i.e. when $\mathcal{U}_{\partial} = \mathcal{U}$, then the equality

$$u = -p/\bar{a}, \ (x,t) \in \gamma_T,\tag{4.6}$$

follows from inequality (4.5).

If the solution $(y,p)^{\mathrm{T}}$ to problem (2.3), (4.4), (4.6) is smooth enough on $\bar{\Omega}_{lT}$, $l = 1,2$, then the differential problem, specified by equalities (1.1), (1.2), (1.5), (1.6), (2.1) and (4.3) and by the condition

$$\left[\sum_{i,j=1}^{n}\left(a_{ij}\frac{\partial^2 y}{\partial x_j \partial t}+k_{ij}\frac{\partial y}{\partial x_j}\right)\cos(v,x_i)\right]=\omega-p/\bar{a}, \ (x,t)\in\gamma_T,$$

corresponds to problem (2.3), (4.4), (4.6), where the optimal control u is found by formula (4.6).

9.5   CONTROL UNDER BOUNDARY CONDITION WITH OBSERVATION ON A THIN INCLUSION

For every control $u \in \mathcal{U} = L_2(\Gamma_T)$, determine a state $y(x,t;u)$ as a generalized solution to the initial boundary-value problem specified by equation (1.1), initial conditions (1.5) and (1.6), conjugation conditions (2.1) and (2.2) (under $u = 0$) and the boundary condition

$$\sum_{i,j=1}^{n}\left(a_{ij}\frac{\partial^2 y}{\partial x_j \partial t}+k_{ij}\frac{\partial y}{\partial x_j}\right)\cos(v,x_i)=-\alpha y+\beta+u, \ (x,t)\in\Gamma_T.\tag{5.1}$$

The cost functional is

$$J(u) = \int\limits_0^T \int\limits_\gamma \left(y(u) - z_g \right)^2 d\gamma dt + \int\limits_0^T \int\limits_\Gamma \bar{a} u^2 d\Gamma dt, \qquad (5.2)$$

where $0 < a_0 \le \bar{a} \le a_1 < \infty$; $a_0, a_1 = \text{const}$, $\bar{a} \in L_2(\Gamma)$, and it may be rewritten as

$$J(u) = \pi(u,u) - 2L(u) + \int\limits_0^T \int\limits_\gamma \left(z_g - y(\cdot,t;0) \right)^2 d\gamma dt; \qquad (5.3)$$

in this case,

$$\pi(u,v) = \int\limits_0^T \left(y(\cdot,t;u) - y(\cdot,t;0), \ y(\cdot,t;v) - y(\cdot,t;0) \right)_{L_2(\gamma)} dt$$

and

$$L(v) = \left(z_g - y(\cdot,t;0), \ y(\cdot,t;v) - y(\cdot,t;0) \right)_{L_2(\gamma) \times L_2}.$$

The generalized problem corresponds to initial boundary-value problem (1.1), (1.5), (1.6), (2.1), (2.2) (under $u = 0$), (5.1) and means to find a function $y(x,t;u) \in W(0,T)$ that satisfies the following equality system $\forall w(x) \in V_0$:

$$\left(\frac{d^2 y}{dt^2}, w \right) + a_0 \left(\frac{dy}{dt}, w \right) + a_1(y,w) =$$

$$= (f,w) - \int\limits_\gamma \omega w d\gamma + \int\limits_\Gamma \beta w d\Gamma + \int\limits_\Gamma u w d\Gamma, \quad t \in (0,T),$$

$$a_0(y,w) = a_0(y_0,w), \quad t = 0, \qquad (5.4)$$

$$\left(\frac{dy}{dt}, w \right) = (y_1, w), \quad t = 0.$$

The forthcoming statement takes place.

Theorem 5.1. *Initial boundary-value problem (1.1), (1.5), (1.6), (2.1), (2.2) (under $u = 0$), (5.1) has a unique generalized solution $y(x,t;u) \in W(0,T)$ $\forall u \in \mathcal{U}$.*

The validity of Theorem 5.1 is stated by analogy with the proof of Theorem 1.1.

Proceed from the first equality of system (5.4), and it is easy to see that $y(u_1) \neq y(u_2)$ under $u_1 \neq u_2$. Let $\tilde{y}' = \tilde{y}(u')$ and $\tilde{y}'' = \tilde{y}(u'')$ be solutions from $W(0,T)$ to problem (5.4) under f, ω and $\beta = 0$ and under a function u that is equal, respectively, to u' and u''. The equality

$$z^2(t) + 2\int_0^t a_0\left(\frac{d(\tilde{y}' - \tilde{y}'')}{dt}, \frac{d(\tilde{y}' - \tilde{y}'')}{dt}\right)d\tau =$$

$$= 2\int_0^t \left((u' - u''), \frac{d}{dt}(\tilde{y}' - \tilde{y}'')\right)_{L_2(\Gamma)} d\tau + z^2(0) \qquad (5.5)$$

is true, where $z^2(t) = \left\|\dfrac{d(\tilde{y}' - \tilde{y}'')}{dt}\right\|^2 + a_1(\tilde{y}' - \tilde{y}'', \tilde{y}' - \tilde{y}'')(t)$.

Since the inequality

$$\int_0^T a_0\left(\frac{d(\tilde{y}' - \tilde{y}'')}{dt}, \frac{d(\tilde{y}' - \tilde{y}'')}{dt}\right)dt \leq$$

$$\leq \frac{1}{4\varepsilon}\|u' - u''\|^2_{L_2(\Gamma)\times L_2} + \varepsilon\, c_0 \left\|\frac{d}{dt}(\tilde{y}' - \tilde{y}'')\right\|^2_{V\times L_2}$$

takes place and the ellipticity condition is taken into account, then the inequality

$$(\mu - \varepsilon c_0)\left\|\frac{d(\tilde{y}' - \tilde{y}'')}{dt}\right\|^2_{V\times L_2} \leq \frac{1}{4\varepsilon}\|u' - u''\|^2_{L_2(\Gamma)\times L_2}, \qquad (5.6)$$

where $\mu - \varepsilon c_0 > 0$ and c_0 is the constant obtained from the inequalities of the embedding theorems, follows from equality (5.5).

Use inequality (5.6), consider the embedding theorems and Friedrichs inequality, and the inequality

$$\mu_1 \|\tilde{y}' - \tilde{y}''\|^2_{L_2(\gamma) \times L_2} \leq c_1 \|u' - u''\|_{L_2(\Gamma) \times L_2},$$

where μ_1, $c_1 = \text{const} > 0$, is obtained from equality (5.5).

Therefore, the linear functional $L(\cdot)$ and bilinear form $\pi(\cdot, \cdot)$ of cost functional (5.3) are continuous on $\mathcal{U}$.

On the basis of [58, Chapter 1, Theorem 1.1], the validity of the following statement is proved.

Theorem 5.2. *If a system state is determined as a solution to problem (5.4), then there exists a unique element u of a convex set $\mathcal{U}_{\partial}$ that is closed in $\mathcal{U}$, and relation (1.37) takes place for u, where the cost functional has the form of expression (5.2).*

As for the control $v \in \mathcal{U}$, the conjugate state $p(v)$ is specified by the equality system

$$\frac{\partial^2 p}{\partial t^2} - A\left(\frac{\partial p}{\partial t}\right) - a\frac{\partial p}{\partial t} + K(p) + bp = 0, \ (x,t) \in \Omega_T,$$

$$\sum_{i,j=1}^{n} \left(-a_{ij}\frac{\partial^2 p}{\partial x_j \partial t} + k_{ij}\frac{\partial p}{\partial x_j}\right)\cos(v, x_i) = -\alpha p, \ (x,t) \in \Gamma_T,$$

$$[p] = 0, \ (x,t) \in \gamma_T, \tag{5.7}$$

$$\left[\sum_{i,j=1}^{n} \left(-a_{ij}\frac{\partial^2 p}{\partial x_j \partial t} + k_{ij}\frac{\partial p}{\partial x_j}\right)\cos(v, x_i)\right] = y(v) - z_g, \ (x,t) \in \gamma_T,$$

$$p(x,T) = 0, \ x \in \bar{\Omega}_1 \cup \bar{\Omega}_2,$$

$$\frac{\partial p}{\partial t}(x,T) = 0, \ x \in \bar{\Omega}_1 \cup \bar{\Omega}_2.$$

Problem (5.7) has the unique generalized solution $p(v) \in W(0,T)$ as the unique one for the following equality system:

$$\left(\frac{d^2p}{dt^2}, w\right) - a_0\left(\frac{dp}{dt}, w\right) + a_1(p, w) =$$

$$= -\left(y(v) - z_g, w\right)_{L_2(\gamma)}, \quad \forall w \in V_0, \quad t \in (0, T),$$

$$a_0(p, w) = 0, \quad \forall w \in V_0, \quad t = T, \tag{5.8}$$

$$\left(\frac{dp}{dt}, w\right) = 0, \quad \forall w \in V_0, \quad t = T \ ;$$

in this case, the spaces V_0 and $W(0, T)$ are specified in point 9.2.

Use the difference $y(v) - y(u)$, instead of w in the first equality of system (5.8), where $v = u$, find the integral from 0 to T of the result, and the equality

$$\int_0^T \left(\frac{d^2p(v)}{dt^2}, y(v) - y(u)\right) dt - \int_0^T a_0\left(\frac{dp}{dt}, y(v) - y(u)\right) dt +$$

$$+ \int_0^T a_1\left(p, y(v) - y(u)\right) dt = -\left(y(u) - z_g, y(v) - y(u)\right)_{L_2(\gamma) \times L_2}$$

is obtained, from which the equality

$$(p, v - u)_{L_2(\Gamma) \times L_2} = -\left(y(u) - z_g, y(v) - y(u)\right)_{L_2(\gamma) \times L_2}$$

follows. Therefore, the optimality condition for the control $u \in \mathcal{U}_\partial$ is

$$(-p + \bar{a}u, v - u)_{L_2(\Gamma) \times L_2} \geq 0, \quad \forall v \in \mathcal{U}_\partial. \tag{5.9}$$

Thus, the optimal control $u \in \mathcal{U}_\partial$ is specified by relations (5.4), (5.8) and (5.9). If the constraints are absent, i.e. when $\mathcal{U}_\partial = \mathcal{U}$, then the equality

$$-p(u) + \bar{a}u = 0, \quad (x, t) \in \Gamma_T, \tag{5.10}$$

follows from inequality (5.9). If the solution $(y, p)^T$ to problem (5.4), (5.8), (5.10) is smooth enough on $\bar{\Omega}_{lT}$, $l = 1, 2$, then the differential problem of finding the vector-function $(y, p)^T$, that satisfies equalities (1.1), (1.5), (1.6), (2.1), (2.2) (under $u = 0$) and (5.7) and the condition

$$\sum_{i,j=1}^{n} \left(a_{ij} \frac{\partial^2 y}{\partial x_j \partial t} + k_{ij} \frac{\partial y}{\partial x_j} \right) \cos(\nu, x_i) = -\alpha y + \beta + p/\bar{a}, \quad (x,t) \in \Gamma_T,$$

corresponds to problem (5.4), (5.8), (5.10), where the optimal control is $u = p/\bar{a}$ under $(x,t) \in \Gamma_T$.

9.6 CONTROL UNDER BOUNDARY CONDITION WITH FINAL OBSERVATION

Assume that equation (1.1) is specified in the domain Ω_T. The initial conditions have the form of expressions (1.5) and (1.6) under $t = 0$, conjugation conditions (2.1) and (2.2) (under $u = 0$) are specified on the section γ_T and initial condition (5.1), where the control is $u \in \mathcal{U} = L_2(\Gamma_T)$ is specified, in their turn, on Γ_T.

The cost functional is

$$J(u) = \int_{\Omega} \left(y(x, T; u) - z_g \right)^2 dx + \int_{0}^{T} \int_{\Gamma} \bar{a} u^2 d\Gamma dt, \qquad (6.1)$$

where $0 < a_0 \le \bar{a} \le a_1 < \infty$; $a_0, a_1 = \text{const}$, and it may be rewritten as

$$J(u) = \pi(u, u) - 2L(u) + \int_{\Omega} \left(z_g(x) - y(x, T; 0) \right)^2 dx;$$

where

$$\pi(u, v) = \left(y(\cdot, T; u) - y(\cdot, T; 0), \; y(\cdot, T; v) - y(\cdot, T; 0) \right) + \int_{0}^{T} \int_{\Gamma} \bar{a} uv d\Gamma dt$$

and

$$L(v) = \left(z_g - y(\cdot, T; 0), \ y(\cdot, T; v) - y(\cdot, T; 0) \right).$$

The generalized problem corresponds to initial boundary-value problem (1.1), (1.5), (1.6), (2.1), (2.2) (under $u = 0$), (5.1) and means to find a function $y(x, t; u) \in W(0, T)$ that satisfies equality system (5.4) $\forall w(x) \in V_0$; the space $W(0, T)$ is specified in point 9.1 and the spaces V and V_0 are specified, in their turn, in point 9.2. Theorem 5.1 takes place. Equality (5.5) is true, from which the inequality

$$a_1 (\tilde{y}' - \tilde{y}'', \ \tilde{y}' - \tilde{y}'')(T) + (\mu - \varepsilon c_0) \left\| \frac{d(\tilde{y}' - \tilde{y}'')}{dt} \right\|_{V \times L_2}^2 \leq \frac{1}{4\varepsilon} \| u' - u'' \|_{L_2(\Gamma) \times L_2}^2$$

is obtained, where $0 < \varepsilon < \mu / c_0$; μ, $c_0 = \text{const} > 0$.

Proceed from it and from the Friedrichs inequality, and the inequality

$$\| \tilde{y}' - \tilde{y}'' \|(T) \leq c_1 \| u' - u'' \|_{L_2(\Gamma) \times L_2}$$

follows. Therefore, the linear functional $L(\cdot)$ and bilinear form $\pi(\cdot, \cdot)$ of cost functional (6.1) are continuous on $\mathcal{U}$.

On the basis of [58, Chapter 1, Theorem 1.1], the validity of the following statement is proved.

Theorem 6.1. *If a system state is determined as a solution to problem (5.4), then there exists a unique element u of a convex set $\mathcal{U}_\partial$ that is closed in $\mathcal{U}$, and relation (1.37) takes place for u, where the cost functional has the form of expression (6.1).*

As for the control $v \in \mathcal{U}$, the conjugate state $p(v)$ is specified by equality system

$$\frac{\partial^2 p}{\partial t^2} - A\left(\frac{\partial p}{\partial t} \right) - a \frac{\partial p}{\partial t} + K(p) + bp = 0, \ (x, t) \in \Omega_T,$$

$$\sum_{i,j=1}^{n} \left(-a_{ij} \frac{\partial^2 p}{\partial x_j \partial t} + k_{ij} \frac{\partial p}{\partial x_j} \right) \cos(\nu, x_i) = -\alpha p, \ (x, t) \in \Gamma_T,$$

$$[p] = 0, \quad (x,t) \in \gamma_T, \tag{6.2}$$

$$\left[\sum_{i,j=1}^{n} \left(-a_{ij} \frac{\partial^2 p}{\partial x_j \partial t} + k_{ij} \frac{\partial p}{\partial x_j} \right) \cos(\nu, x_i) \right] = 0, \quad (x,t) \in \gamma_T,$$

$$p(x,T;v) = 0, \quad x \in \bar{\Omega}_1 \cup \bar{\Omega}_2,$$

$$\frac{\partial p}{\partial t}(x,T;v) = y(x,T;v) - z_g, \quad x \in \bar{\Omega}_1 \cup \bar{\Omega}_2.$$

Problem (6.2) has the unique generalized solution $p(v) \in W(0,T)$ as the unique one to the equality system

$$\left(\frac{d^2 p}{dt^2}, w \right) - a_0 \left(\frac{dp}{dt}, w \right) + a_1(p, w) = 0, \quad \forall w \in V_0, \ t \in (0,T),$$

$$a_0(p, w) = 0, \quad \forall w \in V_0, \ t = T, \tag{6.3}$$

$$\left(\frac{dp}{dt}, w \right) = \left(y(x,T;v) - z_g, w \right), \quad \forall w \in V_0, \ t = T \ ;$$

in this case, the spaces V_0 and $W(0,T)$ are specified in point 9.2.

Use the difference $y(v) - y(u)$, instead of u in the first equality of system (6.3), where $v = u$, find the integral from 0 to T of the result, and the equality

$$\left(y(x,t;u) - z_g, y(v) - y(u) \right)(T) + \int_0^T (p, v - u)_{L_2(\Gamma)} dt = 0$$

is obtained that yields the equality

$$\left(y(x,t;u) - z_g, y(v) - y(u) \right)(T) = - \int_0^T (p, v - u)_{L_2(\Gamma)} dt \ .$$

Therefore, the control $u \in \mathcal{U}_\partial$ is optimal if and only if the following inequality is true:

$$\int\limits_{0}^{T}\int\limits_{\Gamma}(-p+\bar{a}u)(v-u)\,d\Gamma\,dt \geq 0, \quad \forall v \in \mathcal{U}_{\partial}.\qquad(6.4)$$

Thus, the optimal control $u \in \mathcal{U}_{\partial}$ is specified by relations (5.4), (6.3) and (6.4). If the constraints are absent, i.e. when $\mathcal{U}_{\partial} = \mathcal{U}$, then the equality

$$u = p/\bar{a}, \ (x,t) \in \Gamma_T,\qquad(6.5)$$

is obtained from inequality (6.4).

If the solution $(y,p)^{\mathrm{T}}$ to problem (5.4), (6.3), (6.5) is smooth enough on $\bar{\Omega}_{lT}$, $l = 1, 2$, then the differential problem, specified by equalities (1.1), (1.5), (1.6), (2.1), (2.2) (under $u{=}0$), (5.1) (under $u = p/\bar{a}$) and (6.2), corresponds to problem (5.4), (6.3), (6.5), where the optimal control is $u = p/\bar{a}$ under $(x,t) \in \Gamma_T$.

OPTIMAL CONTROL OF A DEFORMED
COMPLICATED SOLID BODY STATE

10.1   DISTRIBUTED CONTROL

Assume that the elastic equilibrium equation system [122]

$$-\sum_{k=1}^{3} \frac{\partial \sigma_{ik}}{\partial x_k} = f_i(x), \quad i = \overline{1,3}, \tag{1.1}$$

is specified in bounded, continuous and strictly Lipschitz domains Ω_1 and $\Omega_2 \in R^3$; in this case, $x = (x_1, x_2, x_3)$; $\sigma_{ki} = \sigma_{ik} = \sigma_{ik}(y) = \sigma_{ik}(x;y) = \sum_{l,m=1}^{3} c_{iklm} \varepsilon_{lm}$; σ_{ik} and ε_{lm} are elements, respectively, of stress and deformation tensors, c_{iklm} are elastic constants; $i,k = \overline{1,3}$, $\varepsilon_{ml} = \varepsilon_{lm} = \varepsilon_{lm}(y) = \varepsilon_{lm}(x;y) = \frac{1}{2}\left(\frac{\partial y_l}{\partial x_m} + \frac{\partial y_m}{\partial x_l}\right)$, $y = (y_1(x), y_2(x), y_3(x))^T$ is a displacement vector, $y_i(x)$ is a projection of this vector on an i-th axis of the Cartesian coordinate system and $f = (f_1(x), f_2(x), f_3(x))^T$ is a mass force vector.

Suppose that the elasticity coefficients have the features of symmetry $c_{iklm} = c_{lmik} = c_{kiml}(x)$ and satisfy the condition [72]

$$\sum_{i,k,l,m=1}^{3} c_{iklm}\, \varepsilon_{ik}\, \varepsilon_{lm} \geq \alpha_0 \sum_{i,k=1}^{3} \varepsilon_{ik}^2, \quad \alpha_0 = \text{const} > 0. \tag{1.1'}$$

The condition

$$y = 0 \tag{1.2}$$

is specified on a boundary $\Gamma = (\partial\Omega_1 \cup \partial\Omega_2) \backslash \gamma$ ($\gamma = \partial\Omega_1 \cap \partial\Omega_2 \neq \varnothing$) of a domain $\overline{\Omega}$ and, on a section $\gamma = \partial\Omega_1 \cap \partial\Omega_2$ of the domain $\overline{\Omega}$, the conjugation conditions for an imperfect contact are [21]

$$[y_n] = 0 \tag{1.3}$$

and

$$[\sigma_n] = 0, \quad [\tau_s] = 0, \quad \{\tau_s\}^{\pm} = r[y_s]. \tag{1.4}$$

Conditions (1.3) and (1.4) illustrate the continuities for normal components of displacement and stress vectors and for a stress vector component tangent and show the proportionality of a stress vector component to jumping of a displacement vector component tangent.

In this case, $[\varphi] = \varphi^+ - \varphi^-$, $\varphi^+ = \{\varphi\}^+ = \varphi(x)$ under $x \in \partial\Omega_2 \cap \gamma$, $\varphi^- = \{\varphi\}^- = \varphi(x)$ under $x \in \partial\Omega_1 \cap \gamma$, $r = r(x)$, $0 \leq r$, $r \in L_2(\gamma)$.

Let there be a control Hilbert space $\mathcal{U}$ and mapping $B \in \mathcal{L}(\mathcal{U}; V')$, where V' is a space dual with respect to a state Hilbert space V. Assume the following: $\mathcal{U} = L_2(\Omega)$.

For every control $u \in \mathcal{U}$, determine a system state y as a generalized solution to the boundary-value problem specified by the equation

$$-\sum_{k=1}^{3} \frac{\partial \sigma_{ik}(y)}{\partial x_k} = f_i(x) + B_i u, \quad i = \overline{1,3}, \ x \in \Omega, \tag{1.5}$$

and by conditions (1.2)–(1.4), where $Bu = (B_1 u, B_2 u, B_3 u)^{\mathrm{T}}$, $u \in L_2(\Omega)$.

Specify the observation

$$Z(u) = C\, y(u), \tag{1.6}$$

where $C \in \mathcal{L}(V; \mathcal{H})$ and $\mathcal{H}$ is some Hilbert space. Assume the following:

$$C\, y(u) \equiv y(u), \quad \mathcal{H} = V \subset L_2(\Omega). \tag{1.7}$$

Bring a value of the cost functional

$$J(u) = \|C\,y(u) - z_g\|_{\mathcal{H}}^2 + (\mathcal{N}u, u)_{\mathcal{U}} \tag{1.8}$$

in correspondence with every control $u \in \mathcal{U}$; in this case, $z_g = \left(z_{g_1}, z_{g_2}, z_{g_3}\right)^{\mathrm{T}}$ is a known element of the space $\mathcal{H}$,

$$\mathcal{N} \in \mathcal{L}(\mathcal{U}; \mathcal{U}), \quad (\mathcal{N}u, u)_{\mathcal{U}} \geq v_0 \|u\|_{\mathcal{U}}^2, \quad v_0 = \text{const} > 0, \quad \forall u \in \mathcal{U}. \tag{1.9}$$

Assume the following: $f \in L_2(\Omega)$, $Bu \equiv u \in L_2(\Omega)$, $\mathcal{N}u = \bar{a}(x)u$, $0 < a_0 \leq \bar{a}(x) \leq a_1 < \infty$, $\bar{a}|_{\Omega_l} \in C(\Omega_l)$, $l = 1, 2$; $a_0, a_1 = \text{const}$, $(\varphi, \psi)_{\mathcal{U}} = (\varphi, \psi) = \int_\Omega \varphi\,\psi\,dx$.

The forthcoming statement is true.

Theorem 1.1. *A unique state, namely, a function* $y = y(u) \in$
$\in V = \left\{ v: v|_{\Omega_l} \in W_2^1(\Omega_l), \; l = 1, 2; \; v = (v_1, v_2, v_3)^{\mathrm{T}}, \; v|_\Gamma = 0, \; [v_n]|_\gamma = 0 \right\}$,
corresponds to every control $u \in \mathcal{U}$, *delivers a minimum to the energy functional* [21]

$$\Phi(v) = a(v, v) - 2l(v) \tag{1.10}$$

on V, and it is a unique solution in V to the weakly stated problem: Find an element $y \in V$ that meets the equation

$$a(y, v) = l(u, v), \quad \forall v \in V, \tag{1.11}$$

where

$$a(y, v) = \int_\Omega \sum_{i,k,l,m=1}^3 c_{iklm}\,\varepsilon_{ik}(y)\,\varepsilon_{lm}(v)\,dx + \int_\gamma r[y_s][v_s]\,d\gamma,$$

$$l(v) = l(u, v) = (f, v) + (u, v). \tag{1.11'}$$

Proof. Proceed from the Cauchy-Bunyakovsky inequalities and embedding theorems [55], and the following inequalities are obtained $\forall v, z \in V$:

$$|a(v, z)| \leq c_1 \|v\|_V \|z\|_V \quad \text{and} \quad |l(v)| \leq c_2 \|v\|_V; \tag{1.12}$$

in this case, $\|v\|_V = \left\{ \sum_{i=1}^{2} \|v\|_{W_2^1(\Omega_i)}^2 \right\}^{1/2}$, $\|\cdot\|_{W_2^1(\Omega_i)}$ is the norm of the Sobolev

space $W_2^1(\Omega_i)$.

Consider inequality (1.1') and the Friedrichs [21] and Korn [16] ones, and the inequality

$$a(v,v) \geq \alpha_0' \|v\|_V^2, \quad \forall v \in V, \tag{1.13}$$

is derived. Inequalities (1.12) provide the continuity [49] of the bilinear form $a(\cdot,\cdot): V \times V \to R^1$ and linear functional $l(\cdot): V \to R^1$ on V and inequality (1.13) provides the V-ellipticity of the form $a(\cdot,\cdot)$ on V.

Therefore, according to the Lax-Milgramm lemma [16] problem (1.11) has the unique solution $y = y(u)$ in V. The equivalence for problems (1.10) and (1.11) is easily stated. Theorem is proved.

Take the aforesaid assumptions into account, and cost functional (1.8) may be rewritten as

$$J(u) = \|y(u) - z_g\|^2 + (\bar{a}u, u) = \pi(u,u) - 2L(u) + \|z_g - y(0)\|^2, \tag{1.14}$$

where $\|\varphi\| = (\varphi,\varphi)^{1/2}$ and the bilinear form $\pi(\cdot,\cdot)$ and linear functional $L(\cdot)$ are specified by the expressions

$$\pi(u,v) = (y(u) - y(0), y(v) - y(0)) + (\bar{a}u, v),$$

$$L(v) = (z_g - y(0), y(v) - y(0)). \tag{1.15}$$

The form $\pi(\cdot,\cdot)$ is coercive on $\mathcal{U}$ since

$$\pi(u,u) \geq a_0(u,u). \tag{1.15'}$$

Let $\tilde{y}' = \tilde{y}(u')$ and $\tilde{y}'' = \tilde{y}(u'')$ be solutions from V to problem (1.11) under $f = 0$ and under a function $u = u(x)$ that is equal, respectively, to u' and u''. Proceed from inequality (1.13), and the inequality

$$\alpha_0' \|\tilde{y}' - \tilde{y}''\|^2 \leq \alpha_0' \|\tilde{y}' - \tilde{y}''\|_V \leq a(\tilde{y}' - \tilde{y}'', \ \tilde{y}' - \tilde{y}'') \leq$$

$$\leq \|u' - u''\| \cdot \|\tilde{y}' - \tilde{y}''\|, \tag{1.16}$$

where $\alpha_0' = \text{const} > 0$, follows from equation (1.11).

The derived inequality provides the continuity of the linear functional $L(\cdot)$ and bilinear form $\pi(\cdot,\cdot)$ on $\mathcal{U}$.

On the basis of [58, Chapter 1, Theorem 1.1], the validity of the following statement is proved.

Theorem 1.2. *Let a system state be determined as a solution to equivalent problems (1.10) and (1.11). Then, there exists a unique element u of a convex set $\mathcal{U}_\partial$ that is closed in $\mathcal{U}$, and*

$$J(u) = \inf_{v \in \mathcal{U}_\partial} J(v) \tag{1.17}$$

takes place for u.

Definition 1.1. If an element $u \in \mathcal{U}_\partial$ meets condition (1.17), it is called an optimal control.

The inequality

$$\pi(u, v - u) \geq L(v - u), \quad \forall v \in \mathcal{U}_\partial, \tag{1.18}$$

is the necessary and sufficient condition for $u \in \mathcal{U}_\partial$ to be the optimal control.

As for the control $v \in \mathcal{U}$, the conjugate state $p(v) \in V^* = V$ is specified as the generalized solution to the problem specified, in its turn, by the following equalities:

$$-\sum_{k=1}^{3} \frac{\partial \sigma_{ik}(p)}{\partial x_k} = y_i(v) - z_{g_i}, \quad i = \overline{1,3}, \quad x \in \Omega,$$

$$p = 0, \quad x \in \Gamma,$$

$$[p_n] = 0, \quad x \in \gamma, \tag{1.19}$$

$$[\sigma_n(p)] = 0, \quad [\tau_s(p)] = 0, \quad \{\tau_s(p)\}^{\pm} = r[p_s], \quad x \in \gamma.$$

The generalized problem corresponds to boundary-value problem (1.19) and means to find a function $p \in V$ that satisfies the equation

$$a(p,z) = l_1(y,z), \quad \forall z \in V, \tag{1.20}$$

where the bilinear form $a(\cdot,\cdot)$ is specified by expression (1.11'), and

$$l_1(y,z) = (y,z) - (z_g,z).$$

Therefore, the necessary and sufficient condition for the existence of the optimal control $u \in \mathcal{U}_\partial$ is the one under which the relations

$$a(y(u),z) = l(u,z), \quad \forall z \in V, \tag{1.21}$$

$$a(p(u),z) = l_1(y(u),z), \quad \forall z \in V, \tag{1.22}$$

and

$$\left(y(u) - z_g, y(v) - y(u)\right) + \left(\bar{a}u, v - u\right) \geq 0, \quad \forall v \in \mathcal{U}_\partial, \tag{1.23}$$

are met.

Choose the difference $y(v) - y(u)$ instead of z in equalities (1.21) and (1.22), and the equality

$$\left(y(u) - z_g, y(v) - y(u)\right) = \left(v - u, p(u)\right), \quad \forall v \in \mathcal{U}_\partial,$$

is obtained. Take it into account, and inequality (1.23) has the form

$$\left(p(u) + \bar{a}u, v - u\right) \geq 0, \quad \forall v \in \mathcal{U}_\partial. \tag{1.24}$$

Therefore, the necessary and sufficient condition for the existence of the optimal control $u \in \mathcal{U}_\partial$ is the one under which the relations (1.21), (1.22) and (1.24) are met.

If the constraints are absent, i.e. when $\mathcal{U}_\partial = \mathcal{U}$, then the equality

$$p(u) + \bar{a}u = 0 \tag{1.25}$$

follows from condition (1.24). Therefore, when the constraints are absent, the control u can be excluded from equality (1.21) by means of equality (1.25). On the basis of equalities (1.21), (1.22) and (1.25), the problem is obtained: Find a vector-function $(y,p)^T \in H = \left\{v = (v_1,v_2,)^T : v_i \in V, \right.$

$i = 1,2\}$ that satisfies the equality system

$$a(y,z) = l(-p/\bar{a},z), \quad \forall z \in V,$$

$$a(p,z) = l_1(y,z), \quad \forall z \in V, \tag{1.26}$$

and the vector solution $(y,p)^{\mathrm{T}}$ is found from this system along with the optimal control

$$u = -p/\bar{a}, \quad x \in \Omega. \tag{1.26'}$$

If the vector solution $(y,p)^{\mathrm{T}}$ to problem (1.26) is smooth enough on $\bar{\Omega}_l$, viz., $y, p\big|_{\bar{\Omega}_l} \in C^1(\bar{\Omega}_l) \cap C^2(\Omega_l)$, $l=1,2$, then the differential problem of finding the vector-function $(y,p)^{\mathrm{T}}$, that satisfies the relations

$$-\sum_{k=1}^{3} \frac{\partial \sigma_{ik}(y)}{\partial x_k} + p_i/\bar{a} = f_i, \quad i = \overline{1,3}, \quad x \in \Omega,$$

$$-\sum_{k=1}^{3} \frac{\partial \sigma_{ik}(p)}{\partial x_k} - y_i = -z_{g_i}, \quad i = \overline{1,3}, \quad x \in \Omega,$$

$$y = 0, \quad p = 0, \quad x \in \Gamma,$$

$$[y_n] = [p_n] = 0, \quad x \in \gamma, \tag{1.27}$$

$$[\sigma_n(y)] = [\sigma_n(p)] = 0, \quad x \in \gamma,$$

$$[\tau_s(y)] = [\tau_s(p)] = 0, \quad \{\tau_s(y)\}^{\pm} = r[y_s], \quad \{\tau_s(p)\}^{\pm} = r[p_s], \quad x \in \gamma,$$

corresponds to problem (1.26).

Definition 1.2. A generalized (weak) solution to boundary-value problem (1.27) is called a vector-function $U = (y,p)^{\mathrm{T}} \in H$ that satisfies the equation

$$a(U,Z) = l(Z), \quad \forall Z \in H, \tag{1.28}$$

where

$$U = (U_1, U_2)^{\mathrm{T}}, \quad Z = (Z_1, Z_2)^{\mathrm{T}}, \quad U_i, Z_i \in V, \quad i = 1,2;$$

$$a(U,Z) = a_0(U,Z) + (p/\bar{a}, Z_1) - (y, Z_2),$$

$$a_0(U,Z) = \int_{\Omega} \sum_{j=1}^{2} \sum_{i,k,l,m=1}^{3} c_{iklm} \, \varepsilon_{ik}(U_j) \, \varepsilon_{lm}(Z_j) \, dx +$$

$$+ \int_\gamma r \sum_{j=1}^{2} [U_{js}][Z_{js}] \, d\gamma; \tag{1.29}$$

$$l(Z) = (f, Z_1) - (z_g, Z_2). $$

Let the constraint

$$\alpha_0 \mu - \frac{1}{2a_0} - \frac{1}{2} > 0 \tag{1.30}$$

be met, where α_0 and μ are the positive constants, respectively, from inequality (1.1') and the Friedrichs one [21].

Proceed from the Cauchy-Bunyakovsky and Friedrichs inequalities and embedding theorems [55], take constraints (1.1') and (1.30) into account, and the inequalities

$$a(Z,Z) \geq \bar{\alpha}_1 \|Z\|_H^2, \quad \forall Z \in H, \quad \bar{\alpha}_1 = \text{const} > 0,$$

$$|a(U,Z)| \leq c_1 \|U\|_H \|Z\|_H, \quad \forall U, Z \in H, \quad c_1 = \text{const} > 0, \tag{1.31}$$

are true for the bilinear form $a(\cdot,\cdot): H \times H \to R^1$, i.e. such form is H-elliptic and continuous [49] on H, where

$$\|Z\|_H = \left\{ \sum_{i,j=1}^{2} \|Z_i\|_{W_2^1(\Omega_j)}^2 \right\}^{1/2}. \tag{1.32}$$

Consider the Cauchy-Bunyakovsky inequality, and the inequality

$$|l(Z)| \leq c_2 \|Z\|_H, \quad c_2 = \text{const}, \tag{1.33}$$

is obtained $\forall Z \in H$.

The following statement is valid.

Theorem 1.3. *Let constraints (1.1') and (1.30) be met. Then, there exists a solution $U \in H$ to problem (1.28).*

The validity of Theorem 1.3 is stated on the basis of the Lax-Milgramm lemma, when inequalities (1.31) and (1.33) are taken into account.

Problem (1.28) can be solved approximately by means of the finite-element method. Divide the domains $\overline{\Omega}_i$ into N_i finite-elements

$\bar{e}_i^j$ $(j = \overline{1, N_i},\ i = 1, 2)$ of the regular family [16]. Specify the subspace $H_k^N \subset H$ $(N = N_1 + N_2)$ of the vector-functions $V_k^N(x)$. The components $v_{lk}^N\big|_{\bar{\Omega}_i} \in C(\bar{\Omega}_i)$, $l = \overline{1, 6}$, $i = 1, 2$, of $V_k^N(x)$ are the complete polynomials of the power k that contain the variables x_1, x_2 and x_3 at every $\bar{e}_i^j$. Then, the linear algebraic equation system

$$A\bar{U} = B \qquad (1.34)$$

follows from equation (1.28), and the solution $\bar{U}$ to system (1.34) exists and such solution is unique. The vector $\bar{U}$ specifies the unique approximate solution $U_k^N \in H_k^N$ to problem (1.28) as the unique one to the equation

$$a\left(U_k^N,\ V_k^N\right) = l\left(V_k^N\right), \quad \forall V_k^N \in H_k^N. \qquad (1.35)$$

If $U = U(x) \in H$ are the solutions to problem (1.28), then:

$$a\left(U - U_k^N,\ V_k^N\right) = 0, \quad \forall V_k^N \in H_k^N.$$

The following relation is true:

$$\bar{\alpha}_1 \left\| U - U_k^N \right\|_H \le a\left(U - U_k^N, U - U_k^N\right) = a\left(U - U_k^N, U - \tilde{U}\right), \forall \tilde{U}_k^N \in H_k^N.$$

Therefore,

$$\left\| U - U_k^N \right\|_H \le \frac{c_1}{\bar{\alpha}_1} \left\| U - \tilde{U} \right\|_H, \qquad (1.36)$$

where $\bar{\alpha}_1$ and c_1 are the positive constants from respective inequalities (1.31). Suppose that $\tilde{U} \in H_k^N$ is a complete interpolation polynomial for the solution U at every $\bar{e}_i^j$.

Take the interpolation estimates [16] into account, assume that every component U_l $(l = \overline{1, 6})$ of the solution U on Ω_j belongs to the Sobolev space $W_2^{k+1}(\Omega_j)$, $j = 1, 2$, and the estimate

$$\left\| U - U_k^N \right\|_H \le ch^k, \qquad (1.37)$$

where h is a maximum diameter of all the finite elements $\bar{e}_i^j$, $c = const$, follows from inequality (1.36).

Take estimate (1.37) into account, and the estimate

$$\left\| u - u_k^N \right\|_V \le c_2 \left\| p - p_k^N \right\|_V \le c_3 h^k \qquad (1.38)$$

takes place for the approximation $u_k^N(x) = -p_k^N / \bar{a}$ of the control $u = u(x)$.

10.2 DISTRIBUTED CONTROL WITH OBSERVATION AND WITH TAKING SIGHT ON JUMPING OF A DISPLACEMENT VECTOR COMPONENT TANGENT AT A DOMAIN SECTION

Assume that elasticity theory equation system (1.5), where $Bu \equiv u$, is specified in the bounded, continuous and strictly Lipschitz domains Ω_1 and $\Omega_2 \in R^3$ for every control $u \in \mathcal{U} = L_2(\Omega)$. Boundary condition (1.2) is specified on the boundary Γ and constraints (1.3) and (1.4) are specified, in their turn, on the section γ. Specify the observation as follows:

$$Cy(u) = [y_s(u)], \quad x \in \gamma.$$

Bring a value of the cost functional

$$J(u) = \int_\gamma \left([y_s(u)] - z_g \right)^2 d\gamma + (\bar{a}u, u) \qquad (2.1)$$

in correspondence with every control $u \in \mathcal{U}$; in this case, z_g is a known scalar function from $L_2(\gamma)$, and the function $\bar{a}(x)$ is specified in point 10.1.

It is shown in point 10.1 that a unique state, namely, a function $y(u) \in V$ corresponds to every control $u \in \mathcal{U}$, minimizes energy functional (1.10) on V and meets equation (1.11), where the bilinear form

$a(\cdot,\cdot): V \times V \to R^1$ and functional $l(u,v)$ are specified by respective expressions (1.11′).

Rewrite cost functional (2.1) as

$$J(u) = \pi(u,u) - 2L(u) + \left\| z_g - [y_s(0)] \right\|^2_{L_2(\gamma)}; \qquad (2.2)$$

in this case, the bilinear form $\pi(\cdot,\cdot)$ and linear functional $L(\cdot)$ are expressed as

$$\pi(u,v) = \left([y_s(u)] - [y_s(0)], \ [y_s(v)] - [y_s(0)] \right)_{L_2(\gamma)} + (\bar{a}u, v)$$

and

$$L(v) = \left(z_g - [y_s(0)], \ [y_s(v)] - [y_s(0)] \right)_{L_2(\gamma)},$$

where $(\varphi, \psi)_{L_2(\gamma)} = \displaystyle\int\limits_{\gamma} \varphi\psi \, d\gamma$.

Inequality (1.15′) is true for the bilinear form $\pi(\cdot,\cdot)$, i.e. such form is coercive on $\mathcal{U}$. Introduce the denotation $\tilde{y}(v) = y(v) - y(0)$. Then:

$$[\tilde{y}_s(\alpha_1 u_1 + \alpha_2 u_2)] = \sum_{i=1}^{2} \alpha_i [\tilde{y}_s(u_i)], \ \forall \alpha_1, \alpha_2 \in R^1, \ \forall u_1, u_2 \in \mathcal{U}. \ (2.3)$$

Proceed from equality (2.3), and the linearity of the functional $L(v)$ and the bilinearity of the form $\pi(u,v)$ are stated.

Let $\tilde{y}' = \tilde{y}(u')$ and $\tilde{y}'' = \tilde{y}(u'')$ be solutions from V to problem (1.11) under $f = 0$ and under a function $u = u(x)$ that is equal, respectively, to u' and u''. Since the inequality

$$\int\limits_{\gamma} \left((\tilde{y}'_s - \tilde{y}''_s)^+ - (\tilde{y}'_s - \tilde{y}''_s)^- \right)^2 d\gamma \leq$$

$$\leq 2 \left(\int\limits_{\gamma} \left((\tilde{y}'_s - \tilde{y}''_s)^+ \right)^2 d\gamma + \int\limits_{\gamma} \left((\tilde{y}'_s - \tilde{y}''_s)^- \right)^2 d\gamma \right) \leq$$

$$\leq c_0 \left\| \tilde{y}'_s - \tilde{y}''_s \right\|^2_V \leq c'_0 \left\| \tilde{y}' - \tilde{y}'' \right\|^2_V, \ c'_0 = \text{const} > 0, \qquad (2.3')$$

is true in accordance with the embedding theorems [55], then:

$$\frac{1}{c_0'}\left\|[\tilde{y}_s']-[\tilde{y}_s'']\right\|_{L_2(\gamma)}^2 \leq \|\tilde{y}'-\tilde{y}''\|_V^2 \leq$$

$$\leq \frac{1}{\alpha_0'} a(\tilde{y}'-\tilde{y}'',\tilde{y}'-\tilde{y}'') \leq \frac{1}{\alpha_0'}\|u'-u''\|\|\tilde{y}'-\tilde{y}''\|_V .$$

Therefore, the inequality

$$\left\|[\tilde{y}_s']-[\tilde{y}_s'']\right\|_{L_2(\gamma)} \leq \frac{\sqrt{c_0'}}{\alpha_0'}\|u'-u''\|$$

is obtained that provides the continuity of the linear functional $L(\cdot)$ and bilinear form $\pi(\cdot,\cdot)$ on $\mathcal{U}$.

On the basis of [58, Chapter 1, Theorem 1.1], the validity of the following statement is proved.

Theorem 2.1. *Let a system state be determined as a solution to problems (1.10) and (1.11). Then, there exists a unique element u of a convex set $\mathcal{U}_\partial$ that is closed in $\mathcal{U}$, and relation (1.17) takes place for u, where the cost functional has the form of expression (2.1).*

As for the control $v \in \mathcal{U}$, the conjugate state $p(v) \in V$ is specified a generalized solution to the problem specified, in its turn, by the equalities

$$\sum_{k=1}^{3}\frac{\partial \sigma_{ik}(p)}{\partial x_k}=0, \quad i=\overline{1,3}, \quad x \in \Omega,$$

$$p=0, \quad x \in \Gamma,$$

$$[p_n]=0, \quad [\sigma_n(p)]=0, \quad x \in \gamma, \tag{2.4}$$

$$[\tau_s(p)]=0, \quad \{\tau_s(p)\}^{\pm}=r[p_s]+[y_s]-z_g, \quad x \in \gamma.$$

The generalized problem corresponds to boundary-value problem (2.4) and means to find a function $p \in V$ that satisfies the equation

$$a(p,z)=l_1(y,z), \quad \forall z \in V, \tag{2.5}$$

where the bilinear form $a(\cdot,\cdot)$ is specified by expression (1.11'), and

$$l_1(y,z) = -\int_\gamma \left([y_s] - z_g\right)[z_s] d\gamma. \tag{2.5'}$$

Therefore, the necessary and sufficient condition for the existence of the optimal control $u \in \mathcal{U}_\partial$ is the one under which the relations

$$a(y(u),z) = l(u,z), \quad \forall z \in V, \tag{2.6}$$

$$a(p(u),z) = l_1(y(u),z), \quad \forall z \in V, \tag{2.7}$$

and

$$\left([y_s(u)] - z_g, \ [y_s(v)] - [y_s(u)]\right)_{L_2(\gamma)} + (au, v-u) \ge 0, \quad \forall v \in \mathcal{U}_\partial, \tag{2.8}$$

are met.

Choose the difference $y(v) - y(u)$ instead of z, obtain the equality

$$a(p(u), y(v) - y(u)) = -\int_\gamma \left([y_s(u)] - z_g\right)\left([y_s(v)] - [y_s(u)]\right) d\gamma$$

from equality (2.7), and the equality

$$a(p(u), y(v) - y(u)) = (v - u, p(u))$$

follows from equality (2.6). Therefore, the equality

$$\int_\gamma \left([y_s(u)] - z_g\right)\left([y_s(v)] - [y_s(u)]\right) d\gamma = -(p(u), v-u) \tag{2.8'}$$

is derived. Take it into account, and inequality (2.8) has the form

$$(-p(u) + au, v-u) \ge 0, \quad \forall v \in \mathcal{U}_\partial. \tag{2.9}$$

If the constraints are absent, i.e. when $\mathcal{U}_\partial = \mathcal{U}$, then the equality

$$-p + au = 0, \quad x \in \Omega, \tag{2.10}$$

follows from inequality (2.9). Therefore, when the constraints are absent, the control u can be excluded from equality (2.6) by means of equality (2.10). On the basis of equalities (2.6), (2.7) and (2.10), the problem is obtained: Find the vector-function $(y,p)^T \in H = \{v = (v_1, v_2)^T : v_i \in V,$ $i = 1, 2\}$ that satisfies the equality system

$$a(y,z) = l(p/\bar{a},z), \ \forall z \in V \ ,$$
$$a(p,z) = l_1(y,z), \ \forall z \in V \ , \tag{2.11}$$

and the vector-solution $(y,p)^{\mathrm{T}}$ is found from this system along with the optimal control

$$u = p/\bar{a}, \ x \in \Omega. \tag{2.12}$$

If the vector solution $(y,p)^{\mathrm{T}}$ to problem (2.11) is smooth enough on $\overline{\Omega}_l$, viz., y, $p|_{\overline{\Omega}_l} \in C^1(\overline{\Omega}_l) \cap C^2(\Omega_l)$, $l = 1,2$, then the differential problem of finding the vector-function $(y,p)^{\mathrm{T}}$, that satisfies system (2.4) and equalities

$$-\sum_{k=1}^{3} \frac{\partial \sigma_{ik}(y)}{\partial x_k} - p_i/\bar{a} = f_i, \ i = \overline{1,3}, \ x \in \Omega,$$

$$y = 0, \ x \in \Gamma,$$

$$[y_n] = 0, \ [\sigma_n(y)] = 0, \ x \in \gamma,$$

$$[\tau_s(y)] = 0, \ \{\tau_s(y)\}^{\pm} = r[y_s], \ x \in \gamma, \tag{2.13}$$

corresponds to problem (2.11).

Definition 2.1. A generalized (weak) solution to boundary-value problem (2.4), (2.13) is called a vector-function $U = (y,p)^{\mathrm{T}} \in H$ that satisfies equation like (1.28), where

$$a(U,Z) = a_0(U,Z) - (p/\bar{a},Z_1) + ([y_s],[Z_{2s}])_{L_2(\gamma)},$$

$$l(Z) = (f,Z_1) + (z_g,[Z_{2s}])_{L_2(\gamma)}. \tag{2.13'}$$

Let the constraint

$$\min\left\{\alpha_0\mu - \frac{1}{a_0}, \ \alpha_0\mu - 1, \ \frac{\alpha_0'}{2} - \frac{c_0'}{2}\right\} > 0 \tag{2.14}$$

be met, where α_0', c_0' and μ are the constants, respectively, from inequalities (1.13) and (2.3') and the Friedrichs one.

Therefore, inequalities like (1.31) and (1.33) take place for the bilinear form $a(\cdot,\cdot)$ and linear functional $l(\cdot)$ specified by respective expressions (2.13′). The following statement is valid.

Theorem 2.2. *Let constraint (2.14) be met. Then, there exists a solution* $U \in H$ *to problem (1.28), where the bilinear form* $a(\cdot,\cdot)$ *and linear functional* $l(\cdot)$ *are specified by respective expressions (2.13′).*

Assume that the classical solution U on Ω_l to problem (2.4), (2.13) belongs to the Sobolev space $W_2^{k+1}(\Omega_l)$, $l = 1,2$. Then, estimate like (1.37) takes place for the approximate solution $U_k^N \in H_k^N$ obtained by means of the finite-element method. Therefore, estimate (1.38) takes place, in its turn, for the approximation $u_k^N(x) = p_k^N / \bar{a}$ of the control $u = p/\bar{a}$.

10.3 DISTRIBUTED CONTROL WITH OBSERVATION AND WITH TAKING SIGHT ON A NORMAL DISPLACEMENT VECTOR COMPONENT AT A THIN INCLUSION

Assume that equation system (1.5), where $Bu \equiv u$, is specified in the bounded, continuous and strictly Lipschitz domains Ω_1 and $\Omega_2 \in R^3$ for every control $u \in \mathcal{U} = L_2(\Omega)$. Boundary condition (1.2) is specified on the boundary Γ and constraints (1.3) and (1.4) are specified, in their turn, on the section γ. Specify the observation as follows:

$$Cy(u) = y_n(u), \quad x \in \gamma.$$

Bring a value of the cost functional

$$J(u) = \int_\gamma \left(y_n(u) - z_g \right)^2 d\gamma + \left(\bar{a}u, u \right) \tag{3.1}$$

in correspondence with every control $u \in \mathcal{U}$; in this case, z_g is a known scalar function from $L_2(\gamma)$, and the function $\bar{a}(x)$ is specified in point 10.1.

It is shown in point 10.1 that a unique state, namely, a function $y(u) \in V$ corresponds to every control $u \in \mathcal{U}$, minimizes functional (1.10) on V and meets equation (1.11), where the bilinear form $a(\cdot,\cdot)$ and functional $l(\cdot,\cdot)$ are specified by respective expressions (1.11′).

Rewrite cost functional (3.1) as

$$J(u) = \pi(u,u) - 2L(u) + \left\| z_g - y_n(0) \right\|^2_{L_2(\gamma)}; \qquad (3.2)$$

in this case, the bilinear form $\pi(\cdot,\cdot)$ and linear functional $L(\cdot)$ are expressed as

$$\pi(u,v) = \left(y_n(u) - y_n(0), \; y_n(v) - y_n(0) \right)_{L_2(\gamma)} + (\bar{a}\, u, v)$$

and

$$L(v) = \left(z_g - y_n(0), \; y_n(v) - y_n(0) \right)_{L_2(\gamma)}.$$

Inequality (1.15′) is true for the bilinear form $\pi(\cdot,\cdot)$, i.e. such form is coercive on $\mathcal{U}$. Let $\tilde{y}' = \tilde{y}(u')$ and $\tilde{y}'' = \tilde{y}(u'')$ be solutions from V to problem (1.11) under $f = 0$ and under a function $u = u(x)$ that is equal, respectively, to u' and u''. Since the inequality

$$\int_\gamma (\tilde{y}'_n - \tilde{y}''_n)^2 d\gamma \le c_0 \left\| \tilde{y}' - \tilde{y}'' \right\|^2_V \qquad (3.2')$$

is true in accordance with the embedding theorems, then:

$$\frac{1}{c_0} \left\| \tilde{y}'_n - \tilde{y}''_n \right\|^2_{L_2(\gamma)} \le \left\| \tilde{y}' - \tilde{y}'' \right\|^2_V \le \frac{1}{\alpha'_0} a\left(\tilde{y}' - \tilde{y}'', \tilde{y}' - \tilde{y}'' \right) \le$$

$$\le \frac{1}{\alpha'_0} \left\| u' - u'' \right\| \left\| \tilde{y}' - \tilde{y}'' \right\|_V.$$

Therefore, the inequality

$$\left\| \tilde{y}'_n - \tilde{y}''_n \right\|_{L_2(\gamma)} \le \frac{\sqrt{c_0}}{\alpha'_0} \left\| u' - u'' \right\|$$

is obtained that provides the continuity of the linear functional $L(\cdot)$ and bilinear form $\pi(\cdot,\cdot)$ on $\mathcal{U}$.

On the basis of [58, Chapter 1, Theorem 1.1], the validity of the following statement is proved.

Theorem 3.1. *Let a system state be determined as a solution to equivalent problems (1.10) and (1.11). Then, there exists a unique element u of a convex set $\mathcal{U}_\partial$ that is closed in $\mathcal{U}$, and relation (1.17) takes place for u, where the cost functional has the form of expression (3.1).*

As for the control $v \in \mathcal{U}$, the conjugate state $p(v) \in V$ is specified as a generalized solution to the problem specified, in its turn, by the following equalities:

$$\sum_{k=1}^{3} \frac{\partial \sigma_{ik}(p)}{\partial x_k} = 0, \quad i = \overline{1,3}, \quad x \in \Omega,$$

$$p = 0, \quad x \in \Gamma,$$

$$[p_n] = 0, \quad [\sigma_n(p)] = y_n - z_g, \quad x \in \gamma, \qquad (3.2'')$$

$$[\tau_s(p)] = 0, \quad \{\tau_s(p)\}^{\pm} = r[p_s], \quad x \in \gamma.$$

The generalized problem corresponds to boundary-value problem (3.2'') and means to find a function $p \in V$ that satisfies equation like (2.5), where the bilinear form $a(\cdot,\cdot)$ is specified by expression (1.11'), and

$$l_1(y,z) = -\int_\gamma \left(y_n - z_g\right) z_n \, d\gamma. \qquad (3.3)$$

Therefore, the necessary and sufficient condition for the existence of the optimal control $u \in \mathcal{U}_\partial$ is the one under which the inequality

$$\left(y_n(u) - z_g, \, y_n(v) - y_n(u)\right)_{L_2(\gamma)} + \left(\bar{a}u, v - u\right) \geq 0, \quad \forall v \in \mathcal{U}_\partial, \qquad (3.4)$$

and equalities like (2.6) and (2.7), where the bilinear form $a(\cdot,\cdot)$ and functional $l(\cdot,\cdot)$ are specified by respective expressions (1.11') and $l_1(\cdot,\cdot)$ is specified, in its turn, by expression (3.3), are met.

Choose the difference $y(v) - y(u)$ instead of z, take expression (3.3) into account, obtain the equality

$$a\left(p(u), y(v) - y(u)\right) = -\int_\gamma \left(y_n(u) - z_g\right)\left(y_n(v) - y_n(u)\right) d\gamma$$

from equality (2.7), and the equality

$$a\big(p(u), y(v) - y(u)\big) = \int_{\Omega}(v - u)p(u)\,dx$$

follows from equality (2.6). Therefore, the equality

$$\big(y_n(u) - z_g,\ y_n(v) - y_n(u)\big)_{L_2(\gamma)} = -\big(p(u), v - u\big) \qquad (3.5)$$

is derived. Take it into account, and inequality (3.4) has the form of inequality (2.9). If the constraints are absent, i.e. when $\mathcal{U}_\partial = \mathcal{U}$, then equality (2.10) follows from inequality (2.9). Therefore, when constraints are absent, the control u can be excluded from equality (2.6) by means of equality (2.10). On the basis of equalities (2.6), (2.7) and (2.10), the problem is obtained: Find a vector-function $(y, p)^{\mathrm{T}} \in H = \{v = (v_1, v_2)^{\mathrm{T}} :$

$v_i \in V,\ i = 1, 2\}$ that satisfies equality system (2.11), where $a(\cdot, \cdot)$ and $l(\cdot, \cdot)$ are specified by expressions (1.11′) and $l_1(\cdot, \cdot)$ is specified, in its turn, by expression (3.3). The vector solution $(y, p)^{\mathrm{T}}$ to this problem is obtained and the optimal control u is found by formula (2.12). If the vector solution $(y, p)^{\mathrm{T}}$ to problem (2.11) is smooth enough on $\overline{\Omega}_l$, then the differential problem of finding the vector-function $(y, p)^{\mathrm{T}}$, that satisfies systems (2.13) and (3.2″), corresponds to problem (2.11).

Definition 3.1. A generalized (weak) solution to boundary-value problem (2.13), (3.2″) is called a vector-function $U = (y, p)^{\mathrm{T}} \in H$ that satisfies equation like (1.28), where

$$a(U, Z) = a_0(U, Z) - \big(p/\bar{a},\, Z_1\big) + \big(y_n, Z_{2n}\big)_{L_2(\gamma)}, \qquad (3.5′)$$

$$l(Z) = \big(f, Z_1\big) + \big(Z_g, Z_{2n}\big)_{L_2(\gamma)}.$$

Let the constraint

$$\min\left\{\alpha_0\mu - \frac{1}{a_0}, \ \alpha_0\mu - 1, \ \alpha_0' - c_0\right\} > 0 \qquad (3.6)$$

be met, where α_0, α_0', $c_0(y,p)^T$ and μ are the constants, respectively, from inequalities (1.1′), (1.13) and (3.2′) and the Friedrichs one. Therefore, inequalities like (1.31) and (1.33) take place for the bilinear form $a(\cdot,\cdot)$ and linear functional $l(\cdot)$. The following statement is valid.

Theorem 3.2. *Let constraint (3.6) be met. Then, there exists a solution $U \in H$ to problem (1.28), where the bilinear form $a(\cdot,\cdot)$ and linear functional $l(\cdot)$ are specified by respective expressions (3.5′).*

Assume that the classical solution U on Ω_l to problem (2.13), (3.2″) belongs to the Sobolev space $W_2^{k+1}(\Omega_l)$, $l = 1,2$. Then, estimate like (1.37) takes place for the approximate solution $U_k^N \in H_k^N$ obtained by means of the finite-element method. Therefore, estimate (1.38) takes place, in its turn, for the approximation $u_k^N(x) = p_k^N / \bar{a}$ of the control $u = p/\bar{a}$.

10.4   DISTRIBUTED CONTROL WITH OBSERVATION AT A DOMAIN BOUNDARY PART

Assume that equation system (1.5), where $Bu \equiv u$, is specified in the bounded, continuous and strictly Lipschitz domains Ω_1 and $\Omega_2 \in R^3$ for every control $u \in \mathcal{U} = L_2(\Omega)$. Conjugation conditions (1.3) and (1.4) are specified, in their turn, on the section γ of the domain $\overline{\Omega}$ and, on the boundary Γ of the domain $\overline{\Omega}$ ($\Gamma = (\partial\Omega_1 \cup \partial\Omega_2) \setminus \gamma$, $\gamma = \partial\Omega_1 \cap \partial\Omega_2$), the boundary conditions are

$$y = 0, \quad x \in \Gamma_1, \qquad (4.1)$$

$$\sigma_n = g_1, \quad x \in \Gamma_2, \qquad (4.2)$$

and

$$\tau_s = g_2, \quad x \in \Gamma_2, \tag{4.3}$$

where $\Gamma = \Gamma_1 \cup \Gamma_2$.

Specify the observation as

$$Cy(u) = y(x; u), \quad x \in \Gamma_2.$$

Bring a value of the cost functional

$$J(u) = \int_{\Gamma_2} |y(u) - z_g|^2 \, d\Gamma_2 + (\bar{a}\, u, u) \tag{4.4}$$

in correspondence with every control $u \in \mathcal{U}$; in this case, z_g is a known element from $L_2(\Gamma_2)$, and the function $\bar{a}(x)$ is specified in point 10.1.

Let the Friedrichs inequality

$$\int_\Omega \sum_{i=1}^3 \left(\frac{\partial v_i}{\partial x_i}\right)^2 dx \geq \mu \int_\Omega v^2 dx, \quad \mu = \text{const} > 0, \tag{4.5}$$

take place $\quad \forall v \in V = \left\{ v = (v_1, v_2, v_3)^{\mathrm{T}} : v_i|_{\Omega_l} \in W_2^1(\Omega_l), \quad l = 1, 2; \right.$

$\left. v|_{\Gamma_1} = 0, \quad [v_n]|_\gamma = 0 \right\}$.

The following statement is valid.

Theorem 4.1. *A unique state, namely, a function* $y = y(u) \in V$ *corresponds to every control* $u \in \mathcal{U}$, *delivers a minimum to energy functional (1.10) on V and it is a unique solution in V to weakly stated problem (1.11), where*

$$a(v, z) = \int_\Omega \sum_{i,k,l,m=1}^3 c_{iklm} \varepsilon_{ik}(v) \varepsilon_{lm}(z) \, dx + \int_\gamma r[v_s][z_s] d\gamma, \tag{4.6}$$

$$l(u, v) = (f, v) + (u, v) + \int_{\Gamma_2} (g_1 v_n + g_2 v_s) \, d\Gamma_2.$$

The validity of Theorem 4.1 is stated on the basis of the Lax-Milgramm lemma.

Rewrite cost functional (4.4) as

$$J(u) = \pi(u, u) - 2L(u) + \left\| z_g - y(0) \right\|^2_{L_2(\Gamma_2)},$$

where

$$\pi(u, v) = \left(y(u) - y(0), \ y(v) - y(0) \right)_{L_2(\Gamma_2)} + (\bar{a}u, v)$$

and

$$L(v) = \left(z_g - y(0), \ y(v) - y(0) \right)_{L_2(\Gamma_2)}.$$

Inequality (1.15′) is true for the form $\pi(\cdot, \cdot)$, i.e. such form is coercive on $\mathcal{U}$.

Let $\tilde{y}' = \tilde{y}(u')$ and $\tilde{y}'' = \tilde{y}(u'')$ be solutions from V to problem (1.11), where $a(\cdot, \cdot)$ and $l(\cdot, \cdot)$ are specified by respective expressions (4.6), under $f = 0$ and $g_1 = g_2 = 0$ and under a function $u = u(x)$ that is equal, respectively, to u' and u''. Then, the inequality

$$\left\| \tilde{y}' - \tilde{y}'' \right\|^2_{L_2(\Gamma_2)} \leq c_0 \left\| \tilde{y}' - \tilde{y}'' \right\|^2_V \leq c_1 a \left(\tilde{y}' - \tilde{y}'', \tilde{y}' - \tilde{y}'' \right) \leq$$

$$\leq c_1 \left\| u' - u'' \right\| \left\| \tilde{y}' - \tilde{y}'' \right\|_V,$$

i.e.

$$\left\| \tilde{y}' - \tilde{y}'' \right\|_{L_2(\Gamma_2)} \leq c_2 \left\| u' - u'' \right\|$$

is obtained that provides the continuity of the bilinear form $\pi(\cdot, \cdot)$ and linear functional $L(\cdot)$ on $\mathcal{U}$.

On the basis of [58, Chapter 1, Theorem 1.1], the validity of the following statement is proved.

Theorem 4.2. *Let a system state be determined as a solution to equivalent problems (1.10) and (1.11), where $a(\cdot, \cdot)$ and $l(\cdot, \cdot)$ have the form of expressions (4.6) and the space V is specified in the present point. Then, there exists a unique element u of a convex set $\mathcal{U}_\partial$ that is closed in $\mathcal{U}$, and relation (1.17) takes place for u, where the cost functional has the form of expression (4.4).*

As for the control $v \in \mathcal{U}$, the conjugate state $p(v) \in V$ is specified as a generalized solution to the problem

$$\sum_{k=1}^{3} \frac{\partial \sigma_{ik}(p)}{\partial x_k} = 0, \quad i = \overline{1,3}, \quad x \in \Omega,$$

$$p = 0, \quad x \in \Gamma_1,$$

$$\tau_s(p) = y_s - z_{gs}, \quad x \in \Gamma_2,$$

$$\sigma_n(p) = y_n - z_{gn}, \quad x \in \Gamma_2,$$

$$[p_n] = 0, \quad [\sigma_n(p)] = 0, \quad x \in \gamma, \tag{4.7}$$

$$[\tau_s(p)] = 0, \quad \{\tau_s(p)\}^{\pm} = r[p_s], \quad x \in \gamma.$$

The generalized problem corresponds to boundary-value problem (4.7) and means to find a function $p \in V$ that satisfies equation like (2.5), where the bilinear form $a(\cdot,\cdot)$ is specified by expression (4.6), and

$$l_1(y,z) = \left(y - z_g, z\right)_{L_2(\Gamma_2)}. \tag{4.8}$$

Therefore, the necessary and sufficient condition for the existence of the optimal control $u \in \mathcal{U}_\partial$ is the one under which the inequality

$$\left(y(u) - z_g, \, y(v) - y(u)\right)_{L_2(\Gamma_2)} + (\bar{a}u, v - u) \geq 0, \quad \forall v \in \mathcal{U}_\partial, \tag{4.9}$$

and equalities like (2.6) and (2.7), where the bilinear form $a(\cdot,\cdot)$ and functional $l(\cdot,\cdot)$ are specified by respective expressions (4.6) and the functional $l_1(\cdot,\cdot)$ is specified, in its turn, by expression (4.8), are met.

Choose the difference $y(v) - y(u)$ instead of z, take expression (4.8) into account, obtain the equality

$$a\left(p(u), y(v) - y(u)\right) = \left(y(u) - z_g, \, y(v) - y(u)\right)_{L_2(\Gamma_2)}$$

from equality (2.7), and the equality

$$a\left(p(u), y(v) - y(u)\right) = \left(p(u), \, v - u\right)$$

follows from equality (2.6). Therefore, the equality

$$\left(y(u) - z_g, \, y(v) - y(u)\right)_{L_2(\Gamma_2)} = \left(p(u), v - u\right)$$

is derived. Take it into account, and inequality (4.9) has the form

$$\left(p(u) + \bar{a}u, v - u\right) \geq 0, \quad \forall v \in \mathcal{U}_\partial. \tag{4.10}$$

If the constraints are absent, i.e. when $\mathcal{U}_\partial = \mathcal{U}$, then the equality

$$p(u) + \bar{a}\, u = 0 \tag{4.11}$$

follows from inequality (4.10). Proceed from equality system (2.6), (2.7), where the bilinear form $a(\cdot,\cdot)$ and functional $l(\cdot,\cdot)$ are specified by respective expressions (4.6) and the functional $l_1(\cdot,\cdot)$ is specified, in its turn, by expression (4.8), take the equality $u = -p/\bar{a}$ into account, and the solution to this system, namely, the vector-function $(y,p)^T \in H$ is found and the optimal control is $u = -p/\bar{a}$.

If such solution is smooth enough on $\overline{\Omega}_l$, $l = 1,2$, then the differential problem of finding the vector-function $(y,p)^T$, that satisfies equality system (1.3), (1.4), (4.1)–(4.3), (4.7) and the equation

$$-\sum_{k=1}^{3} \frac{\partial \sigma_{ik}(y)}{\partial x_k} + p_i/\bar{a} = f_i, \quad i = \overline{1,3}, \quad x \in \Omega, \tag{4.12}$$

corresponds to the considered problem.

Definition 4.1. A generalized (weak) solution to boundary-value problem (1.3), (1.4), (4.1)–(4.3), (4.7), (4.12) is called a vector-function $U = (y,p)^T \in H$ that satisfies equation (1.28), where

$$a(U,Z) = a_0(U,Z) + (p/\bar{a}, Z_1) - \int_{\Gamma_2} y\, Z_2\, d\Gamma_2,$$

$$l(Z) = (f, Z_1) - \int_{\Gamma_2} z_g Z_2\, d\Gamma_2. \tag{4.13}$$

Let the constraint

$$\min\left\{\alpha_0 \mu - \frac{1}{a_0}, \quad \alpha_0 \mu - 1, \quad \alpha_0' - c_0 c_0'\right\} > 0 \tag{4.14}$$

be met, where α_0, α_0', c_0, c_0' and μ are the constants, respectively, from inequalities (1.1'), (1.13) and (3.2'), the embedding theorems and Friedrichs inequality.

Therefore, inequalities like (1.31) and (1.33) take place for the bilinear form $a(\cdot,\cdot)$ and linear functional $l(\cdot)$. The following statement is valid.

Theorem 4.3. *Let constraint (4.14) be met. Then, there exists a solution $U \in H$ to problem (1.28), where the bilinear form $a(\cdot,\cdot)$ and linear functional $l(\cdot)$ are specified by respective expressions (4.13).*

Assume that the classical solution U on Ω_l to problem (1.3), (1.4), (4.1)–(4.3), (4.7), (4.12) belongs to the Sobolev space $W_2^{k+1}(\Omega_l)$, $l = 1,2$. Then estimate like (1.37) takes place for the approximate solution $U_k^N \in H_k^N \subset H$ obtained by means of the finite-element method. Therefore, estimate (1.38) takes place, in its turn, for the approximation $u_k^N(x) = -p_k^N/\bar{a}$ of the control $u = -p/\bar{a}$.

10.5 CONTROL AT A DOMAIN BOUNDARY PART WITH TAKING SIGHT ON JUMPING OF A DISPLACEMENT VECTOR COMPONENT TANGENT AT A DOMAIN SECTION

Assume that elasticity theory equation system (1.1) is specified in the bounded, continuous and strictly Lipschitz domains Ω_1 and $\Omega_2 \in R^3$ for every control $u \in \mathcal{U} = L_2(\Gamma_2)$. On the section γ of the domain $\overline{\Omega}$, conjugation conditions (1.3) and (1.4) are specified and, on the boundary Γ of $\overline{\Omega}$, the boundary conditions are

$$y = 0, \quad x \in \Gamma_1,$$
$$\sigma_n = g_1 + u_n, \quad x \in \Gamma_2,$$
$$\tau_s = g_2 + u_s, \quad x \in \Gamma_2, \tag{5.1}$$

where the sectors Γ_1 and Γ_2 of Γ are specified, in their turn, in the previous point.

Specify the observation as

$$Cy(u) = [y_s(u)], \quad x \in \gamma.$$

Bring a value of the cost functional

$$J(u) = \int_\gamma \left([y_s(u)] - z_g \right)^2 d\gamma + (\bar{a}\,u, u)_{L_2(\Gamma_2)} \qquad (5.2)$$

in correspondence with every control $u \in \mathcal{U}$; in this case, z_g is a known element from $L_2(\gamma)$, and the function $\bar{a}(x)$ is specified on Γ_2, $a(x) \in L_2(\Gamma_2)$, $0 < a_0 \leq a(x)$, $a_0 = \text{const}$.

The following statement is valid.

Theorem 5.1. *A unique state, namely, a function* $y = y(u) \in V$ *corresponds to every control* $u \in \mathcal{U}$, *delivers a minimum to energy functional (1.10) on V, and it is a unique solution in V to weakly stated problem (1.11); in this case, the space V is specified in point 10.4, the bilinear form* $a(\cdot, \cdot)$ *is specified by the first formula of expressions (4.6), and*

$$l(u, v) = (f, v) + (u, v)_{L_2(\Gamma_2)} + \int_{\Gamma_2} (g_1 v_n + g_2 v_s) d\Gamma_2 . \qquad (5.3)$$

Rewrite cost functional (5.2) as

$$J(u) = \pi(u, u) - 2L(u) + \left\| z_g - [y_s(0)] \right\|_{L_2(\gamma)}^2,$$

where

$$\pi(u, v) = \left([y_s(u)] - [y_s(0)],\ [y_s(v)] - [y_s(0)] \right)_{L_2(\gamma)} + (\bar{a}u, v)_{L_2(\Gamma_2)}$$

and

$$L(v) = \left(z_g - [y_s(0)],\ [y_s(v)] - [y_s(0)]_{L_2(\gamma)} \right).$$

Inequality (1.15′) is true for the bilinear form $\pi(\cdot, \cdot)$, i.e. such form is coercive on $\mathcal{U}$. Let $\tilde{y}' = \tilde{y}(u')$ and $\tilde{y}'' = \tilde{y}(u'')$ be solutions from V to problem (1.11) under $f = 0$ and $g_1 = g_2 = 0$ and where $a(\cdot, \cdot)$ and $l(\cdot, \cdot)$ are specified, respectively, by the first formula of expressions (4.6) and by expression (5.3). Then, the inequality

$$\frac{1}{c_0'} \left\| [\tilde{y}_s'] - [\tilde{y}_s''] \right\|_{L_2(\gamma)}^2 \leq \left\| \tilde{y}' - \tilde{y}'' \right\|_V^2 \leq \frac{1}{\alpha_0'}\ a(\tilde{y}' - \tilde{y}'', \tilde{y}' - \tilde{y}'') \leq$$

$$\leq \frac{c_1'}{\alpha_0'} \|u' - u''\|_{L_2(\Gamma_2)} \|\tilde{y}' - \tilde{y}''\|_V$$

is obtained, where c_0', c_1' and $\alpha_0' = \text{const} > 0$ and from which the inequality

$$\left\| [\tilde{y}_s'] - [\tilde{y}_s''] \right\|_{L_2(\gamma)} \leq c_1 \|u' - u''\|_{L_2(\Gamma_2)} \tag{5.4}$$

follows that provides the continuity of the bilinear form $\pi(\cdot,\cdot)$ and linear functional $L(\cdot)$ on $\mathcal{U}$.

On the basis of [58, Chapter 1, Theorem 1.1], the validity of the following statement is proved.

Theorem 5.2. *Let a system state be determined as a solution to equivalent problems (1.10) and (1.11), where $a(\cdot,\cdot)$ is specified by the first formula of expressions (4.6) and $l(\cdot,\cdot)$ has the form of expression (5.3). Then, there exists a unique element u of a convex set $\mathcal{U}_\partial$ that is closed in $\mathcal{U}$, and relation (1.17) takes place for u, where the cost functional has the form of expression (5.2).*

For every control $v \in \mathcal{U}$, the conjugate state $p(v) \in V$ is specified as a generalized solution to the problem specified, in its turn, by the equalities of system (2.4), except the second one, and by the constraints

$$p = 0, \quad x \in \Gamma_1,$$

$$\sigma_n = \tau_s = 0, \quad x \in \Gamma_2. \tag{5.4'}$$

The generalized solution to the considered problem exists and such solution is unique in V.

Therefore, the necessary and sufficient condition for the existence of the optimal control $u \in \mathcal{U}_\partial$ is the one under which the relations

$$a(y(u), z) = l(u, z), \quad \forall z \in V, \tag{5.5}$$

$$a(p(u), z) = l_1(y(u), z), \quad \forall z \in V, \tag{5.6}$$

and

$$\left([y_s(u)] - z_g, \ [y_s(v)] - [y_s(u)] \right)_{L_2(\gamma)} +$$

$$+\left(\bar{a}\,u, v-u\right)_{L_2(\Gamma_2)} \geq 0, \quad \forall v \in \mathcal{U}_\partial, \tag{5.7}$$

where $a(\cdot,\cdot)$, $l(\cdot,\cdot)$ and $l_1(\cdot,\cdot)$ are specified by the first formula of expressions (4.6) and by expressions (5.3) and (2.5′), are met.

Choose the difference $y(v) - y(u)$ instead of z, obtain the equality

$$a\left(p(u), y(v) - y(u)\right) = -\left(\left[y_s(u)\right] - z_g, \left[y_s(v)\right] - \left[y_s(u)\right]\right)_{L_2(\gamma)}$$

from equality (5.6), and the equality

$$a\left(p(u), y(v) - y(u)\right) = \left(p(u), v - u\right)_{L_2(\Gamma_2)}$$

follows from equality (5.5). Therefore, the equality

$$\left(\left[y_s(u)\right] - z_g, \left[y_s(v)\right] - \left[y_s(u)\right]\right)_{L_2(\gamma)} = \left(-p(u), v - u\right)_{L_2(\Gamma_2)} \tag{5.8}$$

is derived. Take it into account, and inequality (5.7) has the form

$$\left(-p(u) + \bar{a}u, v - u\right)_{L_2(\Gamma_2)} \geq 0, \quad \forall v \in \mathcal{U}_\partial. \tag{5.9}$$

If the constraints are absent, i.e. when $\mathcal{U}_\partial = \mathcal{U}$, then the equality

$$-p + \bar{a}u = 0, \quad x \in \Gamma_2, \tag{5.10}$$

follows from inequality (5.9).

Therefore, when the constraints are absent, the control u can be excluded from equality (5.5) by means of equality (5.10). On the basis of equality (5.10), the problem is obtained: Find the vector-function $(y, p)^T \in H = \left\{v = (v_1, v_2)^T : v_i \in V, \ i = 1, 2\right\}$ that satisfies the equality system

$$a(y, z) = l\left(p/\bar{a}, z\right), \quad \forall z \in V, \tag{5.11}$$

$$a(p, z) = l_1(y, z), \quad \forall z \in V, \tag{5.12}$$

from which the vector solution $(y, p)^T$ is found along with the optimal control

$$u = p/\bar{a}, \quad x \in \Gamma_2.$$

If the vector solution $(y,p)^T$ to problem (5.11), (5.12) is smooth enough on $\overline{\Omega}_l$, then the differential problem of finding the vector-function $(y,p)^T$, that satisfies equality system (2.4), except its second equality, equalities (1.1), (1.3), (1.4) and (5.4′) and the constraints

$$y = 0, \quad x \in \Gamma_1,$$
$$\sigma_n(y) = g_1 + p_n/\bar{a}, \quad x \in \Gamma_2, \tag{5.13}$$
$$\tau_s(y) = g_2 + p_s/\bar{a}, \quad x \in \Gamma_2,$$

corresponds to problem (5.11), (5.12).

Definition 5.1. A generalized (weak) solution to the boundary-value problem, specified by equalities (1.1), (1.3), (1.4), (5.4′), (5.13) and (2.4), except the second one of system (2.4), is called a vector function $U = (y,p)^T \in H$ that satisfies equation (1.28), where

$$a(U,Z) = a_0(U,Z) - \left(p/\bar{a}, Z_1\right)_{L_2(\Gamma_2)} + \left([y_s],[Z_{2s}]\right)_{L_2(\gamma)}, \tag{5.13′}$$

$$l(Z) = (f, Z_1) + \left(Z_g, [Z_{2s}]\right)_{L_2(\gamma)}.$$

Let the constraint

$$\alpha_0' - \frac{c_0}{2a_0} - \frac{c_1'}{2} > 0 \tag{5.14}$$

be met, where α_0' and c_0 are the constants, respectively, from inequalities (1.13) and the embedding theorems and the constant c_1' is found from the inequality

$$\left\| [y_s] \right\|_{L_2(\gamma)} \le \sqrt{c_1'} \, \|y\|_V.$$

Therefore, inequalities like (1.31) and (1.33) take place for the bilinear form $a(\cdot,\cdot)$ and linear functional $l(\cdot)$. The following statement is valid.

Theorem 5.3. *Let constraint (5.14) be met. Then, there exists a solution $U \in H$ to problem (1.28), where the bilinear form $a(\cdot,\cdot)$ and linear functional $l(\cdot)$ are specified by respective expressions (5.13′).*

Assume that the classical solution U on Ω_l to the problem, specified by equalities (1.1), (1.3), (1.4), (5.4′), (5.13) and (2.4), except the second equality of system (2.4), belongs to the Sobolev space $W_2^{k+1}(\Omega_l)$, $l = 1, 2$. Then, estimate like (1.37) takes place for the approximate solution $U_k^N \in H_k^N$ obtained by means of the finite-element method. Therefore, the estimate

$$\left\| u - u_k^N \right\|_{L_2(\Gamma_2)} \le ch^k \qquad (5.13'')$$

takes place for the approximation $u_k^N(x) = p_k^N / \bar{a}$ of the control $u = p / \bar{a}$.

10.6 CONTROL AT A DOMAIN BOUNDARY PART WITH TAKING SIGHT ON A NORMAL DISPLACEMENT VECTOR COMPONENT AT A DOMAIN SECTION

Assume that elasticity theory equation system (1.1) is specified in bounded, continuous and strictly Lipschitz domains Ω_1 and $\Omega_2 \in R^3$ for every control $u \in \mathcal{U} = L_2(\Gamma_2)$. Conjugation conditions (1.3) and (1.4) are specified, in their turn, on the section γ of the domain $\overline{\Omega}$ and, on its boundary Γ, the boundary conditions have the form of expression (5.1).

Specify the observation as

$$Cy(u) = y_n(u), \quad x \in \gamma.$$

Bring a value of the cost functional

$$J(u) = \int_\gamma \left(y_n(u) - z_g \right)^2 d\gamma + (\bar{a} u, u)_{L_2(\Gamma_2)} \qquad (6.1)$$

in correspondence with every control $u \in \mathcal{U}$; in this case, z_g is a known scalar function from $L_2(\gamma)$, and the function $\bar{a}(x)$ is specified in the previous point.

It is shown in point 10.5 that a unique state, namely, a function $y = y(u) \in V$ corresponds to every control $u \in \mathcal{U}$, minimizes functional

(1.10) on V and meets equation (1.11), where the bilinear form $a(\cdot,\cdot)$ is specified by the first formula of expressions (4.6) and the functional $l(\cdot,\cdot)$ has the form of expression (5.3).

Represent cost functional (6.1) in the form of expression (3.2), where

$$\pi(u,v) = \left(y_n(u) - y_n(0),\ y_n(v) - y_n(0)\right)_{L_2(\gamma)} + (\bar{a}\,u,v)_{L_2(\Gamma_2)}$$

and

$$L(v) = \left(z_g - y_n(0),\ y_n(v) - y_n(0)\right)_{L_2(\gamma)}.$$

It is easy to see that the form $\pi(\cdot,\cdot)$ is coercive on $\mathcal{U}$.

Let $\tilde{y}' = \tilde{y}(u')$ and $\tilde{y}'' = \tilde{y}(u'')$ be solutions from V to problem like (1.11) under $f = 0$ and $g_1 = g_2 = 0$ and under a function $u = u(x)$ that is equal, respectively, to u' and u''. Proceed from the embedding theorem, and the inequality

$$\left\|\tilde{y}'_n - \tilde{y}''_n\right\|_{L_2(\gamma)} \le c_0 \left\|u' - u''\right\|_{L_2(\Gamma_2)}$$

is obtained that provides the continuity of the bilinear form $\pi(\cdot,\cdot)$ and linear functional $L(\cdot)$ on $\mathcal{U}$.

On the basis of [58, Chapter 1, Theorem 1.1], the validity of the following statement is proved.

Theorem 6.1. *Let a system state be determined as a solution to equivalent problems like (1.10) and (1.11), where the bilinear form $a(\cdot,\cdot)$ is specified by the first formula of expressions (4.6) and the functional $l(\cdot,\cdot)$ is specified by expression (5.3). Then, there exists a unique element u of a convex set $\mathcal{U}_\partial$ that is closed in $\mathcal{U}$, and relation (1.17) takes place for u, where cost functional has the form of expression (6.1).*

As for the control $v \in \mathcal{U}$, the conjugate state $p(v) \in V$ is specified as a generalized solution to the problem specified, in its turn, by system (3.2''), except the second equation, and by constraints (5.4'). It is easy to state that the generalized solution to this system exists and that such solution is unique.

Therefore, the necessary and sufficient condition for the existence of the optimal control $u \in \mathcal{U}_\partial$ is the one under which the inequality

$$\left(y_n(u) - z_g, \, y_n(v) - y_n(u)\right)_{L_2(\gamma)} +$$
$$+\left(\bar{a}\,u, v - u\right)_{L_2(\Gamma_2)} \geq 0, \quad \forall v \in \mathscr{U}_\partial, \tag{6.2}$$

and equalities like (2.6) and (2.7), where the bilinear form $a(\cdot,\cdot)$ is specified by the first formula of expressions (4.6), the functional $l(\cdot,\cdot)$ is specified, in its turn, by expression (5.3) and $l_1(\cdot,\cdot)$ has the form of expression (3.3), are met.

Choose the difference $y(v) - y(u)$ instead of z, take expression (3.3) into account, obtain the equality

$$a\left(p(u), y(v) - y(u)\right) = -\int_\gamma \left(y_n(u) - z_g\right)\left(y_n(v) - y_n(u)\right) d\gamma$$

from equality (2.7), and the equality

$$a\left(p(u), y(v) - y(u)\right) = \left(p(u), \, v - u\right)_{L_2(\Gamma_2)}$$

follows from equality (2.6). Therefore, the equality

$$-\left(y_n(u) - z_g, \, y_n(v) - y_n(u)\right)_{L_2(\gamma)} = \left(p(u), v - u\right)_{L_2(\Gamma_2)}$$

is derived. Take it into account, and inequality (6.2) has the form

$$\left(-p(u) + \bar{a}\,u, v - u\right)_{L_2(\Gamma_2)} \geq 0, \quad \forall v \in \mathscr{U}_\partial. \tag{6.3}$$

If the constraints are absent, i.e. when $\mathscr{U}_\partial = \mathscr{U}$, then the equality

$$-p(u) + \bar{a}\,u = 0, \quad x \in \Gamma_2, \tag{6.4}$$

follows from inequality (6.3). Therefore, when the constraints are absent, the control u can be excluded from condition (5.1) by means of equality (6.4), i.e.:

$$y = 0, \quad x \in \Gamma_1,$$
$$\sigma_n = g_1 + p_n/\bar{a}, \quad x \in \Gamma_2, \tag{6.5}$$
$$\tau_s = g_2 + p_s/\bar{a}, \quad x \in \Gamma_2.$$

Definition 6.1. A generalized (weak) solution to boundary-value problem, specified by equalities (1.1), (1.3), (1.4), (6.5) and (3.2″), except the second one of system (3.2″), and by constraints (5.4′), is called a vector-function $U = (y, p)^T \in H$ that satisfies equation (1.28), where

$$a(U, Z) = a_0(U, Z) - \left(p/\bar{a}, Z_1\right)_{L_2(\Gamma_2)} + \left(y_n, Z_{2n}\right)_{L_2(\gamma)},$$
$$l(Z) = (f, Z_1) + \left(Z_g, Z_{2n}\right)_{L_2(\gamma)}. \qquad (6.6)$$

The following statement is valid.

Theorem 6.2. *Let the constraint be met that provides the H-ellipticity of the bilinear form $a(\cdot, \cdot)$ on H. Then, there exists a solution $U \in H$ to problem (1.28), where the bilinear form $a(\cdot, \cdot)$ and linear functional $l(\cdot)$ are specified by respective expressions (6.6).*

Assume that the classical solution U on Ω_l to the considered problem belongs to the Sobolev space $W_2^{k+1}(\Omega_l)$, $l = 1, 2$. Then, estimate like (1.37) takes place for the approximate solution $U_k^N \in H_k^N \subset H$ obtained by means of the finite-element method. Therefore, estimate (5.13″) takes place, in its turn, for the approximation $u_k^N(x) = p_k^N / \bar{a}$ of the control $u = p/\bar{a}$.

REFERENCES

[1] Barenblatt G.I. (1963) *On Some Boundary Value Problems for Equations of Liquid Filtration in Fissured Rock*, Prikladnaya Matematika i Mekhanika, **27**, №2, 348–350

[2] Barenblatt G.I., Entov V.M., Ryzhik V.M. (1978) *Theory of Non-Stationary Filtration of Liquid and Gas*, Nauka: Moscow (in Russian)

[3] Barenblatt G.I., Zhelnov Yu.P., Kochina I.N. (1960) *On Basic Idea of Theory of Homogeneous Lequid Filtration in Fissured Rocks*, Prikladnaya Matematika i Mekhanika, **24**, №5, 852–864

[4] Bear J., Zaslavsky D., Irmay S. (Ed.) (1968) *Physical Principles of Water Percolation and Seepage*, UNESCO

[5] Bellman R.E., Dreyfus S.E. (1962) *Applied Dynamic Programming*, Princenton University Press: Princenton, New York

[6] Belova G.P., Bodrov V.A., Globus A.M. et al (1984) *Hydrophysical Properties of Ions and Mathematical Simulations of Salt Movement in Soil*, Pochvovedeniye, №4, 113–119

[7] Bensoussan A., Lions J.L. (1982) *Controle Impulsionnel et Inequations Quasivariationnelles*, Dunod, Bordas: Paris

[8] Bogan Yu.A. (1999) *A Two-Dimensional Problem of Elasticity Theory for an Orthotrope Plane with a Crack under a Contact of Banks Like a Viscous Friction*, Journal of Applied Mathematics and

374

Technical Physics, Plenum Publishing Corporation, **40**, №4, 195–197

[9]   Boley B.A., Weiner J.H. (1960) *Theory of Thermal Stresses*, Jonn Wiley and Sons, Inc.: New York, London

[10]   Bolotin V.V., Novichkov Yu.N. (1980) *Mechanics of Multilayered Constructions*, Mashinostroenie: Moscow (in Russian)

[11]   Bublik B.N., Kirichenko N.F. (1975) *Foudations of Control Theory*, Vyshcha Shkola: Kyiv (in Ukrainian)

[12]   Butkovsky A.G. (1965) *Theory of Optimal Control of Systems with Distributed Parameters*, Nauka: Moscow (in Russian)

[13]   Butkovsky A.G. (1975) *Methods of Control of Systems with Distributed Parameters*, Nauka: Moscow (in Russian)

[14]   Butkovsky A.G., Samoylenko Yu.I. (1984) *Control of Quantum-Mechanics Processes*, Nauka: Moscow (in Russian)

[15]   Cherepanov G.M. (1983) *Mechanics of Composition Material Destruction*, Nauka: Moscow (in Russian)

[16]   Ciarlet P. (1978) *The Finite Element Method for Elliptic Problems*, North-Holland Publishing Company: Amsterdam, New York, Oxford

[17]   *Construction Element Contact Problems*, By: Podgornyi A.N., Gontarovsky P.P., Kirkach B.N., Matyukhin Yu.I., Khavin G.L. (1989) Naukova Dumka: Kyiv (in Russian)

[18]   Deineka V.S. (1982) *A Numerical Solution to a Boundary-Value Problem that Admits Solution Discontinuity*, Reports of Academy of Sciences of the Ukrainian SSR, vol. A, №11, 28–31 (in Russian)

[19]   Deineka V.S. (1989) *Determining Discontinuous Characteristics for Longitudinal Oscillations of Complicated Rods*, Reports of Academy of Sciences of the Ukrainian SSR, vol. A, №9, 8–11 (in Russian)

[20]   Deineka V.S., Molchanov I.N. (1981) *Schemes in the Method of Finite Elements with High Order of Accuracy for Problems of Elasticity Theory*, Journal of Computation Mathematics and Mathematical Physics, **21**, №2, 452–469 (in Russian)

[21] Deineka V.S., Sergienko I.V. (2001) *Models and Methods for Problem Solution in Heterogeneous Media*, Naukova Dumka: Kyiv (in Russian)

[22] Deineka V.S., Sergienko I.V., Skopetskii V.V. (1995) *Mathematical Models and Methods Used to Calculate Problems with Discontinuous Solutions*, Naukova Dumka: Kyiv (in Russian)

[23] Deineka V.S., Sergienko I.V., Skopetskii V.V. (1998) *Models and Methods Used to Solve Problems with Conjugation Conditions*, Naukova Dumka: Kyiv (in Russian)

[24] Deineka V.S., Sergienko I.V., Skopetskii V.V. (1999) *Eigenvalue Problems with Discontinuous Eigenfunctions and Their Numerical Solutions*, Ukrainian Mathematical Journal, **51**, №10, 1484–1492 (in Ukrainian)

[25] Drenska N.T. (1979) *On a Problem of Conjugation of Two Parabolic Equations*, Differentsialnye Uravneniya, **19**, №12, 2251–2262

[26] Dulnev G.N., Novikov V.V. (1991) *Transfer Processes in Heterogeneous Media*, Energoatomizdat: Leningrad (in Russian)

[27] Dykhta V.A., Samsonyuk O.N. (2000) *Optimal Impulse Control with Applications*, Fizmatlit: Moscow (in Russian)

[28] Dzerzer Ye.S. (1972) *Generalization of Equation of Groundwater Movement with Free Surface*, Reports of Academy of Sciences of the USSR, **202**, №5, 27–46

[29] Egorov A.I. (1978) *Optimal Control of Thermal and Diffusion Processes*, Nauka: Moscow (in Russian)

[30] Egorov A.I. (1988) *Optimal Control of Linear Systems*, Vyshcha Shkola: Kyiv (in Russian)

[31] Egorov A.I., Kapustyan V.E. (1984) *Synthesis of Pulse Optimal Controls in Distributed Systems*, In: Control and Optimization of Thermal and Diffussion Processes, Preprint 84–34, Kyiv, Inst. of Cybernetics

[32] Ekeland I., Temam R. (1976) *Convex Analysis and Variational Problems*, North-Holland Publishing Company: Amsterdam, Oxford; American Elsevier Publishing Company, Inc.: New York

376

[33]   Ermoliev Yu.M. (1979) *Methods of Stohastic Programming*, Nauka: Moscow (in Russian)

[34]   Ermoliev Yu.M., Gulenko V.P., Tzarenko T.I. (1978) *Finite-Difference Method in Optimal Control Problems*, Naukova Dumka: Kyiv (in Russian)

[35]   Ermoliev Yu.M., Lyashko I.I., Mikhalevich V.S., Tyuptya V.I. (1979) *Mathemetical Methods of Operation Research*, Vyshcha Shkola: Kyiv (in Russian)

[36]   Ermoliev Yu.M., Nurminsky Ye.A. (1973) *Limit Extremal Problems*, Cybernetics, №4, 130–132

[37]   Ermoliev Yu.M., Verchenko P.I. (1976) *On Method of Linearization in Limit Extremal Problems*, Cybernetics, №2, 73–79

[38]   Evtushenko A.A., Ivanik E.G. (1999) *Estimation of a Contact Temperature and Wear of a Composed Friction Lining under Braking*, Journal of Engineering Physics and Thermophysics, **72**, №5, 988–994

[39]   Fedorenko R.P. (1978) *An Approximate Solution to Optimal Control Problems*, Nauka: Moscow (in Russian)

[40]   Feldbaum A.A. (1966) *Basis of Optimal Automatic Systems Theory*, Nauka: Moscow (in Russian)

[41]   Gajewski H., Groger K., Zacharias K. (1974) *Nichtlineare Operatorgleichungen und Operatordifferentialgleichungen*, Akademie-Verlag: Berlin

[42]   Gamkrelidze R.V. (1965) *On Some Extremal Problems in the Theory of Differential Equations with Applications to the Theory of Optimal Control*, SIAM J. Control, №3, 106–128

[43]   Glowinski R., Lions J.L., Tremolieres R. (1976) *Analyse Numerique des Inequations Variationnelles*, Publie avec le Concours du Centre National de la Recherche Scientifique, Dunod: Paris

[44]   Goryacheva I.G., Dobychin M.N. (1988) *Contact Problems in Tribology*, Mashinostroenie: Moscow (in Russian)

[45] Ismatov M. (1984) *About One Mixed Problem for Equation Describing the Sound Propagation in Viscous Gas*, Differentzialniye Uravneniya, **20**, №6, 1023–1024

[46] Ivanenko V.I., Mel'nik V.S. (1988) *Variational Methods in Control Problems of Systems with Distributed Parameters*, Naukova Dumka: Kyiv (in Russian)

[47] Johnson K.L. (1985) *Contact Mechanics*, Cambridge University Press

[48] Kachanov L.M. (1981) *Foundation of Plasticity Theory*, Nauka: Moscow (in Russian)

[49] Kolmogorov A.N., Fomin S.V. (1989) *Elements of Function Theory and Functional Analysis*, Nauka: Moscow (in Russian)

[50] Kozhanov A.I. (1996) *Problem with Directional Derivative for Some Pseudo-Parabolic and Similar Equations*, Sibirian Mathematical J., **37**, №6, 1335–1346

[51] Krasovski N.N. (1968) *The Theory of Motion Control*, Nauka: Moscow (in Russian)

[52] Kudinov V.A., Dikon V.V., Remezentsev A.B. (1998) *Construction of Multilayer Composite Materials by Way of Their Unilayer Models*, Russian Academy of Sciences, Energetics, №3, 140–143 (in Russian)

[53] Kulchitsky-Zhigaylo R.D., Evtushenko A.A. (1998) *Influence of a Thin Lining onto Pressure Distribution in Contact Problems under Friction Heat Generation*, Journal of Applied Mechanics and Technical Physics Plenum Publishing Corporation, **39**, №1, 110–118

[54] Ladyzhenskaya O.A., Solonnikov V.A., Uraltseva N.N. (1967) *Linear and Quasilinear Parabolic-Type Equations*, Nauka: Moscow (in Russian)

[55] Ladyzhenskaya O.A., Uraltseva N.N. (1973) *Linear and Quasilinear Elliptic-Type Equations*, Nauka: Moscow (in Russian)

[56] Lions J.L. (1968) *Controlabilite Exacte des System Distibutes*, Comptes Rendue der Scances de L'Academic des Sciences, Ser. S. 302; №13, 471–475

378

[57] Lions J.L. (1968) *Controle de Systemes Gouvernes par des Equations aux Derivees Patielles*, Dunod Gauthier-Villars: Paris

[58] Lions J.L. (1968) *Controle Optimal de Systemes Gouvernes par des Equations aux Derivees Partielles,* Dunod Gauthier-Villars: Paris

[59] Lions J.L. (1969) *Quelques Methodes de Resolution de Problemes aux Limities Non-Lineaires*, Dunod Gauthier-Villars: Paris

[60] Lions J.L. (1975) *Remarks on Some New Non-Linear Boundary-Value Problem*, Lecture Notes in Math., №446, 301–328

[61] Lions J.L. (1979) *Some Remarks on Free Boundary Problems and Optimal Control*, In: Free Boundary Problems, Proc. Semin., vol. 2, Pavia

[62] Lions J.L. (1985) *Some Aspects of Optimal Control of Distributed Systems*, Successes in Mathematical Sciences, **40**, №4, 55–68 (in Russian)

[63] Lions J.L. (1988) *Exact Controllability, Stabilization and Perturbation for Distributed Systems*, SIAM Review, №1, 1–68

[64] Lions J.L., Magenes E. (1968) *Problemes aux Limites Non-Homogenes et Applications*, Dunod: Paris

[65] Luckner L., Schestakow W.M. (1986) *Migrationsprozesse im Boden-und Grundwasser-Bereich*, Nedra: Moscow (in Russian)

[66] Lurje K.A. (1975) *Optimal Control in Problems of Mathematical Physics*, Nauka: Moscow

[67] Lyashko S.I. (2002) *Generalized Optimal Control of Linear Systems with Distributed Parameters*, Kluwer Academic Publishers: Dordrecht, Boston, London

[68] Lykov A.V. (1978) *Heat and Mass Exchange: A Handbook*, Energiya: Moscow

[69] Lyubonova A.Sh. (1989) *Some Issues of Boundary Problems Theory for Pseudo-Parabolic Equations*, Krasnoyarsk (in Russian)

[70] Malovischko V.A. (1991) *On Boundary Value Problems for Singular Pseudo-Parabolic and Pseudo-Hyperbolic Systems*, Differentzialniye Uravneniya, **27**, №12, 2120–2124

[71] Mikhlin S.G. (1952) *A Quadratic Functional Minimum Problem*, Gostekhteorizdat: Moscow-Leningrad

[72] Mikhlin S.G. (1970) *Variational Methods in Mathematical Physics*, Nauka: Moscow (in Russian)

[73] Mordukhovich B.S. (1988) *Approximation Methods in Problems of Optimization and Control*, Nauka: Moscow (in Russian)

[74] Morozov V.A. (1974) *Regular Methods of Solving Non-Correctly Posed Problems*, Prepr. of Computational Center of Moscow State University

[75] Nao R.C., Ting T.W. (1973) *Initial Value Problems for Pseudoparabolic Differential Equations*, Indian Uniq. Math. J., №25, 348–368

[76] Novikova A.M., Motovilovets I.O., Svyatenko O.T. (1972) *Heat Conduction of Multilayer Walls*, Reports of Academy of Sciences of the Ukrainian SSR, vol. A, №3, 250–253 (in Ukrainian)

[77] Novosel'sky S.N., Shul'gin D.F. (1984) *Computation of Non-Stationary Water Transport under Trickle and Underground Irrigation*, Izvestiya of AN Ussr. Meckhanika Zhidkosti i Gaza, №4, 74–81

[78] Oganesyan L.A., Rivkind V.Ya., Rukhovets L.A. (1974) *Variational and Differential Methods Used to Solve Elliptic Equations*, Differential Equations and Their Application, part II, vol. 8 (in Russian)

[79] Oganesyan L.A., Rukhovets L.A. (1979) *Variational and Differential Methods Used to Solve Elliptic Equations*, Erevan, Academy of Sciences of the Armenian SSR (in Russian)

[80] Osipov Yu.S., Suyetov A.P. (1984) *On One Lions Problem*, Reports of the Academy of Sciences of the USSR, **276**, №2, 288–291 (in Russian)

[81] Pelekh B.L., Maksimchuk A.V., Korovaychuk I.M. (1988) *Contact Problems for Layered Elements of Constructions and Bodies with Coverings*, Naukova Dumka: Kyiv (in Russian)

[82] Plumb O.A., Whitaker S. (1988) *Dispersion in Heterogeneous Porous Media*, Water Resour. Res., №7, 913–938

380

[83]   Podstrigach Ya.S., Lomakin V.A., Kolyano Yu.M. (1984) *Thermo-Elasticity of Bodies with a Heterogeneous Structure*, Nauka: Moscow (in Russian)

[84]   Polubarinova-Kochin P.Ya. (1967) *Theory of Ground Water Motion*, Nauka: Moscow (in Russian)

[85]   Pontryagin L.S., Boltyansky V.G., Gamkrelidze R.V., Mishchenko Ye.A. (1969) *Mathematical Theory of Optimal Processes*, Nauka: Moscow (in Russian)

[86]   Pschenichnyi B.N. (1972) *Necessary Conditions of Extremum*, Marcel Decker: New York

[87]   Pschenichnyi B.N. (1980) *Convex Analysis and Extremum Problems*, Nauka: Moscow (in Russian)

[88]   Pschenichnyi B.N., Danilin Yu.M. (1975) *Numerical Methods in Extremum Problems*, Nauka: Moscow (in Russian)

[89]   Rubinstein L.M. (1948) *On Issue about Heat Transport in Heterogeneous Media*, Izvestiya AN SSSR, Ser. Geograph. i Geophys., **12**, №1, 27–46

[90]   Rumynin V.G. (1999) *A Study of the Processes of Ground Water Migration in Stratified Systems with Infiltration Recharge: Case Study of Southeastern Tatarstan*, Water Resources, Kluwer Academic / Plenum Publishers, **26**, №6, 686–702 (in Russian)

[91]   Samarsky A.A. (1977) *Difference Scheme Theory*, Nauka: Moscow (in Russian)

[92]   Samarsky A.A, Andreev V.B. (1976) *Difference Schemes for Elliptic Equations*, Nauka: Moscow (in Russian)

[93]   Samoilenko A.M. (1978) *Realization of Fast-Acting Negative Feedback on the Basis of Dynamic System with Accumulated Energy*, Avtomatika i Telemekhanika, №12, 12–23

[94]   Sanchez-Palencia E. (1980) *Non-Homogeneous Media and Vibration Theory*, Springer-Verlag: New York

[95]   Sergienko I.V. (1999) *Information Science in Ukraine. Formation, Development, Problems*, Naukova Dumka: Kyiv (in Ukrainian)

[96]   Sergienko I.V., Deineka V.S. (1999) *On Determination of the Axisymmetric Stressed State of a Compound Body under Disjoining Pressure*, Int. Appl. Mech., **35**, №1, 46–53

[97]   Sergienko I.V., Deineka V.S. (1999) *Problems with Conjugation Conditions and Their High-Accuracy Computational Discretization Algorithms*, Cybernetics and Systems Analysis, **35**, №6, 930–950

[98]   Sergienko I.V., Deineka V.S. (2000) *On Models with Conjugation Conditions and High-Precision Methods of Their Discretization*, Cybernetics and Systems Analysis, **36**, №1, 83–101

[99]   Sergienko I.V., Deineka V.S. (2001) *High-Precision Algorithms of Solution of Spectral Problems with Eigenvalues in Conjugated Conditions and Boundary Conditions*, Reports of the National Academy of Sciences of Ukraine, №2, 74–80 (in Ukrainian)

[100]   Sergienko I.V., Deineka V.S. (2001) *The Dirichlet and Neumann Problems for Elliptical Equations with Conjugation Conditions and High-Precision Algorithms of Their Discretization*, Cybernetics and Systems Analysis, **37**, №3, 323–347

[101]   Sergienko I.V., Deineka V.S. (2002) *Computational Finite-Element Schemes for Optimal Control of an Elliptic System with Conjugation Conditions*, Cybernetics and Systems Analysis, **38**, №1, 60–75

[102]   Sergienko I.V., Deineka V.S. (2002) *On the Existence of the Unique Generalized Solution of a Quasilinear Elliptic Equation with Conjugation Conditions*, Reports of the National Academy of Sciences of Ukraine, №4, 72–82 (in Ukrainian)

[103]   Sergienko I.V., Deineka V.S. (2002) *Optimal Control of a Conditionally Correct System with Conjugation Conditions*, Cybernetics and Systems Analysis, **38**, №4, 509–527 (in Russian)

[104]   Sergienko I.V., Deineka V.S. (2002) *Optimal Control of a System Described by a Pseudohyperbolic Equation with Conjugation Conditions*, Journal of Automation and Information Sciences, №5, 23–44 (in Russian)

[105]   Sergienko I.V., Deineka V.S. (2002) *Optimal Control of a System Described by a Pseudoparabolic Equation with Conjugation*

Conditions, System Research & Information Technologies, №3, 7–31 (in Russian)

[106] Sergienko I.V., Deineka V.S. (2002) *Optimal Control of an Elliptic System with Complex Thin Inclusion Conditions*, Journal of Automation and Information Sciences, **34**, №1, 47–67

[107] Sergienko I.V., Deineka V.S. (2002) *Optimal Control of System Described by Paraboic Equation with Conjugation Conditions*, Journal of Automation and Information Sciences, №4, 37–56 (in Russian)

[108] Sergienko I.V., Deineka V.S. (2002) *Optimal System Control Described by Quartic Equation with Conjugation Conditions*, Journal of Automation and Information Sciences, №2, 13–33 (in Russian)

[109] Sergienko I.V., Deineka V.S. (2002) *Transmission Problems with Inhomogeneous Principal Conditions of Conjugacy and High-Exact Numerical Algorithms of Their Digitization*, Ukrainian Mathematical Journal, **54**, №2, 258–275 (in Ukrainian)

[110] Sergienko I.V., Deineka V.S. (2003) *Optimal Control of a System Described by a Hyperbolic Equation with Conjugation Conditions*, Cybernetics and Systems Analysis, № 1, 55–74 (in Russian)

[111] Sergienko I.V., Deineka V.S. (2003) *Optimal Control of a System Described by a Two-Dimensional Quartic Equation with Conjugation Conditions*, Cybernetics and Systems Analysis, №2, 112–133 (in Russian)

[112] Sergienko I.V., Deineka V.S. (2003) *Optimal Deformed Condition Control of Compound Body*, Journal of Automation and Information Sciences, №2, 58–79 (in Russian)

[113] Sergienko I.V., Skopetskii V.V., Deineka V.S. (1991) *Mathematical Simulation and Examination of Processes in Heterogeneous Media*, Naukova Dumka: Kyiv (in Russian)

[114] Seymov V.M., Trofimchuk A.N., Savitsky O.A. (1990) *Oscillations and Waves in Layered Media*, Naukova Dumka: Kyiv (in Russian)

[115] Shemiakin E.I. (1955) *Propaganion of Non-Stationary Perturbation in Viscoelastic Medium*, Reports of Academy of Sciences of the USSR, **104**, №1, 34–37

[116] Showalter R.E., Ting T.W. (1970) *Pseudo-Parabolic Partial Differetial Equations*, SIAM J. Math. Anal., №1, 1–26

[117] Shtaerman I.Ya. (1949) *Contact Problem of Elasticity Theory*, Gostekhizdat: Moscow (in Russian)

[118] Sirazetdinov T.K. (1977) *Optimization of Systems with Distributed Parameters*, Nauka: Moscow (in Russian)

[119] Strang G., Fix G.J. (1973) *An Analysis of the Finite Element Method*, Prentice-Hall, Inc., Englewood Cliffs: N. J.

[120] Tikhomirov V.M. (1976) *Some Issues of Approximation Theory*, Moscow: Moscow State University (in Russian)

[121] Tikhonov A.N., Samarsky A.A. (1977) *Mathematical Physics Equations*, Nauka: Moscow (in Russian)

[122] Timoshenko S.P. (1972) *Course of Elasticity Theory*, Naukova Dumka: Kyiv (in Russian)

[123] Tkachenko E.A. (1999) *Influence of a Surface Loading on Stability of Layered Coverings at Increased Temperatures*, Reports of the National Academy of Sciences of Ukraine, №8, 89–93 (in Ukrainian)

[124] Varga R.S. (1971) *Functional Analysis and Approximation Theory in Numerical Analysis*, Society for Industrial and Applied Mathematics: Philadelphia, Pennsylvania

[125] Vasilyev A.G. (1981) *Methods to Solve Extremal Problems*, Nauka: Moscow (in Russian)

[126] Warga J. (1974) *Optimal Control of Differential and Functional Equations*, Academic Press: New York, London